Frontiers in Nanomedicine

(Volume 4)

Applications of Nanomaterials in Medical Procedures and Treatments

Edited by

Felipe López-Saucedo
Department of Radiation Chemistry and Radiochemistry
Institute of Nuclear Sciences
Universidad Nacional Autónoma de México
Mexico

Frontiers in Nanomedicine

(Volume 4)

Applications of Nanomaterials in Medical Procedures and Treatments

Editors: Felipe López-Saucedo

ISSN (Online): 2405-9137

ISSN (Print): 2405-9129

ISBN (Online): 978-981-5136-95-1

ISBN (Print): 978-981-5136-96-8

ISBN (Paperback): 978-981-5136-97-5

First published in 2023.

need for a court order if at any point you breach any terms of this License Agreement. In no event will any delay or failure by Bentham Science Publishers in enforcing your compliance with this License Agreement constitute a waiver of any of its rights.

3. You acknowledge that you have read this License Agreement, and agree to be bound by its terms and conditions. To the extent that any other terms and conditions presented on any website of Bentham Science Publishers conflict with, or are inconsistent with, the terms and conditions set out in this License Agreement, you acknowledge that the terms and conditions set out in this License Agreement shall prevail.

Bentham Science Publishers Pte. Ltd.
80 Robinson Road #02-00
Singapore 068898
Singapore
Email: subscriptions@benthamscience.net

BENTHAM SCIENCE

CONTENTS

PREFACE

Frontiers in Nanomedicine Vol. 4, Applications of Nanomaterials in Medical Procedures and Treatments continues describing relevant topics in nanomaterials for biomedical applications and regulations. Contents of chapters 1 to 7 are organized as follows, molecular imaging and contrast agents; tissue engineering; prosthetics and implants; ophthalmic and cancer therapy; and processing techniques. While last chapter 8 completes the book with a motif of international regulations and standards.

Multidisciplinary research is quickly proliferating in modern times. In this context, nanomedicine intertwines the new wave of high-performance devices with standard methods and pharmaceutics to find final options against deadly diseases.

Once again, I expect you can use this book as a guide to delve into the fundamental issues surrounding nanoscience applied to medicine.

Felipe López-Saucedo
Department of Radiation Chemistry and Radiochemistry
Institute of Nuclear Sciences
Universidad Nacional Autónoma de México
Mexico

List of Contributors

Ajay K. Potbhare — Department of Chemistry, Seth Kesarimal Porwal College of Arts and Scienceand Commerce, Kamptee, India

Ayça Aslan — Department of Bioengineering, Faculty of Chemical and Metallurgical Engineering, Yıldız Technical University, Istanbul, Turkey

Betül Mutlu — Department of Bioengineering, Faculty of Chemical and Metallurgical Engineering, Yıldız Technical University, Istanbul, Turkey

Dimitri Stanicki — General, Organic and Biomedical Chemistry Unit, NMR and Molecular Imaging Laboratory, University of Mons, Mons, Belgium

Erhan Akdoğan — Department of Mechatronics Engineering, Faculty of Mechanical Engineering, Yildiz Technical University, İstanbul, Turkey
Health Institutes of Turkey, İstanbul, Turkey

Felipe López-Saucedo — Institute of Nuclear Sciences, Department of Radiation Chemistry and Radiochemistry, National Autonomous University of Mexico, Mexico City, Mexico

Hatice Feyzan Ay — Department of Bioengineering, Faculty of Chemical and Metallurgical Engineering, Yıldız Technical University, Istanbul, Turkey

Hilal Calik — Department of Bioengineering, Faculty of Chemical and Metallurgical Engineering, Yıldız Technical University, Istanbul, Turkey

Lionel Larbanoix — Center for Microscopy and Molecular Imaging, Gosselies, Belgium

Manar Ezzelarab Ramadan — Department of Parasitology, Faculty of Medicine, Suez University, Suez, Egypt

Manjiri S. Nagmote — Department of Chemistry, Seth Kesarimal Porwal College of Arts and Scienceand Commerce, Kamptee, India

Nagham Gamal Masoud — Faculty of Medicine, Ain Shams University, Cairo, Egypt

Nagwa Mostafa El-Sayed — Department of Medical Parasitology, Research Institute of Ophthalmology, Giza, Egypt

Nakshatra B. Singh — Department of Chemistry and Biochemistry and RDC, Sharda University, Greater Noida, India

Rabia Çakır-Koç — Department of Bioengineering, Faculty of Chemical and Metallurgical Engineering, Yıldız Technical University, Istanbul, Turkey
Biotechnology Institute of Turkey, Health Institutes of Turkey, İstanbul, Turkey

Rabia Yilmaz-Ozturk — Department of Bioengineering, Faculty of Chemical and Metallurgical Engineering, Yıldız Technical University, Istanbul, Turkey

Ratiram G. Chaudhary — Department of Chemistry, Seth Kesarimal Porwal College of Arts and Scienceand Commerce, Kamptee, India

Robert N. Muller — General, Organic and Biomedical Chemistry Unit, NMR and Molecular Imaging Laboratory, University of Mons, Mons, Belgium
Center for Microscopy and Molecular Imaging, Gosselies, Belgium

Sébastien Boutry — Center for Microscopy and Molecular Imaging, Gosselies, Belgium

Selcen Arı-Yuka   Department of Bioengineering, Faculty of Chemical and Metallurgical Engineering, Yıldız Technical University, Istanbul, Turkey

Shirin B. Hanaei   College of Engineering and Physical Sciences, Aston University, Birmingham, United Kingdom

Sıtkı Kocaoğlu   Department of Biomedical Engineering, Faculty of Engineering and Natural Sciences, Ankara Yıldırım Beyazıt University, Ankara, Turkey

Smita S. Bhuyar-Kharkhale   Department of Chemistry, Lemdeo Patil College, Mandhal, India

Sophie Laurent   General, Organic and Biomedical Chemistry Unit, NMR and Molecular Imaging Laboratory, University of Mons, Mons, Belgium
Center for Microscopy and Molecular Imaging, Gosselies, Belgium

Sudhir S. Bhuyar   Department of Chemistry, Seth Kesarimal Porwal College of Arts and Scienceand Commerce, Kamptee, India

Umut Beylik   Department of Health Management, Faculty of Gulhane Health Sciences, University of Health Sciences, Turkey

Yvonne Reinwald   Department of Engineering, School of Science and Technology, Nottingham Trent University, Nottingham, United Kingdom

Zeynep Karavelioglu   Department of Bioengineering, Faculty of Chemical and Metallurgical Engineering, Yıldız Technical University, Istanbul, Turkey

CHAPTER 1

Molecular Imaging and Contrast Agents

Dimitri Stanicki[1], Lionel Larbanoix[2], Sébastien Boutry[2], Robert N. Muller[1,2] and Sophie Laurent[1,2,*]

[1] *General, Organic and Biomedical Chemistry Unit, NMR and Molecular Imaging Laboratory, University of Mons, Mons, Belgium*

[2] *Center for Microscopy and Molecular Imaging, Gosselies, Belgium*

Abstract: As an emerging technology, molecular imaging combines advanced imaging technology with cellular and molecular biology to highlight physiological or pathological processes in living organisms at the cellular level. The main advantage of *in vivo* molecular imaging is its ability to characterize pathologies of diseased tissues without invasive biopsies or surgical procedures. Such technology provides great hope for personalized medicine and drug development, as it can potentially detect diseases in early stages (screening), identify the extent of a disease/anomaly, help to apply directed therapy, or measure the molecular-specific effects of a given treatment. Molecular imaging requires the combination of high-resolution/sensitive instruments with targeted imaging agents that correlate the signal with a given molecular event. In ongoing preclinical studies, new molecular targets, which are characteristic of given diseases, have been identified, and as a consequence, sophisticated multifunctional probes are in perpetual development. In this context, the discovery of new emerging chemical technologies and nanotechnology has stimulated the discovery of innovative compounds, such as multimodal molecular imaging probes, which are multiplex systems that combine targeting moieties with molecules detectable by different imaging modalities.

Keywords: Contrast agents, Diagnostic, Drug delivery, Imaging, Magnetic resonance imaging, Molecular imaging, MRI, Nanoparticles, Nuclear medicine, Optical imaging, PET, Polymers, SPECT, Targeting, Theragnostic, Therapy, Ultrasounds.

INTRODUCTION

Molecular imaging is a technique that allows the characterization of biochemical processes at the cellular and molecular levels in living organisms. This method

---

[*] **Corresponding author Sophie Laurent:** General, Organic and Biomedical Chemistry Unit, NMR and Molecular Imaging Laboratory, University of Mons, Mons, Belgium; and Center for Microscopy and Molecular Imaging, Gosselies, Belgium; E-mail: sophie.laurent@umons.ac.be

Felipe López-Saucedo (Ed.)

helps to understand biological phenomena and reactions involved in different physiological and pathological processes at the nanoscopic scale. By allowing the visualization of a characteristic change at the molecular level, a rapid and accurate diagnosis (by assessing the presence or absence of metastases), a follow-up of therapy (by early assessment of response or resistance to a treatment), or detection of disease recurrence are possible. Oncology, neurology, and cardiology are the three principal areas of bioimaging applications [1 - 3]. One of the main advantages of this technique is its ability to characterize diseased tissue noninvasively, allowing for more personalized treatment planning.

The implementation of an active targeting strategy implies the development of a probe resulting from the combination of a specific vector (*i.e.,* a moiety able to specifically recognize the target of interest) and an imaging agent. The use of an imaging probe involves, among other things, the choice of the imaging modality by considering the strengths and limitations of each technique. Many imaging techniques are available for preclinical and clinical studies. Among the most used are X-ray imaging, ultrasound, magnetic resonance imaging, nuclear imaging, and optical imaging [4 - 6]. It is important to choose the appropriate imaging technique based on the desired application. Imaging modalities differ in the equipment used and instrumental properties: sensitivity, precision, spatial and temporal resolutions, tissue penetration, quantification, acquisition time, and cost (Table **1**). Another important parameter is the toxicity induced by certain techniques using ionizing radiation.

Table 1. Characteristics of noninvasive imaging modalities [7].

Imaging Technique	Positron Emission Tomography (PET)	Single Photon Emission Computed Tomography (SPECT)	Magnetic Resonance Imaging (MRI)	Optical Imaging	Ultrasound (US)
Detection	High energy γ rays	Lower energy γ rays	Radio waves (magnetic field)	Visible light and near-infrared	High-frequency sound
Spatial resolution	1-2 mm	1-2 mm	25 -100 μm	2-5 mm	50-500 μm
Depth	No limit	No limit	No limit	1-2 cm	mm to cm
Temporal resolution	10 s to min	min	10 s to min	s to min	s to min
Sensitivity	10^{-11}-10^{-12} M	10^{-10}-10^{-11} M	10^{-3}-10^{-5} M	10^{-9}-10^{-12} M	Not well characterized

(Table 1) cont.....

Imaging Technique	Positron Emission Tomography (PET)	Single Photon Emission Computed Tomography (SPECT)	Magnetic Resonance Imaging (MRI)	Optical Imaging	Ultrasound (US)
Types of probes used common CA	Radiolabeled ^{19}FDG, ^{3}H$_2$O, ^{68}Ga-DOTA	Radiolabeled ^{99m}Tc-HMPO, ^{111}In octreotide	Gd-complexes, iron oxide nanoparticles	Fluorophores, quantum dots, rhodamine	Micro-bubbles Sonovue®, Acusphere
Some examples of applications	Cerebral/blood flow, degenerative diseases, brain development	Cerebral infarction, ischemia, dementia, cardiac imaging	Angiography, cell labeling	Gene expression, cell tracking	Echography morphological studies, liver lesions

IMAGING TECHNOLOGIES

Nuclear Imaging

Nuclear medicine, which is based on the disintegration/detection of radioactive atoms injected in the patient, is a functional molecular imaging technique because it allows the visualization and localization of accumulated radiomolecules. In recent decades, the use of radionuclides in medicine has been considerably developed to become a clinical specialty integrating both imaging (diagnosis) and the treatment of pathologies (therapy). Radionuclides exhibit different properties (half-life time, type of emission, energy, scope of action, production, availability, cost, *etc.*). All these parameters influence the choice of radionuclide to be used according to the application envisaged (Table **2**).

Table 2. Main radionuclides used in imaging and therapy and their properties (half-life, mode of disintegration, application).

Radionuclides	Half-life	Disintegration mode	Applications
^{11}C	20.3 min	β^+	PET
^{18}F	109.8 min	β^+	PET
^{68}Ga	67.8 min	β^+	PET
^{67}Ga	78.3 h	γ	SPECT
^{124}I	99.6 h	β^+	PET
^{123}I	13.2 h	γ	SPECT
^{131}I	8 d	β^-	Therapy
^{64}Cu	12.7 h	β^+	PET/therapy
^{67}Cu	61.9 h	γ	SPECT

(Table 2) cont.....

Radionuclides	Half-life	Disintegration mode	Applications
^{99m}Tc	6 h	γ	SPECT
^{111}In	67.9 h	γ	SPECT
^{177}Lu	6.7 d	β⁻	SPECT/therapy
^{90}Y	64.8 h	β⁻	Therapy
^{225}Ac	9.9 d	α	Therapy
^{212}Pb	10.64 h	α	Therapy
^{89}Zr	78.42 h	β⁺	PET

Nonmetallic elements, such as carbon-11, fluorine-18 or iodine-131, can be covalently attached to the molecule of interest (Figs. **1** and **2**). The small (or absent) structural modification allows us to maintain their pharmacological properties. However, direct covalent radiolabeling requires the development of a new radiosynthesis protocol for each new molecule [8]. Moreover, radiolabeling conditions are often harsh (high temperature, acidic or basic pH), which makes them incompatible with sensitive systems. For metallic elements, such as gallium, copper, indium, technetium or lutecium, attachment to the carrier molecule generally occurs through complexation. This requires the introduction of a chelating agent, which, due to its large size, may modify the pharmacokinetics and biodistribution properties of the molecule of interest in a non-negligible way. As a counterpart, radiolabeling is generally carried out under milder conditions and is therefore compatible with sensitive moieties, such as biomolecules. Many chelating agents are available, and the choice of a chelating agent is based on the radionuclide. DOTA (1,4,7,10-tetraazacyclododecane-1,4,7,10-tetraacetic acid) is the most widely used chelate [9 - 12].

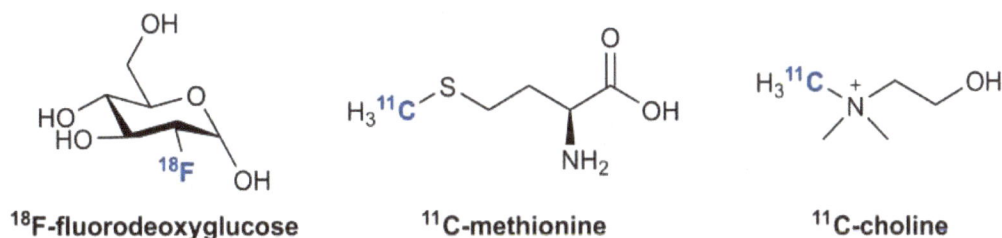

^{18}F-fluorodeoxyglucose   **^{11}C-methionine**   **^{11}C-choline**

Fig. (1). Examples of some radiomolecules: ^{18}F-fluorodeoxyglucose, ^{11}C-methionine, and ^{11}C-choline.

Two technologies based on different radioelements are currently used: positron emission tomography (PET) and single photon emission computed tomography (SPECT). They differ in the mode of disintegration of the injected radionuclide and the instrumentation used. These two types of imaging systems are based on the detection of γ photons from the decay of a radionuclide.

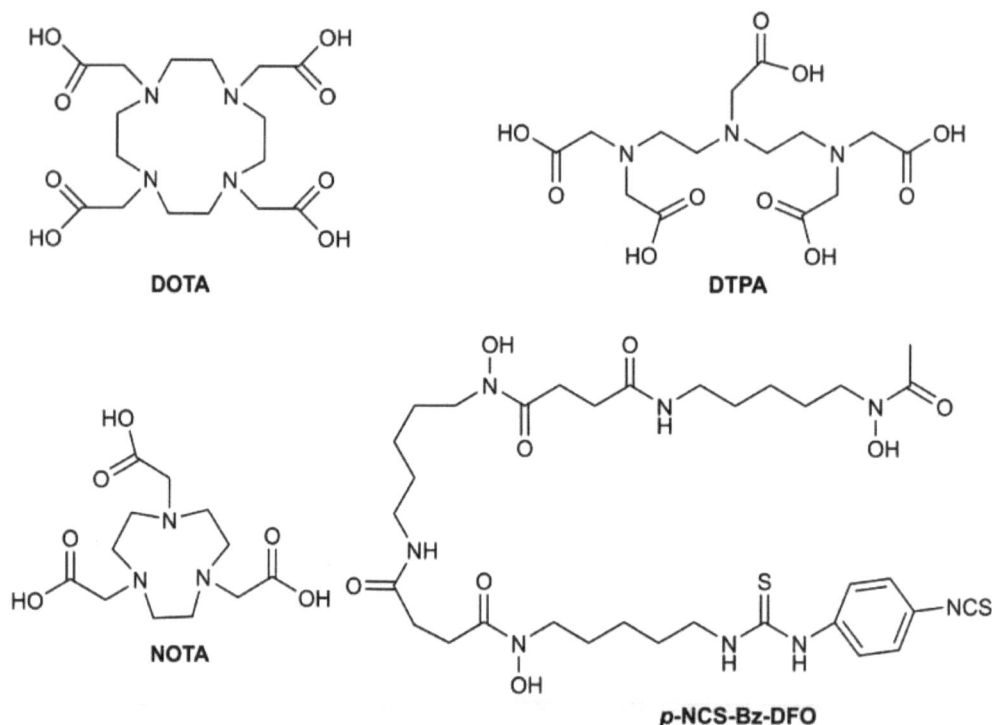

Fig. (2). Examples of chelating agents used for the complexation of metal radionuclides. N and O ligands: DTPA, DTPA NOTA, and (*p*-NCS-Bz-DFO).

SPECT Imaging

Injected radioisotopes emit γ photons detected by a network of detectors capable of 360° rotation around the patient. Collimators are implemented in the detectors to exclusively collect the γ rays arriving perpendicularly. By identifying the location of radionuclides, a three-dimensional image can be reconstructed. The collimator improves resolution, but due to the loss of many signals, a decrease in sensitivity is observed. Despite many advances, the spatial resolution of SPECT is still relatively low (<1 mm in preclinical and 8 to 12 mm in clinical applications). If the acquisition times are relatively long, which can be a source of discomfort for the patient, the sensitivity is high (10^{-11} M), and tissue penetration is unlimited. SPECT is particularly used in cardiology but also in oncology and neurology [13 - 15] (Fig. **3**). This imaging modality is, therefore, currently widely used in preclinical and clinical applications despite poorer resolution than PET.

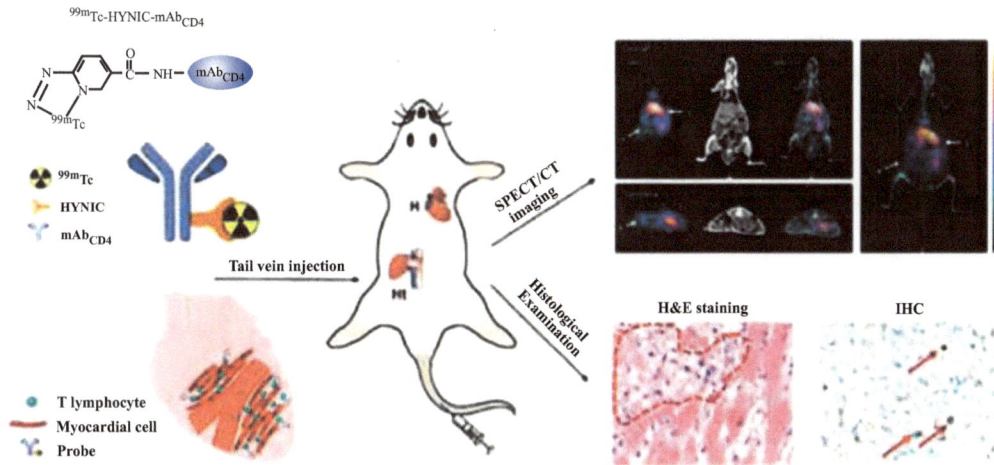

Fig. (3). The ^{99m}Tc-HYNIC-mAb$_{CD4}$ probe achieved high affinity and specificity of binding to CD4+ T lymphocytes and accumulation in the transplanted heart. (Reprinted with permission from [15]. Copyright (2021) American Chemical Society).

PET Imaging

PET requires the injection of a positron-emitting radionuclide. Each positron is annihilated with an electron and emits two γ photons that are detected simultaneously by detectors located on a ring (crown) around the patient. The detection of the two coincidences leads to a reduction of background noise and helps to identify the site of annihilation precisely. The acquisition of a large number of coincident events then makes it possible to reconstruct a 3D image. This technique allows better resolution (1 mm in preclinical and 3 to 4 mm in clinical applications) as well as better sensitivity (10^{-11} to 10^{-12} M). The concentrations of injected radionuclides are very low (pM to nM), and quantification of radioactive signals is possible. Because of these many advantages, PET is widely used preclinically and clinically. The precision of this technique makes it possible to follow processes at the molecular level, such as the interaction of a ligand with its receptor, which makes PET very useful for the diagnosis and monitoring of numerous pathologies. This imaging tool is widely used in oncology for diagnosis, tumor characterization, staging, and therapeutic monitoring of patients [16 - 19].

PET is also used for the detection of inflammation and infection or to study the metabolism and viability of the myocardium as well as to monitor myocardial perfusion or atherosclerosis in cardiology [20, 21]. Neurology also uses this technique extensively for the diagnosis and monitoring of different pathologies, such as Alzheimer's disease, Parkinson's disease, or cerebral vascular pathologies

[22, 23]. The technique also facilitates the development and evaluation of new drugs [24]. Two limitations are observed: (i) exposure to ionizing radiation, and (ii) short half-lives of radionuclides, which implies that synthesis of the radiopharmaceutical, injection into the patient, and imaging examination should be carried out in a limited time.

Optical Imaging

Optical imaging also provides functional information and is based on the detection of an optical signal obtained by luminescence, such as fluorescence, bioluminescence or chemiluminescence. Chemiluminescence and biolumi-nescence describe the emission of light after a chemical or biochemical reaction, respectively. Fluorescence requires the injection of an exogenous fluorescent molecule. This imaging technique allows us to obtain images with very good sensitivity (10^{-9} at 10^{-16} M) and good spatial resolution (2-3 mm). Acquisition times are short enough to follow real-time processes. The major limitation is the low penetrability into tissues (a few mm to 1 cm). Therefore, optical imaging is mainly used in preclinical studies but still very seldom in the clinic except for endoscopy [25], surgery guided by fluorescence [26], analysis of biopsy samples, or examination of surface tissues, such as skin or the eye. In addition, some molecules in the human body can fluoresce, which can lead to high background noise in imaging (autofluorescence). To improve the quality of the images, the use of fluorophores with excitation and emission maxima in a wavelength range where the endogenous compounds absorb very little light is needed. These wavelengths are in the therapeutic window, which is situated in the near-infrared range (NIR). Depending on the wavelength range, two windows can be distinguished: NIR-I between 650 and 950 nm and NIR-II-III (also called SWIR) between 950 and 1700 nm. In these 2 windows, the absorbance of photons, scattering of photons, and autofluorescence by tissues are significantly reduced. Consequently, the depth of penetration, sensitivity, and spatial resolution (especially in the SWIR window) are far better. Many fluorescent probes are currently available. These molecules are characterized by several photophysical properties:

i. The excitation (λ_{exc}) and emission (λ_{em}) wavelengths; a good fluorescent probe should have excitation and emission wavelengths in the therapeutic window (between 650 and 1700 nm) to obtain the best possible signal.
ii. The Stokes displacement (the difference between the maxima of excitation and emission peaks in the spectrum); the Stokes displacement must be as high as possible to reduce interference.
iii. The molar extinction coefficient (the capacity to absorb light).

iv. The fluorescence quantum yield j (the emission efficiency of the fluorophore); the molar extinction coefficient and the quantum efficiency must be as high as possible to obtain good sensitivity.

v. Brightness (the intensity of fluorescence).

vi. Photobleaching (the loss of molecular fluorescence).

In addition to these parameters, the probe must be stable under physiological conditions; it must be water-soluble to avoid aggregation, and finally, it should have a grafting function to affix to the targeting moiety.

There are two types of fluorophores: inorganic fluorophores (lanthanides, semiconductor nanoparticles (quantum dots), transition metal-ligand complexes, *etc.*) and organic fluorophores [27, 28]. Lanthanides can be used as chelates emitting in the near infrared-II (NIR-II), which makes it possible to obtain images of good quality with a high signal-to-noise ratio [29, 30]. Most organic fluorophores emit near the infrared-I (NIR-I) region. Many families exist, such as those having a xanthene core (fluoresceins, rhodamines, and eosins), BODIPYs, cyanines, porphyrins, *etc.* [31 - 34] (Fig. **4**). Only indocyanine green (ICG), methylene blue, fluorescein, and 5-ALA and its derivatives (Metvixia and Hexvix) are approved by the FDA (Food and Drug Administration) [35 - 38]. Indocyanine green is the fluorophore most widely used in the medical field (*e.g.,* in ophthalmic angiography, as a marker in the evaluation of perfusion of tissues and organs, or in assistance for the biopsy of sentinel lymph nodes with tumors).

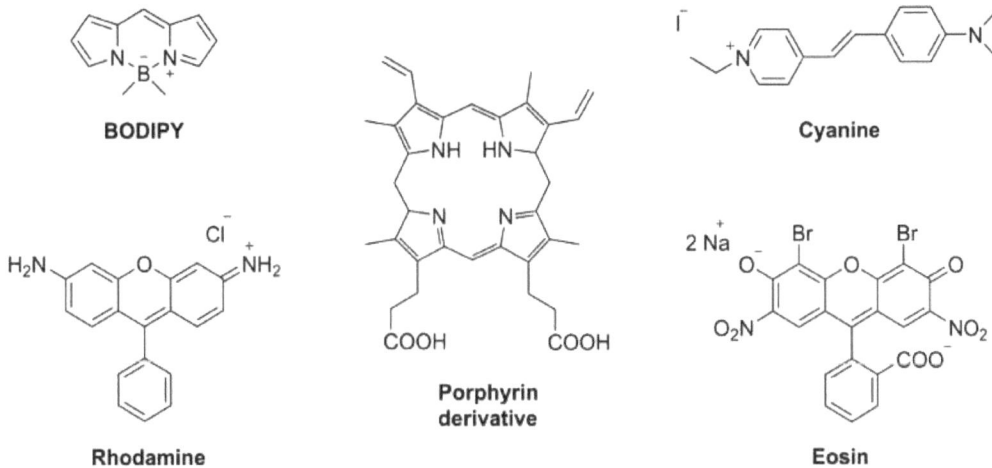

Fig. (4). Examples of fluorophores used for optical imaging. BODIPY, a porphyrin derivative, cyanine, rhodamine, and eosin are typical families of molecules with this characteristic.

Fluorescent probes can be nonvectorized. For example, indocyanine green (ICG) has been widely used in cardiology, ophthalmology, neurosurgery, and cancer [39 - 41] (Figs. **5** and **6**).

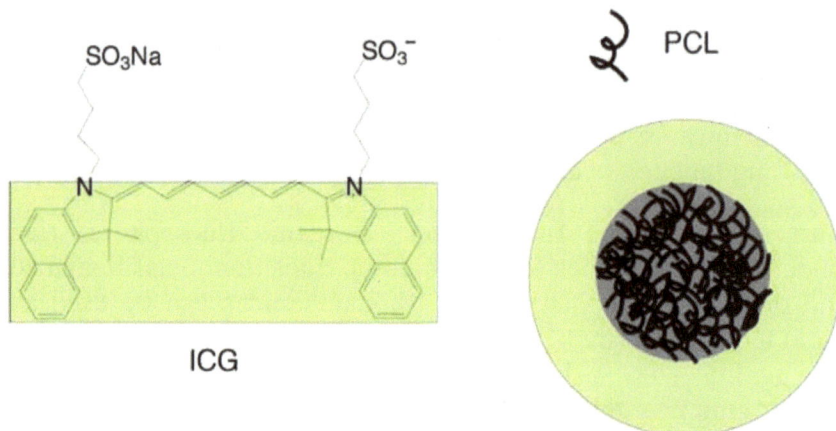

Fig. (5). Scheme of indocyanine green (ICG)-coated polycaprolactone (PCL) micelles. Micelles were formed through the coassembly of PCL and ICG. (Reprinted with permission from [41]. Copyright (2020) American Chemical Society).

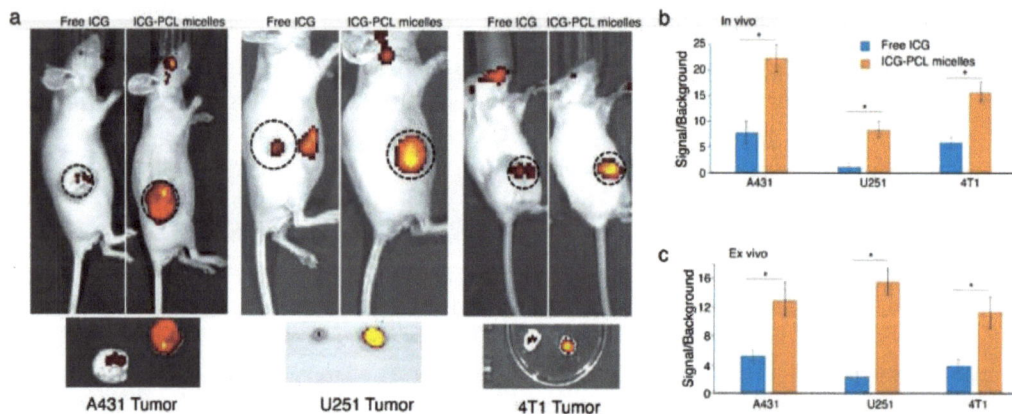

Fig. (6). (**a**) Fluorescent images of mice with A431, U251, and 4T1 tumors 24 h after injection of ICG or ICG-PCL micelles at an ICG concentration of 5 mg kg^{-1} body weight. Tumor location is indicated by a dotted black circle (top row). Some fluorescence can also be seen in the liver and at the injection site (eye). Bottom row: Fluorescent images of excised tumors 24 h post administration of free ICG or the ICG-PCL micelles (bottom row). (**b**) Semiquantitative analysis of tumor fluorescence from *in vivo* studies. (**c**) Semiquantitative analysis of tumor fluorescence from excised tumors. (Reprinted with permission from [41]. Copyright (2020) American Chemical Society).

Magnetic Resonance Imaging

In comparison to nuclear or optical imaging modalities, MRI has many advantages: (i) high spatial resolution in the micrometer range (on research devices), (ii) no emission of ionizing radiation, and (iii) no limitation for tissue penetration. As a counterpart, MRI exhibits a low sensitivity (mM range), is expensive, and is slower than the other techniques described above. MRI experiments [42] are essentially based on the study of hydrogen atom nuclei present in water molecules, given that hydrogen has very favorable NMR properties and that water represents the most abundant molecule in living organisms (>65%). Nuclear magnetic resonance consists of the study of an atomic nucleus present in a given sample, which is subjected to a fixed magnetic field (B_0, applied along an Oz axis) and an oscillating electromagnetic field (electromagnetic wave, radio frequency, B_1). In the absence of a magnetic field, the magnetic moments of hydrogen atoms are randomly oriented. When placed in a magnetic field, B_0, these magnetic moments align themselves according to the direction of the field, which is the steady state. These magnetic moments are in fact precessing around the field B_0 according to a frequency depending on the nature of the nucleus and described by the Larmor equation, Eq. **1**, (with v_0 corresponding to the precession frequency of the nucleus, which is proportional to the field B_0, and the gyromagnetic ratio γ, which is specific to each nucleus).

$$v_0 = \gamma\, 2\pi\, B_0 \tag{1}$$

When the equilibrium state is perturbed by the application of the radio frequency wave (B_1, applied in the xOy plane along Ox), the magnetic moments resonate with this field and precess about B_1. When this RF wave is turned off, the magnetic moments return to their equilibrium state, which constitutes a relaxation phenomenon. During relaxation, the system emits an electromagnetic wave called a signal or FID (Free Induction Decay), to which a Fourier transform is applied to derive the spectrum.

The relaxation of magnetic moments can be longitudinal (T_1 relaxation) or transverse (T_2 relaxation): the longitudinal relaxation time T_1 is characteristic of the time for the magnetization (or magnetic moments) to return to equilibrium along the z-axis (M_z). It is also called spin-lattice relaxation because, during the return of protons from the high energy level to a lower energy level, there is the emission of the energy absorbed during the excitation by interaction with the surrounding medium (lattice) (Eq. **2**).

$$M_z = M_0(1 - e^{-t/T_1}) \tag{2}$$

T_2 varies with the molecular structure of the sample under study and is higher in solutions than in solids. The transverse relaxation time T_2 is characteristic of the time for the return to equilibrium of the magnetization in the xy plane (Mxy), as presented in Eq. **3**. Transverse relaxation is also called spin-spin relaxation because it is the consequence of proton spins interacting with each other and does not involve energy transfer.

$$M_{xy} = M_0 e^{-t/T_2} \tag{3}$$

In media, such as biological tissues, the evolution of T_1 and T_2 mainly depends on "how fast" spins can move (motion frequency represented by correlation time τ_c or "tumbling rate"). This motion frequency can be related to the sizes of spin-containing molecules (or the presence of associated water spins) and to the freedom with which water molecules move within tissues (*i.e.,* viscosity), including a solid, a normally structured (soft) tissue, or a fluid-containing compartment (*e.g.,* brain ventricles). T_1 has an optimum (or peak) value at a certain frequency of molecular motion naturally occurring around the proton spin Larmor frequency (fat represents this optimum) and allowing for efficient energy exchanges (short T_1 relaxation). T_2 is more closely related to how local B_0 field inhomogeneities induced by spins are rather fixed due to slow motion (short T_2) or averaged due to fast motions (long T_2) (Fig. **7**).

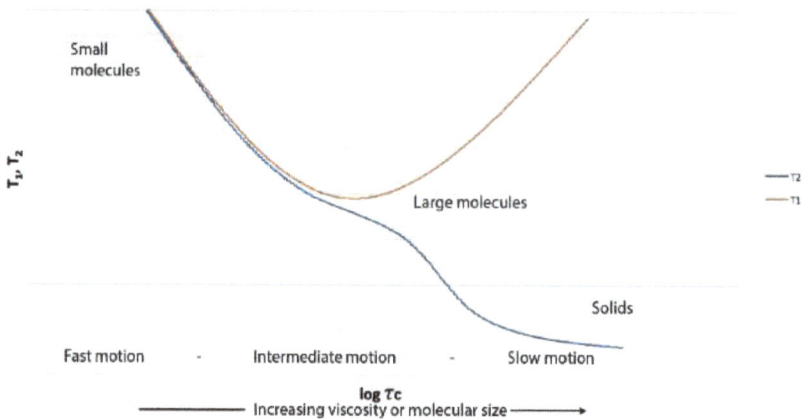

Fig. (7). Dependence of the relaxation time (T_1, T_2) on the molecular motion correlation time (τ_c).

MRI experiments require spatial encoding of the sample under study. Localization of the signal in space is possible using field gradients. The gradient corresponds to a linear variation of the magnetic field as a function of the position in space. The Larmor relation is then modified according to Eq. **4**, where Gx, Gy, and Gz correspond to the field gradients in each of the directions in space (x, y, and z).

$$v = \gamma \, 2\pi \, (B_0 + xG_x + yG_y + zG_z) \tag{4}$$

The contrast in MRI is expressed as different gray levels in the image. In the human body, differences in contrast appear according to the different biological tissues. The contrast will vary mainly according to the relaxation times (T_1, T_2) and the spin density of the different tissues; the higher the spin density is, the more intense the signal. Some sequence parameters will allow the contrast in the image to vary according to the intrinsic properties of the sample (T_1, T_2, T_2^*).

All spin echo sequences are built according to the same scheme: application of a first RF pulse at 90°, followed by a second RF pulse at 180°, which avoids spin dephasing due to B_0 field inhomogeneities (Fig. **8**).

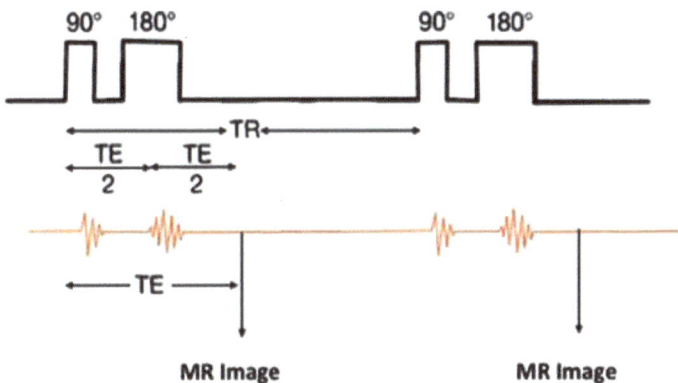

Fig. (8). Schematic illustration of a spin-echo pulse sequence.

T_1- and T_2-weighted sequences differ mainly in the parameters of repetition time (TR) and echo time (TE) after 90° RF excitation. TR corresponds to the time between two successive repetitions of the 90° RF pulse, TE corresponds to the time interval between the 90° RF pulse and a spin echo, and it determines the moment when the signal is measured. T_1-weighted sequences have short TE and TR. In this way, total magnetizations of tissues with short T_1 longitudinal relaxation times will have time to return to equilibrium (bright signal on the MR image), while those with longer T_1 will not have time to return to their equilibrium position (dark signal on the MR image). Thus, media with long T_1 (aqueous media) will appear dark on the image, while media with shorter T_1 (fat) will appear brighter. T_2-weighted sequences require long TE and TR. A long TR minimizes the influence of longitudinal T_1 relaxation, while a long TE favors the influence of transverse T_2 relaxation on the image contrast. As a result, media with a long T_2 appear bright on the image (aqueous media), while media with a shorter T_2 appear darker (fat) (Fig. **9**). The contrast in the image strongly depends on the relaxation time of the protons of the water molecules in the tissue or sample observed. In some cases, the relaxation difference involved in pathology is limited. MRI does not then allow differentiation of the signals for the different tissues and thus the establishment of a specific diagnosis. To increase the contrast in MRI, superparamagnetic substances (such as iron oxides) or exogenous paramagnetic metal complexes called "contrast agents" have been introduced.

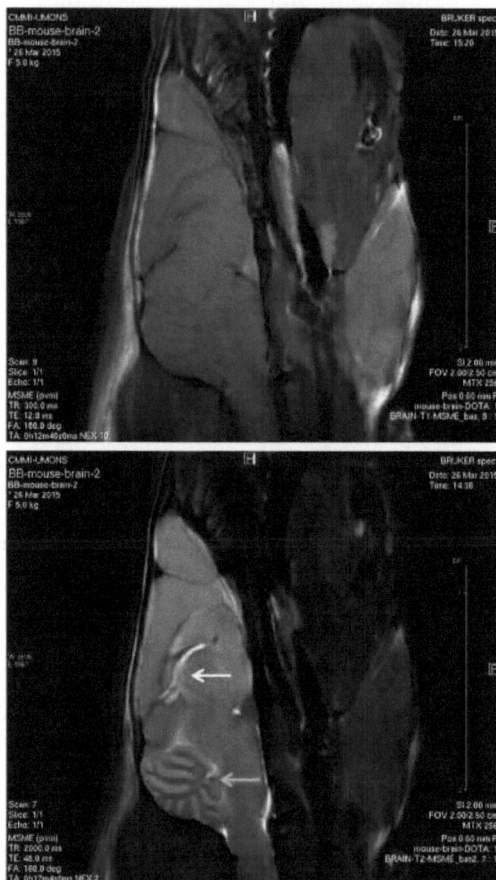

Fig. (9). T_1- (top) and T_2 (bottom)-weighted images acquired from a mouse head (main contrasts can be seen in the T2-weighted image (bottom), especially in the brain (cerebellum (dark and light gray, blue arrow) and ventricles (bright, white arrow)).

Gadolinium complexes (Fig. **10a**) decrease T_1 and thus increase the proton relaxation rate, which leads to brighter areas in images. These are called positive contrast agents [43 - 45]. Another contrast agent used, superparamagnetic iron oxide nanoparticles (Fig. **10b**), decreases T_2 and leads to darker areas in the images [46, 47]. These are called negative contrast agents.

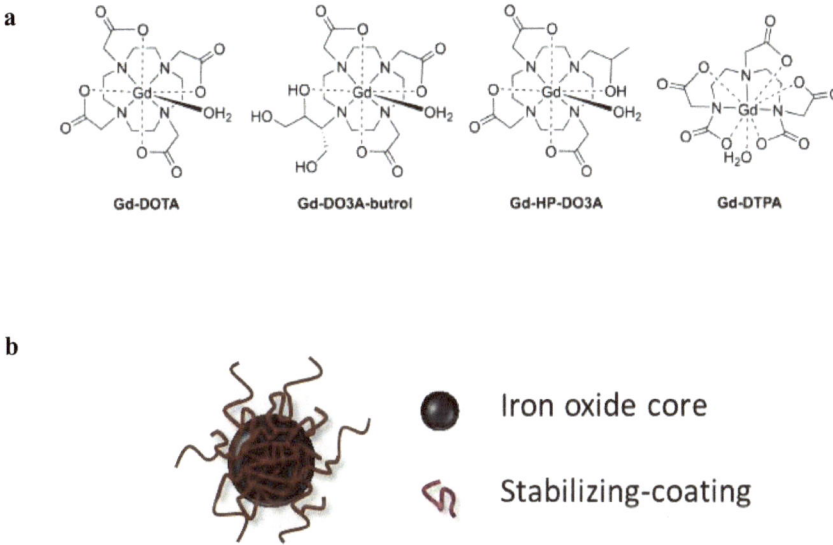

Fig. (10). Examples of Gd complexes (**a**) and iron oxide nanoparticles (**b**) used as contrast agents.

The injection of non-specific contrast agents only provides structural information. Many teams are interested in the use of targeted contrast agents that provide information on biochemical events. For example, iron oxide nanoparticles grafted with biovectors, such as peptides, allow specific targeting of cancerous cells (Fig. 11) [48].

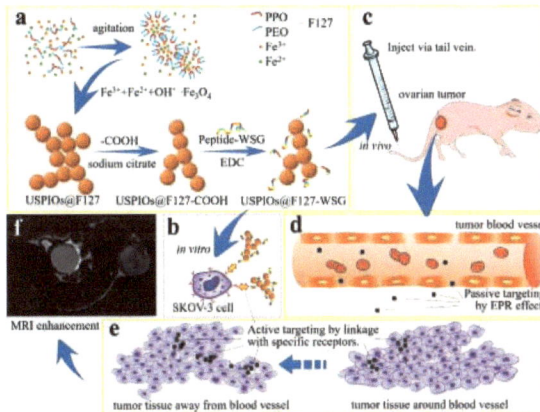

Fig. (11).   Schematic illustration of the rational design of USPIOs@F127-WSG for MRI contrast enhancement (**a-c**), with passive targeting through the vascular interspace by the EPR effect during blood circulation (**d**), by entering the tumor and actively binding to SKOV-3 cells (**e**). Finally, MRI contrast enhancement between the tumor and surrounding tissues was obtained after the injection of USPIOs@F127-WSG (**f**). (Reprinted with permission from [48]. Copyright (2019) American Chemical Society).

Ultrasound

Ultrasound is an imaging technique based on the reflection of ultrasound. A transducer converts an electrical signal into sound waves that penetrate the body and are reflected by various biological tissues. These signals are analyzed and processed to build two-dimensional morphological images. The distance, intensity, and direction of the sound waves are calculated and allow for image transcription [49]. To improve the signal/noise ratio, it is possible to inject air microbubbles on the order of one micrometer covered with a layer of lipids or polymers [50 - 53] (Fig. **12**). This process is efficient but does not provide molecular information. By grafting a targeting vector, it is possible to observe processes, such as angiogenesis or inflammation [53, 54] (Figs. **12 - 15**).

The medium resolution can be improved by increasing the frequencies of the sound waves. However, this results in shorter wavelengths and, thus, limited tissue penetration. The main advantages of ultrasound imaging are the availability, low cost, and portability of the ultrasound device. In addition, the injection of air microbubbles provides good sensitivity (10^{-12} M). However, this tool mainly allows to obtain two-dimensional morphological information by detecting only soft tissues. This technique is widely used in obstetrics and, when coupled with Doppler, in vascular imaging. Its infrequent use in oncology is due to its resolution and the difficulty of interpreting images.

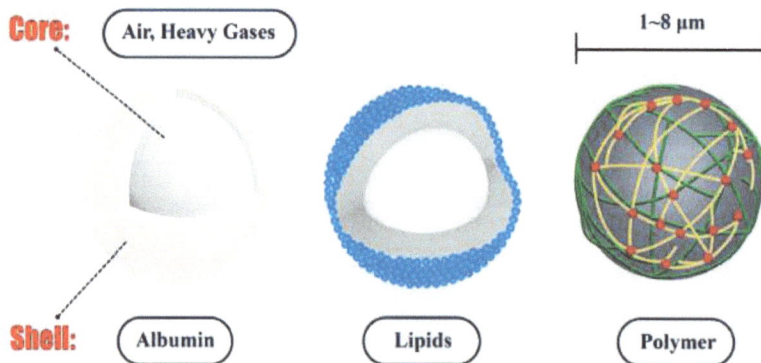

Fig. (12). Schematic illustration of conventional ultrasound MB contrast agents. MBs are typically 1−8 μm in diameter and composed of an inner gas core stabilized by various shell materials, such as proteins, lipids, or polymers, to prolong the lifetime of the gas in the bloodstream. The gas core usually consists of single air or bioinert heavy gases, such as perfluorocarbons (PFCs) or sulfur hexafluoride. (Reprinted with permission from [53]. Copyright (2018) American Chemical Society).

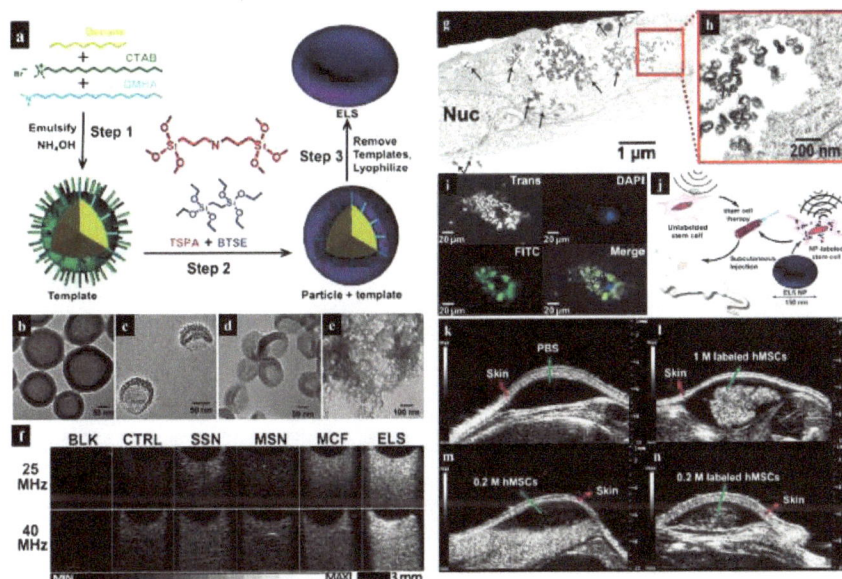

Fig. (13). Preparation of exosome-like silica nanoparticles and ultrasound images and quantification of cell echogenicity *in vivo*. (**a**) Schematic of ELS nanoparticle fabrication and morphology. TSPA (red) changed the overall stiffness of the silica shells and rendered them more elastic to allow the formation of ELS nanoparticles. (b–e) TEM images of silica products made with (**b**) 0%, (**c**) 20%, (**d**) 40%, and (**e**) 100% TSPA (red). (**b**) Hollow spheres were obtained when no TSPA was added; (**e**) a silica gel was formed with only TSPA. (**f**) Ultrasound intensity analysis of ELS with other silica nanoparticles. SSNs, MSNs, and MCFs also increased the echogenicity of hMSCs but not as strongly as the ELS nanoparticles. (**g**) TEM images of ELS-labeled hMSCs indicated aggregation of ELS inside the cells. ELS was located both inside and on the cells. Arrows indicate ELS nanoparticles, and Nuc indicates the nucleus. (**h**) This higher-magnification TEM image indicated that the ELS retained the unique curvature after entering the hMSCs. (**i**) Epifluorescence microscopy with hMSC nuclei in blue and ELS nanoparticles fluorescently tagged in green. (**j**) ELS-labeled hMSCs were subcutaneously injected with a Matrigel carrier into nude mice. The majority of the ELS was specifically bound to hMSCs. *In vivo* ultrasound images of (**k**) PBS, (**l**) 1 million ELS-labeled hMSCs, (**m**) 0.2 million unlabeled hMSCs, and (**n**) 0.2 million ELS-labeled hMSCs. (Reprinted with permission from [53]. Copyright (2018) American Chemical Society).

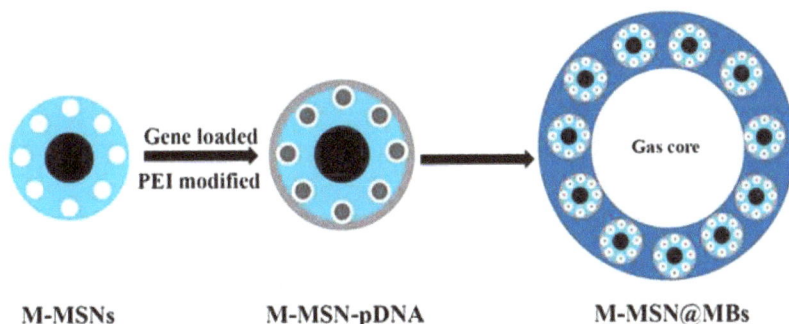

Fig. (14). Schematic representation of tumor-targeting gene delivery by M-MSN@MBs combined with ultrasound and magnetic attraction. (Reprinted with permission from [54]. Copyright (2020) American Chemical Society).

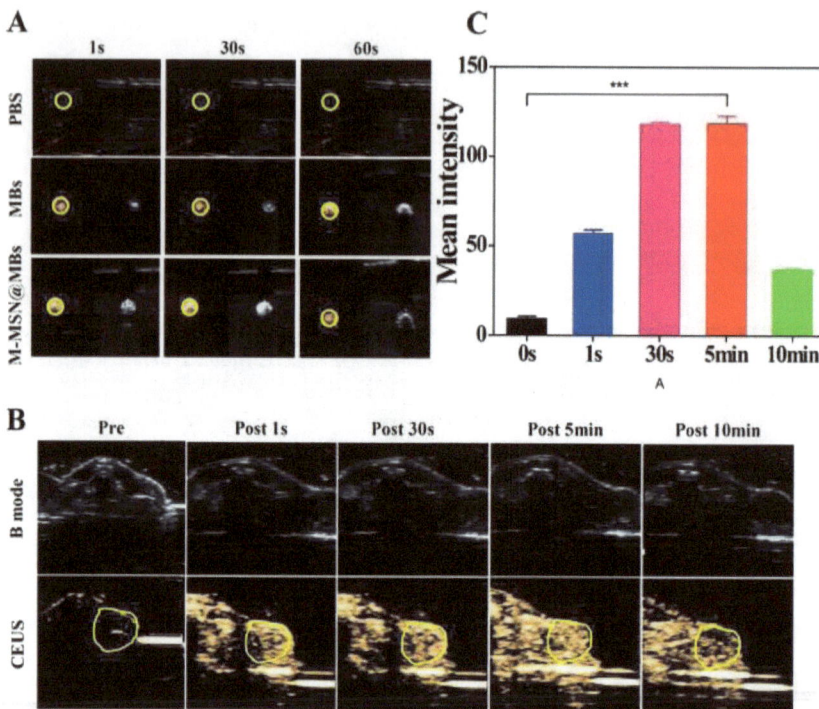

Fig. (15). Ultrasound imaging performance of M-MSN@MBs. (**A**) Ultrasound imaging of PBS, MBs, and M-MSN@MBs in phantom. (**B**) Ultrasound imaging of MBs or M-MSN@MBs at different times after tail vein injection in tumor-bearing mice (n = 3). (**C**) Mean intensity of *in vivo* ultrasound imaging at different times after injection. ***P < 0.001. (Reprinted with permission from [54]. Copyright (2020) American Chemical Society).

MULTIMODAL IMAGING

Each imaging technique has its own advantages and limitations (Table **3**). To benefit from the strengths and overcome the limitations of one imaging technique, it is possible to combine several modalities. This is called multimodal imaging [55, 56]. Multimodality, such as PET/MRI, PET/SPECT, or PET/CT, can be used to obtain anatomical and molecular information while providing enough information for clinical diagnosis.

Table 3. Advantages and limitations of different imaging modalities [7].

Imaging Technique	Disadvantages	Advantages
PET-SPECT	Low spatial resolution, radiation risks, high cost (for PET), cyclotron or generator is needed.	High sensitivity, quantitative, no penetration limit.

(Table 3) cont.....

Imaging Technique	Disadvantages	Advantages
MRI	Low sensitivity, high cost, time-consuming scan and processing.	Morphological and functional imaging, no penetration limit, high spatial resolution.
Optical imaging	Photobleaching, limited penetration, low spatial resolution, autofluorescence disturbing.	Low cost, easy manipulation, high sensitivity, and detection of fluorochrome in live and dead cells.
Ultrasounds	Limited resolution and sensitivity, and low data reproducibility.	Safety, low cost, wide availability, and real-time.

These combinations require the use of at least two kinds of imaging probes. There are two possible approaches: two molecules, each containing a modality, can be coinjected, or both modalities can be introduced directly with the same molecule/particle injected. This last strategy may appear better because the bimodal system has the same pharmacokinetic behavior for both modalities. This allows a perfect correlation between the two imaging methods. In addition, the use of multimodal probes allows more information to be obtained in less time and with just one injection. However, not all techniques have the same sensitivity, which can be problematic when developing a bimodal compound. As an example, optical and nuclear imaging are much more sensitive than CT and MRI. In this sense, the use of nanosystems, such as metal oxide nanoparticles [57], liposomes or micelles [58], dendrimers [59], or quantum dots [60], appears to be a promising approach to modulate the amount of detectable probe owing to the sensitivity of the related modality. Several studies have also reported the combination of optical imaging and MRI. Again, the difference in sensitivity obtained despite the addition of contrast agents must be compensated for by the introduction of a higher number of MRI contrast agents compared to the fluorophore [61, 62]. Another problem that can be encountered is the incompatibility of the physical properties of fluorophores and MRI contrast agents (for example, iron oxide is a quencher of fluorescence). However, it has been shown that fluorescence quenching depends on the size of the nanoparticles. By increasing the size of the nanoparticles, fluorescence can be observed. It is, therefore, possible to combine iron oxide nanoparticles and fluorescent probes [63] and to use other types of nanoparticles (quantum dots, for example) [64]. For example, fluorescence and MR images of HepG2 tumor-bearing mice after injection of enzyme-responsive polymeric nanoparticles are given in (Figs. **16 - 18**) [65].

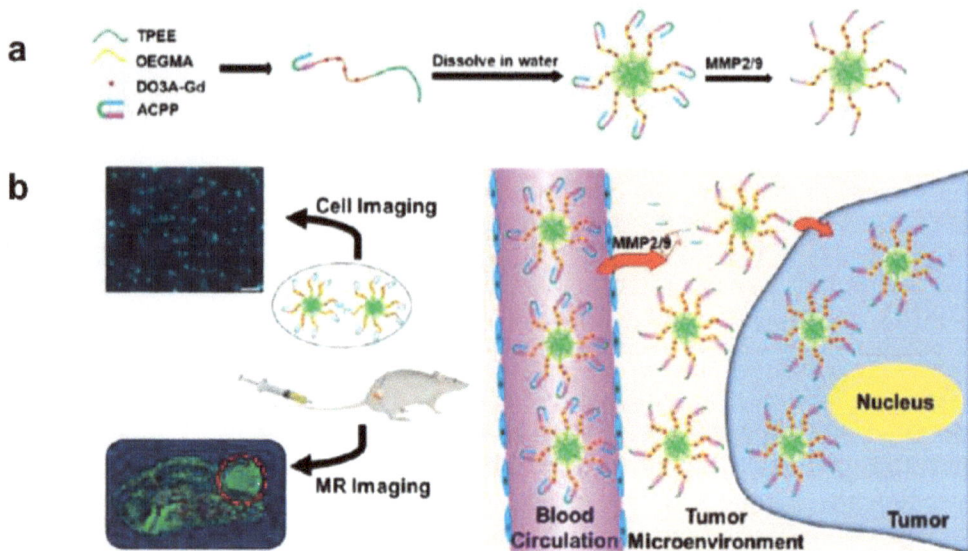

Fig. (16). Preparation of enzyme-responsive polymeric nanoparticles integrating MRI and fluorescence imaging (**a**) and their *in vivo* MRI applications (**b**). (Reprinted with permission from [65]. Copyright (2020) American Chemical Society).

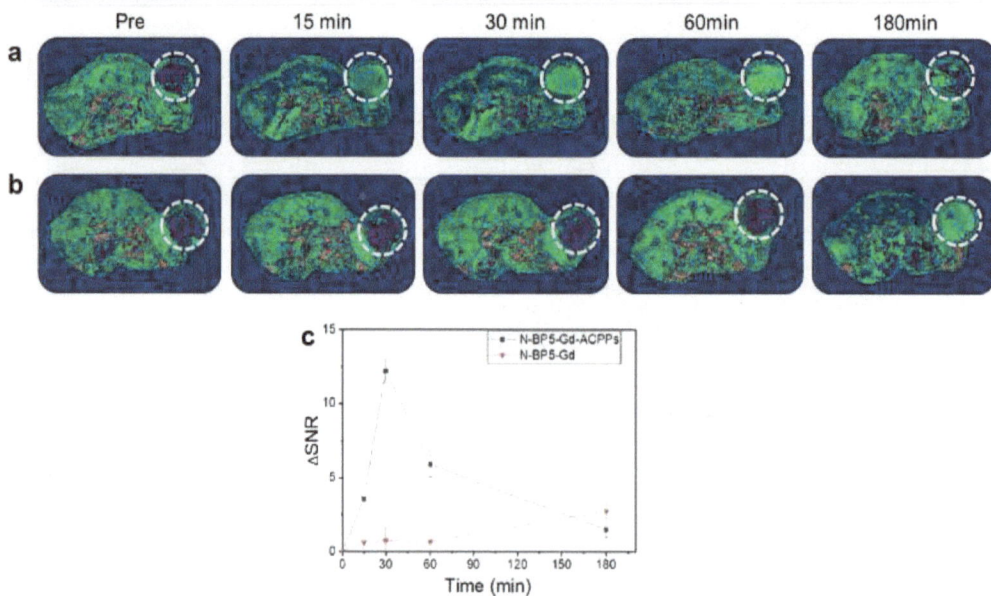

Fig. (17). MR images of HepG2 tumor-bearing mice after injection of (**a**) N-BP5-Gd-ACPPs and (**b**) N-BP--Gd at various times. (**c**) ΔSNR in tumors produced by N-BP5-Gd-ACPPs and N-BP5-Gd. (Reprinted with permission from [65]. Copyright (2020) American Chemical Society).

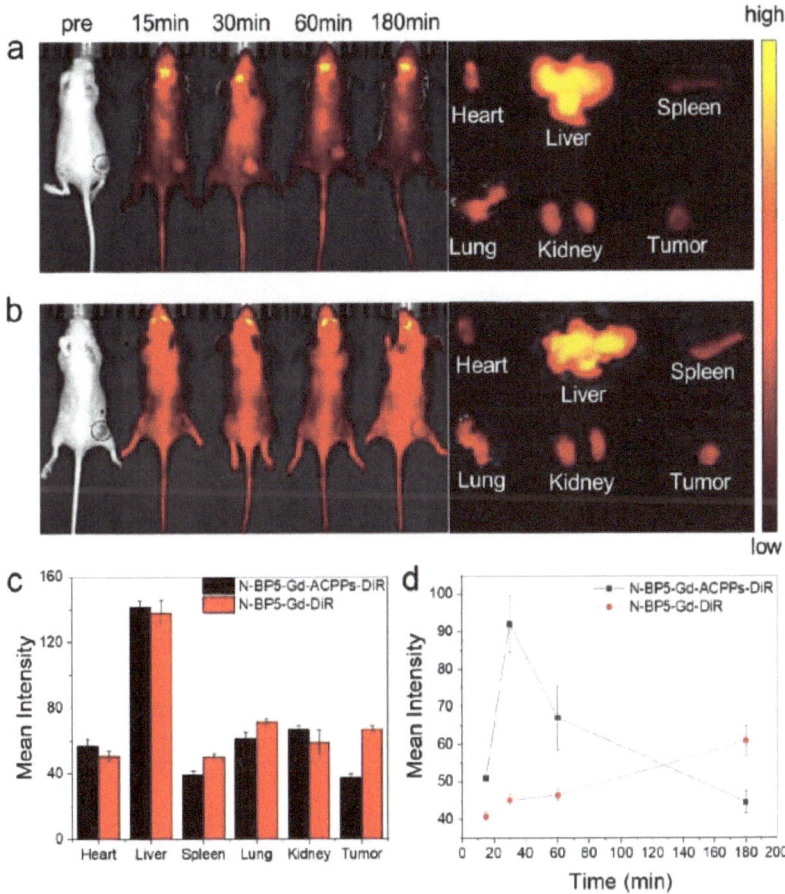

Fig. (18). Fluorescence images of HepG2 tumor-bearing mice after injection of (**a**) N-BP5-Gd-ACPPs loaded with DiR and (**b**) N-BP5-Gd loaded with DiR at various times, and fluorescence images of tumors and organs harvested at 180 min; (**c**) Mean intensity values (DiR fluorescence) for tumors and major organs harvested at 3h post-injection; (**d**) Mean intensity for DiR fluorescence from N-BP5-Gd-ACPPs and N-BP- -Gd loaded with DiR after injection at various times (n = 3). (Reprinted with permission from [65]. Copyright (2020) American Chemical Society).

CONCLUSION

Molecular imaging appears to be promising for modern medical imaging. It can help with the early detection/screening of pathologies as well as with treatment follow-up. It is hoped that new strategies involving early diagnosis and immediate treatment monitoring will improve success rates for curing diseases with high mortality rates, such as cardiovascular diseases or cancers. In recent years, this technology has witnessed certain progress (*e.g.,* in the identification of many molecular targets and the development of novel molecular imaging contrast agents), but some key problems related to molecular imaging probes and imaging equipment have not yet been solved.

Although several imaging instruments are available, they have their limitations. For example, MRI is a useful clinical diagnostic tool, but it suffers from poor sensitivity. Therefore, attention should be paid to improving existing instruments and developing systems combining different modalities. Indeed, if multimodal imaging presents many advantages, problems, such as the difficulty of designing a PET/MRI system suited for the entire body, cost increases, and the development of an "all-in-one" multimodal contrast agent, still exist and need to be solved. With the development of technologies and the emergence of innovative chemical processes, it is hoped that significant advancements will be achieved in the field and lead to targeted multifunctional molecular imaging probes for clinical applications.

CONSENT FOR PUBLICATION

Not applicable.

CONFLICT OF INTEREST

The author declares no conflict of interest, financial or otherwise.

ACKNOWLEDGEMENTS

This work was performed with the financial support of the FNRS, FEDER, ARC, Walloon Region, COST, and CMMI Fund Wallonia.

REFERENCES

[1]   Vaz SC, Oliveira F, Herrmann K, Veit-Haibach P. Nuclear medicine and molecular imaging advances in the 21st century. Br J Radiol 2020; 93(1110): 20200095.
[http://dx.doi.org/10.1259/bjr.20200095] [PMID: 32401541]

[2]   Engel J Jr, Pitkänen A. Biomarkers for epileptogenesis and its treatment. Neuropharmacology 2020; 167: 107735.
[http://dx.doi.org/10.1016/j.neuropharm.2019.107735] [PMID: 31377200]

[3]   Farber G, Boczar KE, Wiefels CC, *et al.* The future of cardiac molecular imaging. Semin Nucl Med 2020; 50(4): 367-85.
[http://dx.doi.org/10.1053/j.semnuclmed.2020.02.005] [PMID: 32540033]

[4]   Bogner W, Otazo R, Henning A. Accelerated MR spectroscopic imaging—a review of current and emerging techniques. NMR Biomed 2021; 34(5): e4314.
[http://dx.doi.org/10.1002/nbm.4314] [PMID: 32399974]

[5]   Alberti P. A review of novel biomarkers and imaging techniques for assessing the severity of chemotherapy-induced peripheral neuropathy. Expert Opin Drug Metab Toxicol 2020; 16(12): 1147-58.
[http://dx.doi.org/10.1080/17425255.2021.1842873] [PMID: 33103947]

[6]   Haj-Mirzaian A, Kadivar A, Kamel IR, Zaheer A. Updates on imaging of liver tumors. Curr Oncol Rep 2020; 22(5): 46.
[http://dx.doi.org/10.1007/s11912-020-00907-w] [PMID: 32296952]

[7]   Mahmoudi M, Serpooshan V, Laurent S. Engineered superparamagnetic iron oxide nanoparticles for biomolecular imaging. Nanoscale 2011; 3(8): 3007-26.
[http://dx.doi.org/10.1039/c1nr10326a] [PMID: 21717012]

[8]   Yang X, Wang F, Zhu H, Yang Z, Chu T. Synthesis and bioevaluation of novel [^{18}F]FDG-conjugated 2-nitroimidazole derivatives for tumor hypoxia imaging. Mol Pharm 2019; 16(5): 2118-28.
[http://dx.doi.org/10.1021/acs.molpharmaceut.9b00075] [PMID: 30964298]

[9]   Evangelista L, Ravelli I, Bignotto A, Cecchin D, Zucchetta P. Ga-68 DOTA-peptides and F-18 FDG PET/CT in patients with neuroendocrine tumor: A review. Clin Imaging 2020; 67: 113-6.
[http://dx.doi.org/10.1016/j.clinimag.2020.05.035] [PMID: 32559681]

[10]   McElroy KM, Binkovitz LA, Trout AT, *et al.* Pediatric applications of Dotatate: early diagnostic and therapeutic experience. Pediatr Radiol 2020; 50(7): 882-97.
[http://dx.doi.org/10.1007/s00247-020-04688-z] [PMID: 32495176]

[11]   Ambrosini V, Kunikowska J, Baudin E, *et al.* Consensus on molecular imaging and theranostics in neuroendocrine neoplasms. Eur J Cancer 2021; 146: 56-73.
[http://dx.doi.org/10.1016/j.ejca.2021.01.008] [PMID: 33588146]

[12]   Wong P, Li L, Chea J, *et al.* Antibody Targeted PET imaging of ^{64}Cu-DOTA-anti-CEA PEGylated lipid nanodiscs in CEA positive tumors. Bioconjug Chem 2020; 31(3): 743-53.
[http://dx.doi.org/10.1021/acs.bioconjchem.9b00854] [PMID: 31961138]

[13]   Zhou X, Yan N, Cornel EJ, *et al.* Bone-targeting polymer vesicles for simultaneous imaging and effective malignant bone tumor treatment. Biomaterials 2021; 269: 120345.
[http://dx.doi.org/10.1016/j.biomaterials.2020.120345] [PMID: 33172607]

[14]   Liu J, Zhang Y, Li Q, *et al.* Development of injectable thermosensitive polypeptide hydrogel as facile radioisotope and radiosensitizer hotspot for synergistic brachytherapy. Acta Biomater 2020; 114: 133-45.
[http://dx.doi.org/10.1016/j.actbio.2020.07.032] [PMID: 32688087]

[15]   Li H, Chen Y, Jin Q, *et al.* Noninvasive radionuclide molecular imaging of the CD4-Positive T Lymphocytes in acute cardiac rejection. Mol Pharm 2021; 18(3): 1317-26.
[http://dx.doi.org/10.1021/acs.molpharmaceut.0c01155] [PMID: 33506680]

[16]   Ventura M, Bernards N, De Souza R, *et al.* Zheng. Longitudinal PET imaging to monitor treatment efficacy by liposomal irinotecan in orthotopic patient-derived pancreatic tumor models of high and low hypoxia. Mol Imaging Biol 2020; 22(3): 653-64.
[http://dx.doi.org/10.1007/s11307-019-01374-x] [PMID: 31482415]

[17]   Li T, Hu X, Fan Q, Chen Z, Zheng Z, Zhang R. The Novel DPP-BDT nanoparticles as efficient photoacoustic imaging and positron emission tomography agents in living mice. Int J Nanomed 2020; 15: 5017-26.
[http://dx.doi.org/10.2147/IJN.S238679] [PMID: 32764933]

[18]   Shi Y, Fu Q, Li J, *et al.* Covalent Organic Polymer as a carborane carrier for imaging-facilitated boron neutron capture therapy. ACS Appl Mater Interfaces 2020; 12(50): 55564-73.
[http://dx.doi.org/10.1021/acsami.0c15251] [PMID: 33327054]

[19]   Goud NS, Bhattacharya A, Joshi RK, Nagaraj C, Bharath RD, Kumar P. Carbon-11: Radiochemistry and target-based PET molecular imaging. Applications in oncology, cardiology, and Neurology. J Med Chem 2021; 64(3): 1223-59.
[http://dx.doi.org/10.1021/acs.jmedchem.0c01053] [PMID: 33499603]

[20]   Cardoso R, Leucker TM. Applications of PET-MR imaging in cardiovascular disorders. PET Clin 2020; 15(4): 509-20.

[http://dx.doi.org/10.1016/j.cpet.2020.06.007] [PMID: 32888548]

[21]   van der Bijl P, Knuuti J, Delgado V, Bax JJ. Cardiac sympathetic innervation imaging with PET radiotracers. Curr Cardiol Rep 2021; 23(1): 4.
        [http://dx.doi.org/10.1007/s11886-020-01432-9] [PMID: 33245510]

[22]   Schweitzer JS, Song B, Herrington TM, *et al.* Personalized iPSC-derived dopamine progenitor cells for Parkinson's disease. N Engl J Med 2020; 382(20): 1926-32.
        [http://dx.doi.org/10.1056/NEJMoa1915872] [PMID: 32402162]

[23]   Jain P, Chaney AM, Carlson ML, Jackson IM, Rao A, James ML. Neuroinflammation PET imaging: Current opinion and future directions. J Nucl Med 2020; 61(8): 1107-12.
        [http://dx.doi.org/10.2967/jnumed.119.229443] [PMID: 32620705]

[24]   Mukai H, Watanabe Y. Review: PET imaging with macro- and middle-sized molecular probes. Nucl Med Biol 2021; 92: 156-70.
        [http://dx.doi.org/10.1016/j.nucmedbio.2020.06.007] [PMID: 32660789]

[25]   Tang Y, Anandasabapathy S, Richards-Kortum R. Advances in optical gastrointestinal endoscopy: a technical review. Mol Oncol 2020.
        [http://dx.doi.org/10.1002/1878-0261.12792] [PMID: 32915503]

[26]   Achterberg FB, Deken MM, Meijer RPJ, *et al.* Clinical translation and implementation of optical imaging agents for precision image-guided cancer surgery. Eur J Nucl Med Mol Imaging 2021; 48(2): 332-9.
        [http://dx.doi.org/10.1007/s00259-020-04970-0] [PMID: 32783112]

[27]   Wang S, Li B, Zhang F. Molecular fluorophores for deep-tissue bioimaging. ACS Cent Sci 2020; 6(8): 1302-16.
        [http://dx.doi.org/10.1021/acscentsci.0c00544] [PMID: 32875073]

[28]   Wilson AJ, Devasia D, Jain PK. Nanoscale optical imaging in chemistry. Chem Soc Rev 2020; 49(16): 6087-112.
        [http://dx.doi.org/10.1039/D0CS00338G] [PMID: 32700702]

[29]   Zhou H, Zeng X, Li A, *et al.* Upconversion NIR-II fluorophores for mitochondria-targeted cancer imaging and photothermal therapy. Nat Commun 2020; 11(1): 6183.
        [http://dx.doi.org/10.1038/s41467-020-19945-w] [PMID: 33273452]

[30]   Yu H, Ji M. Recent advances of organic near-infrared II fluorophores in optical properties and imaging functions. Mol Imaging Biol 2021; 23(2): 160-72.
        [http://dx.doi.org/10.1007/s11307-020-01545-1] [PMID: 33030708]

[31]   Kwon YD, Byun Y, Kim HK. ^{18}F-labelled BODIPY dye as a dual imaging agent: Radiofluorination and applications in PET and optical imaging. Nucl Med Biol 2021; 93: 22-36.
        [http://dx.doi.org/10.1016/j.nucmedbio.2020.11.004] [PMID: 33276283]

[32]   Li Y, Zhou Y, Yue X, Dai Z. Cyanine conjugate-based biomedical imaging probes. Adv Healthc Mater 2020; 9(22): 2001327.
        [http://dx.doi.org/10.1002/adhm.202001327] [PMID: 33000915]

[33]   Feng J, Chen S, Yu YL, Wang JH. Red-emission hydrophobic porphyrin structure carbon dots linked with transferrin for cell imaging. Talanta 2020; 217: 121014.
        [http://dx.doi.org/10.1016/j.talanta.2020.121014] [PMID: 32498886]

[34]   Wang L, Du W, Hu Z, Uvdal K, Li L, Huang W. Hybrid rhodamine fluorophores in the visible/NIR region for biological imaging. Angew Chem Int Ed 2019; 58(40): 14026-43.
        [http://dx.doi.org/10.1002/anie.201901061] [PMID: 30843646]

[35]   van Manen L, Handgraaf HJM, Diana M, *et al.* A practical guide for the use of indocyanine green and methylene blue in fluorescence-guided abdominal surgery. J Surg Oncol 2018; 118(2): 283-300.
        [http://dx.doi.org/10.1002/jso.25105] [PMID: 29938401]

[36]   Forgách L, Hegedűs N, Horváth I, *et al.* Fluorescent. Prussian blue-based biocompatible nanoparticle system for multimodal imaging contrast. Nanomaterials (Basel) 2020; 10(9): 1732.
[http://dx.doi.org/10.3390/nano10091732] [PMID: 32878344]

[37]   Hadjipanayis CG, Stummer W, Sheehan JP. 5-ALA fluorescence-guided surgery of CNS tumors. J Neurooncol 2019; 141(3): 477-8.
[http://dx.doi.org/10.1007/s11060-019-03109-y] [PMID: 30671710]

[38]   Hadjipanayis CG, Stummer W. 5-ALA and FDA approval for glioma surgery. J Neurooncol 2019; 141(3): 479-86.
[http://dx.doi.org/10.1007/s11060-019-03098-y] [PMID: 30644008]

[39]   Duprée A, Rieß HC, von Kroge PH, *et al.* Validation of quantitative assessment of indocyanine green fluorescent imaging in a one-vessel model. PLoS One 2020; 15(11): e0240188.
[http://dx.doi.org/10.1371/journal.pone.0240188] [PMID: 33206647]

[40]   Thongvitokomarn S, Polchai N. Indocyanine green fluorescence versus blue dye or radioisotope regarding detection rate of sentinel lymph node biopsy and nodes removed in breast cancer: A systematic review and meta-analysis. Asian Pac J Cancer Prev 2020; 21(5): 1187-95.
[http://dx.doi.org/10.31557/APJCP.2020.21.5.1187] [PMID: 32458621]

[41]   Luo L, Yan L, Amirshaghaghi A, *et al.* Indocyanine green-coated polycaprolactone micelles for fluorescence imaging of tumors. ACS Appl Bio Mater 2020; 3(4): 2344-9.
[http://dx.doi.org/10.1021/acsabm.0c00091] [PMID: 32455339]

[42]   Rinck P. Magnetic Resonance in Medicine: A Critical Introduction, The Basic Textbook of the European Magnetic Resonance Forum. 12th edition.. Published by BoD, Germany 2018.

[43]   Avasthi A, Caro C, Pozo-Torres E, Leal MP, García-Martín ML. Magnetic nanoparticles as MRI contrast agents. Top Curr Chem (Cham) 2020; 378(3): 40.
[http://dx.doi.org/10.1007/s41061-020-00302-w] [PMID: 32382832]

[44]   Anani T, Rahmati S, Sultana N, David AE. MRI-traceable theranostic nanoparticles for targeted cancer treatment. Theranostics 2021; 11(2): 579-601.
[http://dx.doi.org/10.7150/thno.48811] [PMID: 33391494]

[45]   Yuan D, Ellis CM, Davis JJ. Mesoporous silica nanoparticles in bioimaging. Materials (Basel) 2020; 13(17): 3795.
[http://dx.doi.org/10.3390/ma13173795] [PMID: 32867401]

[46]   Zhao S, Yu X, Qian Y, Chen W, Shen J. Multifunctional magnetic iron oxide nanoparticles: an advanced platform for cancer theranostics. Theranostics 2020; 10(14): 6278-309.
[http://dx.doi.org/10.7150/thno.42564] [PMID: 32483453]

[47]   Hou Z, Liu Y, Xu J, Zhu J. Surface engineering of magnetic iron oxide nanoparticles by polymer grafting: synthesis progress and biomedical applications. Nanoscale 2020; 12(28): 14957-75.
[http://dx.doi.org/10.1039/D0NR03346D] [PMID: 32648868]

[48]   Yin J, Yao D, Yin G, Huang Z, Pu X. Peptide-decorated ultrasmall superparamagnetic nanoparticles as active targeting MRI contrast agents for ovarian tumors. ACS Appl Mater Interfaces 2019; 11(44): 41038-50.
[http://dx.doi.org/10.1021/acsami.9b14394] [PMID: 31618000]

[49]   Sjöstrand S, Evertsson M, Jansson T. Magnetomotive ultrasound imaging systems: Basic principles and first applications. Ultrasound Med Biol 2020; 46(10): 2636-50.
[http://dx.doi.org/10.1016/j.ultrasmedbio.2020.06.014] [PMID: 32753288]

[50]   Son S, Kim JH, Wang X, *et al.* Multifunctional sonosensitizers in sonodynamic cancer therapy. Chem Soc Rev 2020; 49(11): 3244-61.
[http://dx.doi.org/10.1039/C9CS00648F] [PMID: 32337527]

[51]   Siddique S, Chow JCL. Application of nanomaterials in biomedical imaging and cancer therapy.

Nanomaterials (Basel) 2020; 10(9): 1700.
[http://dx.doi.org/10.3390/nano10091700] [PMID: 32872399]

[52]   Chowdhury SM, Abou-Elkacem L, Lee T, Dahl J, Lutz AM. Ultrasound and microbubble mediated therapeutic delivery: Underlying mechanisms and future outlook. J Control Release 2020; 326: 75-90.
[http://dx.doi.org/10.1016/j.jconrel.2020.06.008] [PMID: 32554041]

[53]   Li Y, Chen Y, Du M, Chen ZY. Ultrasound technology for molecular imaging: From contrast agents to multimodal imaging. ACS Biomater Sci Eng 2018; 4(8): 2716-28.
[http://dx.doi.org/10.1021/acsbiomaterials.8b00421] [PMID: 33434997]

[54]   Du M, Chen Y, Tu J, *et al.* Ultrasound responsive magnetic mesoporous silica nanoparticle-loaded microbubbles for efficient gene delivery. ACS Biomater Sci Eng 2020; 6(5): 2904-12.
[http://dx.doi.org/10.1021/acsbiomaterials.0c00014] [PMID: 33463299]

[55]   Burke BP, Cawthorne C, Archibald SJ. Multimodal nanoparticle imaging agents: design and applications. Philos Trans- Royal Soc, Math Phys Eng Sci 2017; 375(2107): 20170261.
[http://dx.doi.org/10.1098/rsta.2017.0261] [PMID: 29038384]

[56]   Wu M, Shu J. Multimodal molecular imaging: Current status and future directions. Contrast Media Mol Imaging 2018; 2018: 1-12.
[http://dx.doi.org/10.1155/2018/1382183] [PMID: 29967571]

[57]   Huang Y, Li L, Zhang D, *et al.* Gadolinium-doped carbon quantum dots loaded magnetite nanoparticles as a bimodal nanoprobe for both fluorescence and magnetic resonance imaging. Magn Reson Imaging 2020; 68: 113-20.
[http://dx.doi.org/10.1016/j.mri.2020.02.003] [PMID: 32032662]

[58]   Prasad R, Jain NK, Yadav AS, *et al.* Liposomal nanotheranostics for multimode targeted *in vivo* bioimaging and near-infrared light mediated cancer therapy. Commun Biol 2020; 3(1): 284.
[http://dx.doi.org/10.1038/s42003-020-1016-z] [PMID: 32504032]

[59]   Chen JS, Chen J, Bhattacharjee S, *et al.* Functionalized nanoparticles with targeted antibody to enhance imaging of breast cancer *in vivo*. J Nanobiotechnology 2020; 18(1): 135.
[http://dx.doi.org/10.1186/s12951-020-00695-2] [PMID: 32948179]

[60]   Albuquerque GM, Souza-Sobrinha I, Coiado SD, *et al.* Quantum dots and $Gd^{(3+)}$ chelates: Advances and challenges towards bimodal nanoprobes for magnetic resonance and optical imaging. Top Curr Chem (Cham) 2021; 379(2): 12.
[http://dx.doi.org/10.1007/s41061-021-00325-x] [PMID: 33550491]

[61]   Yang CT, Hattiholi A, Selvan ST, *et al.* Gadolinium-based bimodal probes to enhance T1-Weighted magnetic resonance/optical imaging. Acta Biomater 2020; 110: 15-36.
[http://dx.doi.org/10.1016/j.actbio.2020.03.047] [PMID: 32335310]

[62]   Singh G, Ddungu JLZ, Licciardello N, Bergmann R, De Cola L, Stephan H. Ultrasmall silicon nanoparticles as a promising platform for multimodal imaging. Faraday Discuss 2020; 222(0): 362-83.
[http://dx.doi.org/10.1039/C9FD00091G] [PMID: 32108214]

[63]   Ding N, Sano K, Kanazaki K, *et al.* Sensitive photoacoustic/magnetic resonance dual imaging probe for detection of malignant tumors. J Pharm Sci 2020; 109(10): 3153-9.
[http://dx.doi.org/10.1016/j.xphs.2020.07.010] [PMID: 32679213]

[64]   Sharifianjazi F, Jafari Rad A, Bakhtiari A, *et al.* Biosensors and nanotechnology for cancer diagnosis (lung and bronchus, breast, prostate, and colon): a systematic review. Biomed Mater 2022; 17(1): 012002.
[http://dx.doi.org/10.1088/1748-605X/ac41fd] [PMID: 34891145]

[65]   Xia B, Yan X, Fang WW, *et al.* Activatable cell-penetrating peptide conjugated polymeric nanoparticles with Gd-chelation and aggregation-induced emission for bimodal MR and fluorescence imaging of tumors. ACS Appl Bio Mater 2020; 3(3): 1394-405.
[http://dx.doi.org/10.1021/acsabm.9b01049] [PMID: 35021632]

CHAPTER 2

Synthetic Biology and Tissue Engineering

Betül Mutlu[1,*], **Selcen Arı-Yuka**[1], **Ayça Aslan**[1] and **Rabia Çakır-Koç**[1,2]

[1] *Department of Bioengineering, Faculty of Chemical and Metallurgical Engineering, Yıldız Technical University, Istanbul, Turkey*

[2] *Biotechnology Institute of Turkey, Health Institutes of Turkey, İstanbul, Turkey*

Abstract: Advanced approaches that can mimic the structure and function of natural tissue in tissue engineering applications that use multidisciplinary engineering approaches to repair damaged or dysfunctional tissues are fed forward by current engineering applications. Manipulating cells or cell groups in an integrated manner into the scaffold, similar to the native tissue composition, is the main challenge in these approaches. Synthetic biology approaches, originating from genetic engineering, based on the use of advanced tools in the manipulation of cells at the molecular level, are one of the most current issues in tissue engineering that shed light on the programming of cells. Synthetic biology tools allow the reprogramming of cells whose transcriptional, translational, or post-translational molecular mechanisms have been engineered by stimulating them with intrinsic or extrinsic signals. Combining these advanced and excellent tools from synthetic biology with materials engineering applications of tissue engineering is the latest fashion. This chapter discusses going beyond conventional tissue engineering applications, synthetic biological molecular tools, circuit designs that allow the complex behavior of cells to be manipulated with these tools, and approaches that enable the integration of these tools into the material component of tissue engineering.

Keywords: Biomaterials, Biotechnology, CRISPR, Genetic circuit, Genetic engineering, Genome editing, Hydrogels, Laci, Molecular tools, Reprogramming of cells, Saps, Spytag-spycatcher, Synthetic biology, Talens, Teto, Tetr, Tissue engineering, Transcription factors, Virus-like particles, Zfns.

INTRODUCTION

Tissue engineering is a multidisciplinary science that aims to regenerate and restore the structural and functional properties of biological tissues damaged for any reason [1]. Tissue engineering includes various approaches in which cells are used in combination with biochemical factors and biomaterials to repair damaged tissue and form new tissue [2, 3].

---

[*] **Corresponding author Betül Mutlu:** Department of Bioengineering, Faculty of Chemical and Metallurgical Engineering, Yıldız Technical University, Istanbul, Turkey; E-mail: betlmutl@gmail.com

Felipe López-Saucedo (Ed.)

Cell-material-biochemical factor-based approaches have been investigated for many years to direct cell differentiation and promote tissue formation. Early tissue engineering applications put a lot of effort into promoting cell behaviors by creating many different classes of materials in different scaffold configurations and topographies to mimic the natural microenvironment [4]. Despite the use of a range of physical, chemical, and mechanical cues to support cell proliferation and lineage-specific differentiation in all these studies, it has still not been possible to develop tissue engineering products for clinical applications, except for a few applications [5].

A current limitation of this situation is the inability to maintain the cell phenotype during synthetic new tissue formation, the inability to control the spatial arrangement, and the dynamic interaction of cells and material properties [6]. Tissue development is a very complex process and depends on the appropriate scaffold and a large number of different signaling molecules, as well as precisely tuned cellular responses [7 - 9]. To restore the function of naturally varied tissues more appropriately, high-resolution spatial control over the cell is essential [10]. Biomaterials and bioactive molecules to guide cell behavior indirectly affect the properties of cells, such as survival, proliferation, differentiation and morphogenesis, through the manipulation of cellular events at the molecular level [11, 12]. All these cellular processes are highly related to various receptor-protein interactions, genetic factors, and the expression of genes. For this reason, in recent years, researchers have focused on the development of new tissue engineering strategies that directly involve molecular and genetic processes [13, 14]. In addition to these great advances with the integration of developments in molecular biology and genetic engineering into tissue engineering, the emergence of the concept of synthetic biology in the past decade has paved the way for a new perspective in tissue engineering. Synthetic biology is a new field in which engineering principles are applied to biological components to understand, manipulate, and modify cell function by designing or regulating organisms or devices [15, 16].

Unlike genetic engineering, synthetic biologists bring together various natural biological parts with different functions from an integrated perspective to redesign cells and create new genetic architectures that perform complex functions, using various molecular biology tools for this purpose [17]. Thanks to this bottom-up approach that allows design and manipulation at the molecular level, different biological functional parts have been combined with endogenous pathways in the organism with various synthetic biology tools to create gene circuits for dynamic gene expression [15]. From the point of view of tissue engineering, the creation of genetic circuits and new gene expression patterns that can control gene expression for tissue formation processes in mammalian cells is a remarkable field of reprogramming of cells. In addition, these techniques can allow the control of cells at the molecular level to better mimic natural cellularity and tissue structure, thereby enabling spatial and temporal manipulation of the artificial tissue formation process with high precision [18]. Moreover, by adapting these modules in synthetic biology to biomaterials, cellular processes can be directly controlled by biomaterials [19].

In this chapter, we aim to discuss synthetic biology techniques and applications within the framework of tissue engineering. In this context, we explain the current synthetic biology tools used in tissue engineering and review studies on how these tools can be applied in cell and biomaterial-targeted tissue engineering strategies.

REPROGRAMMING OF CELLS WITH SYNTHETIC BIOLOGY TOOLS

Transcription Factors

Transcription factors (TFs) are a complex molecular system that directs the expression of various genes by recognizing specific DNA sequences in the genome of living organisms [20]. TFs are key proteins that have important roles in turning genes on and off during the transcription process, which is the first step for gene expression that makes the DNA sequence functional [21]. They function as master regulators for many cellular processes, such as the determination of cell types and developmental patterns, specific cellular mechanisms, and metabolic activities [22, 23]. TFs are distinguished by their sequence specificity, which allows them to activate or inhibit target gene transcription. Recognizing these specific binding sites in the genome and understanding how to control transcription and control gene expression, and determining their specific functional properties of genomes in various species have been very important topics in molecular biology and cell biology [21]. Today, it is thought that many diseases, such as cancers and various metabolic disorders, may be associated with gene expression changes affected by disruptions of TFs and their binding site genes [24 - 26]. TFs are employed as crucial building blocks and regulatory tools in genetic engineering and synthetic biology, in addition to their usual biological and physiological roles in cells [27]. The transfer of transcriptional regulatory parts obtained from different species in mammalian cells is one of the first examples of synthetic biology.

The lactose repressor system (LacI) and tetracycline-controlled transcriptional activation (TetR, TetO) mediated gene regulation, which are bacterial gene expression regulatory systems, are the first examples of synthetic biology and have paved the way for many studies for the controlled expression of endogenous and exogenous genes [28, 29]. In the lac repressor system, lactose and its molecular analog isopropyl β-D-1-thiogalactopyranoside (IPTG) block LacI from binding to the lac operon, allowing transcription to occur [30]. In the Tet system, antibiotics, such as tetracycline and doxycycline (Dox), prevent TetR from binding to the Tet operator [31]. Transcription control can be achieved by the presence/absence or concentration of these small molecules with these transcription regions added to the upstream regions of target genes [32]. It has been shown that the binding domains of LacI and TetR can be inserted into promoter regions to control gene expression in mammalian cells, thereby regulating the expression of beta-galactosidase and luciferase reporter genes [33]. Control of gene expression with these natural transcription factors is an effective system for the control of transcription in stem cells.

For example, a strong and dose-dependent IPTG-regulated expression of transiently transfected reporter genes in LacI-expressing mouse embryonic stem cells (ESCs) with the addition of LacI binding sites to the β-actin promoter has been demonstrated [34].

These transcriptional control systems, which serve to selectively and highly control gene expression, are very important in terms of tissue engineering. Cell-based approaches in tissue engineering and cell modifications with genetic engineering tools are often necessary for cell proliferation, differentiation, and obtaining cells with different characteristics [35]. Many previous studies have shown that the expression of many endogenous or exogenous genes can be successfully realized in mammalian cells and their applications in tissue engineering [36]. Overexpression of these genes, which typically function devoid of endogenous feedback mechanisms, can cause undesirable side effects [37]. For example, high expression of Oct-4, which is an important transcription factor for embryogenesis and generation of induced pluripotent stem cells (iPSCs), increases malignant potential with an aggressive tumor phenotype [38]. The use of synthetic biology tools to control the expression of specific genes for cellular differentiation in tissue engineering provides a major advantage for cell fate determination. An example of this approach is the TetR system reported by Ueblacker *et al.*, developed for the controlled expression of transgenes in growth factor-expressing chondrocytes for the treatment of osteochondral defects [39]. They have designed an expression plasmid containing the lacZ gene and reverse transactivator (rtTA2(s)-M2) under the control of the minimal cytomegalovirus (CMV) promoter fused to the Tet-repressor element (TRE) and used it to transfect chondrocytes. *In vivo* mouse model results showed that Dox-inducible growth factor expression can be used in chondrocytes. In another study, the tetracycline-inducible (tet-off) promoter was used to control the expression of Runx2 osteoblastic transcription factor in primary skeletal myoblasts for the controlled expression of osteogenic factors [40]. The results showed that control of ontogenesis can be achieved by the expression of Runx2 regulated by anhydrotetracycline concentration. TetR inducible gene expression control has been used to control the gene expression of important growth factors in tissue engineering, such as vascular endothelial growth factor (VEGF) [41], bone morfogenetic protein 2 (BMP-2) [42], and glial cell line-derived neurotrophic factor (GDNF) [43].

In addition to TetR and Lac operons, the galactose regulatory transcription factor Gal4 and its effector domain (GAL4/UAS) found in yeasts [44], pancreatic duodenal homeobox-1 [45], which plays a role in regulating the expression of beta-cell gene and somatostatin, the tryptophan operon of *Escherichia coli (E. coli)*, TPR [46], are widely used synthetic transcription factor domains in mammalian cells.

In gene expression regulation by natural TFs, activation or repression is typically mediated by effector domains that allow them to be targeted to specific promoters [47, 48]. Combining these effector domains with transcription factors can increase or decrease gene expression. Examples of the most commonly used transcription rate-enhancing protein activators in mammalian cells include p65 isolated from human NF-κβ [49], viral-derived protein 16 (VP16) [50] or VP64 [51] (4 repeats of VP16), R transactivation (Rta) [52]. Also, the Krüppel-associated box (KRAB) domain is one of the most used transcription repressor domains [53]. Modifications of DNA-binding domains in transcription factors that allow target genes to activate or repress their transcription have provided developing genetic circuits and new cellular gene expression patterns that can synthetically control gene expression [27, 54]. Regulation of gene expression with these synthetic gene circuits can be designed to strongly control important cellular processes, such as proliferation, apoptosis, and differentiation by manipulating the expression of genes in "on" or "off" states [55]. The design of synthetic gene circuits will be mentioned in section Genetic Circuits in Tissue Engineering, in detail.

Genome Editing with Engineered Transcription Factors

Although natural transcription factors are highly effective in a variety of therapeutic and research applications in synthetic biology, there are some disadvantages to using them for clinical applications, such as their limited design features in targeting, exogenously expressed factors, or integrative viral delivery [56, 57]. To integrate the required DNA sequence into the stem cell genome in mammalian cell genome editing, viral vectors, such as retroviruses, adeno-associated viruses, and lentiviruses, are extensively utilized [58]. These widely used techniques have disadvantages affecting stem cells' viability and potencies, such as limited transgene size capacity, toxicity, and immunogenicity in transfected cells or tissue [59]. Also, genetic modification of mammalian cells with these methods may require long periods of time, intensive labor and cost [56].

Significant advances in synthetic biology over the past 10 years have provided valuable molecular tools that can address the problems of these classical genome manipulation techniques. Recent discoveries regarding the understanding of protein-DNA interactions in various organisms have inspired the engineering of designer proteins targetable for any DNA region [60]. These specially designed proteins can act as a template for specific genome editing systems that can alter the DNA sequence, transcriptional regulation, or epigenetic state at any region in the mammalian genome [61]. Zinc finger nucleases (ZFNs), transcription activator-like effectors (TALEs), and clustered regularly spaced short palindromic repeats (CRISPR)/CRISPR-associated protein (Cas) systems are the main programmable genome editing systems used to target novel DNA sequences and manipulate gene expression. These systems are based on protein-based systems for nuclease-directed double-strand breaks that effectively induce non-homologous end joining or homology-directed repair-inducing breaks at specified genomic locations [62].

Zinc Finger Nucleases

Zinc finger nucleases (ZFNs) are the first sequence-specific nucleases developed for site-specific genome editing. ZFNs consist of two parts, a nuclease (cleavage) domain and modular zinc finger domains that specifically bind to DNA by recognizing a specific 3-base pair (bp) in DNA. The engineering of synthetic zinc finger (ZF) domains that target each triplet of DNA base pairs enables the creation of a ZFN that can bind practically any DNA sequence [63, 64].

Transcription Activator-Like Effector Nucleases

Transcription activator-like effectors nucleases (TALENs) are DNA binding proteins that modulate gene transcription identified in plant pathogenic bacteria (*Xanthomonas* and *Ralstonia*) and consist of two parts, a DNA binding domain and a nuclease as ZFNs [65]. These pathogenic bacteria produce TALEN to modulate host gene expression as a defense mechanism, and these proteins recognize and bind DNA sequences depending on a variable number of tandem repeats [66]. DNA-binding domains that recognize a single base pair are sequential sequences of 33-35 amino acid repeats. These sequences recognize a single nucleotide in the main groove of the DNA helix by means of two central hypervariable amino acids, called repeat variable di-residues [67]. The modular structures of TALEN domains allow for generating a binding domain that can usually identify specific 20 bp DNA sequences, and these domains are often combined with a Fok1 nuclease domain to form a TALEN monomer [68].

Although TALENs are similar systems to ZFNs in having a designable DNA binding domain and a nuclease domain, they make it easier to design and manufacture to target new sequences compared to ZFNs [69].

Clustered Regularly Interspaced Palindromic Repeats (CRISPR) and CRISPR-Associated Protein (Cas9)

The latest example of programmable nucleases for DNA-targeted transcriptional control was the discovery of RNA/protein complexes as clustered regularly interspaced palindromic repeats (CRISPR) and associated proteins. In the CRISPR/Cas system, which is a nucleic acid-based adaptive immune response mechanism created against viral parasites in bacteria and archaea, the host incorporates foreign short nucleic acid fragments called protospacer into CRISPR genomic loci [70]. These parasite DNA fragments, which serve as molecular memory, are copied into RNAs and processed to form CRISPR RNAs (crRNAs). The crRNAs containing complementarity sequence against a previous nucleic acid invader recognize the target and combine with transactivating crRNAs (tracrRNAs) and Cas9 endonuclease. In the Type 2 system, crRNAs localize the Cas9 complex to the foreign DNA sequence to promote double-strand break [71]. In order to state the system more simply, gRNAs (guide RNA) that summarize the function of both crRNA and tracrRNA were designed. gRNA is a short synthetic RNA and is composed of three parts: (1) a 20 bp protospacer that provides targeting specificity *via* complementary base pairing with the desired DNA target, (2) a hairpin nucleotide sequence that provides the function of the crRNA/tracrRNA complex necessary for Cas9 nuclease binding, and (3) a transcriptional termination region. In this way, the genomic target sequence of the Cas protein can be altered easily by changing the target sequence present in the gRNA [72].

In parallel with the development of synthetic biology and genetic engineering, ZFN, TALEN and CRISPR/Cas systems have been used in tissue engineering and regenerative medicine in different fields, such as the formation of *in vitro* artificial tissue by directing the differentiation of stem cells into specific lineages, the cellular treatment of various hereditary diseases by obtaining induced pluripotent stem cells, and the design of synthetic genetic circuits [73].

Generation of Induced Pluripotent Stem Cells with Genome Engineering Tools

In 2006, Takahashi and Yamanaka *et al.* demonstrated that adding various genes (Oct-4, Sox-2, Klf-4, c-Myc) to mouse fibroblast cells could lead to achieving stem cells with similar properties to ESCs [74, 75]. A new class of stem cells named iPSCs has been identified as an alternative to ESCs without limitation in terms of clinical and ethical reasons [76]. In tissue engineering and regenerative medicine, iPSCs are promising resources for cell transplantation in damaged tissues, the study of developmental processes [77] and disease mechanisms [78], stem cell-based drug screening [79], and gene therapies [80]. However, genome engineering in human iPSCs is challenging owing to low transfection/transduction efficiency and high apoptosis, and for iPSCs to be used in clinical applications, reprogrammed cells that can remain karyotypically stable for a long time should be created, and effective differentiation of these cells towards the desired tissue type should be ensured [81]. Moreover, genetic mutations in iPSCs must be corrected before treating hereditary diseases [82]. Recent studies suggest that ZF, TALE, and CRISPR/Cas-mediated transcriptional modulators are good tools to direct targeted gene modifications in human embryonic stem cells and iPSCs [80].

The first programmable nucleases used for the generation of iPCSs by genome editing were ZFNs and TALEN. Ji *et al.* constructed ZF-TFs for Oct-4, Sox-2, Klf-4 and c-Myc, four factors important in cellular reprogramming [83]. The engineered ZFs were able to show expression with each of the different activation sites in HEK293T cells and fibroblast cells. Similarly, engineered TALE TFs have been used to reprogram mouse embryonic fibroblasts into iPSCs by inducing OCT4 expression [84].

The successful production of iPSCs cells with engineered transcription factors has paved the way for the use of these tools in cellular treatments of various genetic diseases. Hoher *et al.* demonstrated that ZFN-mediated genome editing provides a highly efficient method of deletion of specific alleles in keratinocyte stem cells without impairing the stem cell potential [85]. Similarly, the successful generation of transgene-free iPSCs carrying homozygous α-thalassemia deletions and using ZFN has been reported [82]. As with ZNFs, TALENs have been investigated for the generation of corrected iPSCs. Ma *et al.* successfully generated non-integration and gene-edited patient-specific pluripotent and normal karyotype iPSCs with the TALEN system from two patients with different types of homozygous mutations for the treatment of β-Thalassemia [86]. Another study reported the successful use of TALE-mediated regulation for restoration of mutation in the transmembrane conductance regulator (CFTR) gene (p.F508del) for iPSCs cellular therapy of cystic fibrosis [87].

Besides TALEN-mediated genome editing studies for the treatment of various genetic diseases, Wook Kwon *et al.* reported that they generated universally compatible immune-responsive human iPSCs *via* TALEN gene editing to prevent mismatches between donor and recipient human leukocyte antigen (HLA) in allogeneic iPSCs transplantation [88]. Dendritic cells differentiated from iPSCs of human dermal fibroblast cells obtained owing to selectively abolishing HLA expression were found to reduce CD4$^+$ T cell activation. This study is promising that engineered iPSCs can solve the problem of immune rejection in allogeneic stem cell therapy.

Although ZF and TALE-mediated genome editing techniques are effective tools for modulating transcription in mammalian cells, their integrity and specificity are lower than CRISPR/Cas9. In addition, the disadvantages of these systems, such as the requirement of protein design, synthesis, and validation that do not allow their routine use, and the difficulties in using modulation of gene expression at multiple loci, have increased the interest in CRISPR/Cas9 [69].

Like ZFN and TALEN, the use of CRISPR/Cas system has also been shown in many studies to generate human embryonic stem and iPSCs with high targeting efficiency. Liu *et al.* have reported that β-Thalassemia patient-specific iPSCs is a common deletion mutant β-41/42 (TCTT) in the HBB gene corrected using the CRISPR/Cas9 system [89]. In addition to these studies, it has been reported that CRISPR/Cas9 mediated genome editing can be successful for iPSCs treatments of diseases, such as Duchenne muscular dystrophy [90], severe combined immunodeficiency [91], fanconi anemia (FAC) [92], hereditary tyrosinemia type I [93], and amyotrophic lateral sclerosis (ALS) [94].

Another remarkable CRISPR-mediated study, in which gene editing and tissue engineering techniques are used together for the treatment of genetic diseases, is on recessive dystrophic epidermolysis bullosa (RDEB), a serious inherited skin disease caused by mutations in the COL7A1 gene. Jackòw *et al.* differentiated iPSCs derived from RDBE patients and corrected the mutation in the COL7A1 gene with the CRISPR/Cas9 system into keratinocytes and fibroblasts, and then tested these human skin equivalents (HSEs) in immunocompromised mice [95]. They reported that HSEs grafted into mice maintained skin integrity and showed restoration of type VII collagen. These studies are notable for the clinical translation of CRISPR-based genome editing applications in tissue engineering.

The ability of synthetic transcription factors to be engineered with nuclease domains, such as transcriptional activators, repressors or epigenetic enzymes, can allow the development of gene controllers with the potential to target any DNA sequence in the genome, thereby regulating the expression of any endogenous gene of interest [48, 53]. As an example, Moreno *et al.* used CRISPR-based transcriptional repression for the treatment of retinitis pigmentosa phenotypes *in vivo* [96]. Researchers have established a direct reprogramming strategy to convert rod-like photoreceptors to cone-like photoreceptors to inhibit further vision loss in patients through suppression of the rod master regulator (Nrl). The efficiency of the developed dCas9-KRAB mediated Nrl repression transcriptional regulation system was tested in a mouse model of retinitis pigmentosa. Mice treated at postnatal day 7 and followed up to day 50 were reported to have improved sight compared to diseased controls. In addition, concomitantly using various activation or repressor sites in activating endogenous coding and non-coding genes and targeting several genes at the same time has also been investigated. For example, Chavez *et al.* described an advanced transcriptional regulator obtained through the rational design of a three-part activator called VPR (VP64-p65-Rta) [97]. They generated Dox-inducible cell lines targeting activation of NGN2 (NEUROG2) and NEUROD1 genes for direct differentiation of fibroblast to neuron (iNeuron) *via* dCas9-VP64 and dCas9-VPR and monitored for phenotypic changes. The results showed that NGN2 and NEUROD1 mRNA expression levels in dCas9-VPR cells were increased 10- and 18-fold, respectively, compared to dCas9-VP64 cells. It has been reported that cells transfected with dCas9 fused VPR (dCas9-VPR) enable rapid and robust differentiation of iPSCs into a neuronal phenotype compared to dCas9-VP64.

GENETIC CIRCUITS IN TISSUE ENGINEERING

The subunits of the gene expression mechanism responsible for the emergence of a phenotype are categorized and considered as modules; separately, the molecular mechanisms of flexibility and strictness of cells in adapting to changing intrinsic and extrinsic conditions are explained by illuminating natural genetic circuits [98, 99].

Due to the complexity of the mechanisms from the genome to the phenotype in the cell, the gene regions involved in these mechanisms are the most ideal approaches to programming cells for tissue engineering [100, 101].

Converting complex mechanisms into simplified discrete modules and their genetic modification is a potent tool for manipulating the responses of programmed cells to both internal and external stimuli. Cell programming approaches and tools can be developed to design cells sensitive to specific molecules by modifying genetic parts, generate molecular responses through activation of stimulus-specific gene expression, or use as delivery platforms (Fig. 1) [73, 100, 102].

In order for the programmed cells to execute the desired functions, the first step is to identify the gene region(s) responsible for the function to be changed or improved. In order for the designed synthetic circuits to acquire predetermined properties in the cell, promoters, transcription factors, activators, repressors, complexity, and control mechanisms of gene expression must be well defined [73].

For optimal system design, the signaling pathways of the cells to be programmed, the biological process and molecular function of the biomolecule (*i.e.,* output) to be expressed by programming the cell, and the cell-cell and cell-extracellular matrix (ECM) interactions in their microenvironment should be well defined [103, 104]. Briefly, the functions and interactions of modules to be modified in the existing biological networks are defined, and a genetic circuit is designed that interferes with the desired function apart from the inherent functions of cells.

The primary challenges in synthetic biological circuit designs are the complication of particular modifications in the native content and function of the cell, and the inconsistency between components in multi-layered module modifications [99, 101]. The regulatory mechanisms of the molecular modules to be used in the functions desired to be achieved to the cell by particle changes should be able to operate in isolation from other mechanisms [105, 106]. In synthetic circuit design, orthogonality is defined as the fact that the signal and regulatory components that stimulate the regulator are independent (*i.e.,* not allowing cross-talk) in a way that does not affect other molecular mechanisms in the native functioning of the cell [55, 99, 101, 105].

The complexity and intentions of the designed synthetic biological circuit can change the degree of orthogonality. For instance, in a circuit designed with multiple gene manipulation, preferring orthogonal mechanisms that do not allow crosstalk in the process from gene to phenotype may ensure that the predetermined function is achieved.

On the other hand, the implementation of Boolean Logic gates in the design of complex multi-component biological circuits allows for specific functions and even multiple signal input-specific cellular responses. Switch and logic operations equipped with electronic circuits enable the circuit to acquire decision-making features [107 - 109].

In synthetic biology tools alike, the advanced techniques to give the cell the ability to decide in conformity with a cellular stimulus (input) draw attention. Basically, in the presence of an exclusive stimulus signal, the switch is ON, and the cell provides gene expression according to the input signal, while in the OFF switch, the stimulus acts as a suppressor [17, 107, 108, 110].

Further, it is feasible to equip biological circuits with Boolean logic operations, such as AND, OR, NOR, NOT, XOR and NAND, in the presence of multiple inputs (Fig. **1**) [111]. A genetic circuit with AND logic operates to exhibit a predetermined function in the cell in the presence of two stimuli; it is sufficient to have at least one of the two stimuli in the OR operation [112].

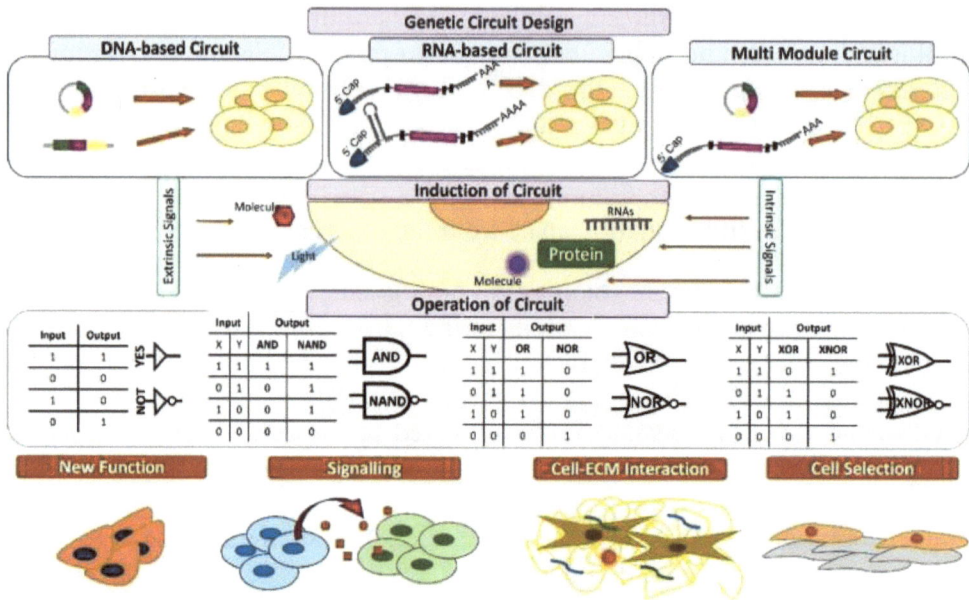

Fig. (1). Synthetic biology tools and approaches in cell programming. Genetic circuits, DNA and RNA, based on a combination of both intrinsic (molecules, nucleotides, or proteins) or extrinsic (such as molecules or light) inducers are activated or deactivated by logic operations. Synthetic biological tools enable cells to reach new functions, generate signals for cell communication, manipulate cell-ECM interactions, or select specific cell lines.

Logic gate is activated (*i.e.,* ON) when two input signals are not present at the same time, which is called NOR [113]. The NAND logic gate is inactive (OFF) in the presence of both signals, and active in other cases (lacking at least one of the two inputs). Circuits activated by the presence of only one of the two signal inputs or OFF of both signal inputs are defined as XOR and XNOR, respectively [114, 115]. In order for the synthetic biological circuit to trigger a predetermined cell function, the mechanism to be triggered by input stimuli and the orthogonality must be well-described during the operation of these gates.

DNA-based Genetic Circuit Design

The controllability of cell differentiation *via* synthetic genetic circuits contributes to the construction of complex cell-cell interactions and to the emergence of more rational approaches in tissue engineering. The changed expression profiles of cells in the differentiation stages are the adjusting factors in the regulation of cell function, cell-cell interactions, and the heterogeneous construction of the tissue [116 - 119]. Synthetic biological tools, on the other hand, afford robust tools for the control, self-organization, and manipulation of multicellular systems [120, 121]. One of the distinguished examples of this is the study of Guye *et al.,* who examined the role of the GATA-binding protein 6 (GATA6) transcription factor in cell function [122]. They have used Dox-inducible lentivirus genetic circuits modified with transgenes encoding GATA6. It was observed that different germline cells could be differentiated from human iPSCs depending on Dox concentration and time. Thus, it has been reported that it is achievable to control complex tissue construction containing cells from different germ layers from single human iPSCs induced by external stimuli.

The most notable challenge in cartilage tissue engineering is the decreased secretion of matrix proteins. This is due to the regulation of the Sox-9 transcription factor in chondrocyte cells [123] and results in the loss of biomechanical properties of cartilage tissue [124]. Yao *et al.* suggested improving the biomechanical properties of cartilage tissue through chondrocytes by regulating the expression of Sox-9 with the approach of induction of tetracycline [125]. In chondrocytes transfected with the vector used to induce the expression of the Sox9 transcription factor, it was reported that the expression of cartilage tissue-specific ECM proteins was increased by Dox-induced activation (ON) of the tetracycline switch. One of the most crucial parameters in the achievement of tissue engineering approaches is cell-cell interactions. In tissue remodeling, the function of the tissue is promoted by the interaction between cells.

Bacchus *et al.* employed one of the advanced techniques provided by synthetic biology for the manipulation of metabolic signals in mammalian cells [126]. In this study, HEK-293 cells were transfected with pWB32 vector containing human cytomegalovirus immediate early promoter (P_{hCMV}) and β-subunit of *E. coli* tryptophan synthase (TrpB26) to convert indole to l-tryptophan to generate the input signal for another cell. The signal produced by this cell (*i.e.,* l-tryptophan) activates the l-tryptophan-dependent transactivator (TRT), which is controlled by the simian virus 40 promoter (P_{SV40}), and human placental secreted alkaline phosphatase (SEAP) is transcribed through the l-tryptophan-inducible promoter (P_{TRT}). In a further step, the authors were able to manipulate the permeability of vascular endothelial cells by implementing engineered cell approaches to create multi-step, two-way communication that responds to the l-tryptophan-producing cell with acetaldehyde production. They suggested a rational and outstanding example in terms of high tissue remodeling needed in tissue engineering.

Approaches tuned with switches provide sophisticated methods for understanding tissue cell differentiation and manipulation. Programmed cell differentiation approaches that will ensure glucose-sensitive insulin secretion for patients with Type-1 diabetes mellitus (T1DM) are quite remarkable. In the study conducted with human iPSCs, it was possible to control cell differentiation programmed to be active or inactive depending on the vanillic acid concentration [127].

In a genetic circuit design, while the codon-modified transcription factor pancreatic and duodenal homeobox 1 ($Pdx1_{cm}$) was active (ON) due to the lack of vanillic acid, codon-modified Neurogenin 3 ($Ngn3_{cm}$) and $Pdx1_{cm}$ was ON and OFF at moderate vanillic acid concentrations, respectively. $Pdx1_{cm}$ and codon-modified V-maf musculoaponeurotic fibrosarcoma oncogene homologue A ($MafA_{cm}$) transcription factors are ON at high vanillic acid concentrations [127]. It has been reported that the progenitor pancreatic cells lacking vanillic acid differentiated into endocrine progenitor cells with the transcription factor $Ngn3_{cm}$ switching from OFF to ON and $Pdx1_{cm}$ switching from ON to OFF at moderate vanillic acid concentration. With increasing vanillic acid concentration, differentiation into β-like cells was observed with the transition of $MafA_{cm}$ from OFF to ON, $Ngn3_{cm}$ from ON to OFF, and $Pdx1_{cm}$ from OFF to ON. This genetic circuit design demonstrates the promising approach of timely controllable and programmable tissue engineering applications *via* generating separate responses to one simple input with different modules. It was reported that the progenitor pancreatic cells lacking vanillic acid differentiated into endocrine progenitor cells with the transcription factor $Ngn3_{cm}$ switching from OFF to ON and $Pdx1_{cm}$ switching from ON to OFF at moderate vanillic acid concentration. With increasing vanillic acid concentration, differentiation into β-like cells was observed with the transition of $MafA_{cm}$ from OFF to ON, $Ngn3_{cm}$ from ON to OFF, and $Pdx1_{cm}$ from OFF to ON. This genetic circuit design demonstrates the promising approach of timely controllable and programmable tissue engineering applications *via* generating separate responses to one simple input with different modules.

Programming intestinal cells as insulin-secreting cells for the generation of β-cells has also revealed an *in vivo* approach to glycemic control. A bacterial artificial chromosome (BAC) vector containing the TetO promoter was used for polycistronic expression of NPM (Ngn3, Pdx1, and MafA) reprogramming factors. It was found that insulin+ cells were induced in the gastrointestinal tract when Dox induction was used for Tet activation. It was reported that blood glucose levels return to normal rapidly when animals use streptozotocin (STZ), which regulates hyperglycemia [128]. With the accomplishment of the genetic circuit with the Dox-triggered TetO promoter, the researchers harvested these cells from animals, loaded them onto the poly glycolic acid (PGA) scaffold and implanted them in the omental flap of NOD SCID gamma (NSG) animals. It was observed that after the induction of hyperglycemia after implantation, blood glycerol levels can be controlled *via* Dox, but an increase in blood glycerol occurs with the removal of the implant [128].

In tissue engineering applications, it is critical that there are superior ECM interactions for the tissue to be implanted without an immune response. In the study carried out by Glass *et al.*, as a cartilage tissue engineering application, they suggested chondrocytes containing an antagonist region to Interleukin-1 (IL-1) gene that reduces the expression of primary ECM components (such as type II collagen and aggrecan) [129]. The Dox-inducible lentiviral Tet-ON vector carrying the IL-1 receptor antagonist (IL-1Ra) transgene region immobilized on 3D polycaprolactone scaffold was used. They induced mesenchymal stem cells to differentiate into mature chondrocytes. Immune system responses can be controlled thanks to the Dox-induced Tet-ON genetic circuit, and they advanced engineered cartilage tissue, which is superior in terms of ECM structuring [129]. Immune system responses can be controlled thanks to the Dox-induced Tet-ON genetic circuit, and they have advanced engineered cartilage tissue, which is superior in terms of ECM structuring [129].

Recent studies on DNA-based gene expression regulation have also suggested advanced technologies for synthetic circuit induction with various input signals. A transient blue light-induced optogenetic "ON" switch permanent gene expression mechanism was suggested for cell differentiation and tissue remodeling [130]. Besides inducing the expression of the master myogenic factor MyoD by blue light, the researchers demonstrated that the same system could control the onset of angiogenesis and tissue morphogenesis through the control of VEGF and angiopoietin-1 expression. However, more rational approaches need to be considered because of the variability of gene expression profiles at the stages of tissue differentiation. The TetON/OFF system was integrated with the 2-cryptochrome-interacting basic helix-loop-helix 1 (Cry2-CIB1) light-inducible binding switch, stimulating dynamic gene expression with two inputs (*i.e.,* light and Dox) [131]. After the transfection of photoactive Tet-OFF adeno-associated viral vectors to hippocampal neurons of transgenic mice, they were stimulated with blue light, and it was stated that gene expression control was successfully performed. In their more recent study, the researchers evaluated 10 candidate photoactivatable transcription activators for optogenetic gene expression regulation mechanisms. It was reported that gene expression could be controlled by changing the activity, sensitivity and patterns of the transcription activator depending on the design [132].

Although the use of DNA-integrating vectors in DNA-based genetic circuit designs is advantageous in terms of efficiency, the dilemma of genomic integration may arise [133]. Especially, despite the high efficiency of genetic circuits designed with vectors, such as retrovirus or lentivirus, the problem of undesired DNA integrations may arise. Methods based on the manipulation of translational control mechanisms (like modRNA) rather than genomic integration may lead to the development of reliable and efficient methods [133].

RNA-based Genetic Circuit Design

In addition to the promoter-transgene complexes in genetic circuit designs, some perspectives allow manipulation or programming of cell function through regulation during or after transcription [134]. After mRNA transcription in the cell, as a result of the translation processes that start with the help of ribosomal subunits and initiation factors, the translation of the gene region existing in the mRNA into the polypeptide sequence occurs through tRNA (Fig. **2A**) [135]. The mainstay of post-transcriptional approaches is the inhibition of the translation of intracellular mRNA content into protein products. mRNA-based approaches rely on adding an aptamer to the end regions of the mRNA transcript that can interact with a particular endogenous biomolecule (non-coding RNA (Fig. **2B**) or protein (Fig. **2C**) [136, 137]. Thus, when the designed mRNA circuit is transfected into the target cell, it will not be translated [133, 137].

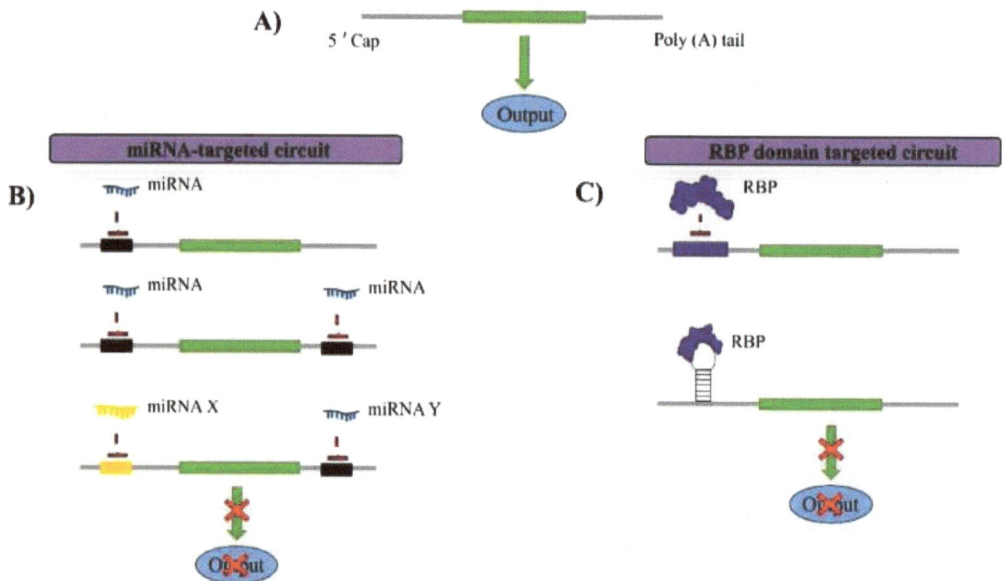

Fig. (2). mRNA translation mechanism and design of RNA-based genetic circuits. **A)** Demonstration of native mRNA translation in central dogma, **B)** Translation manipulation through miRNA-targeted genetic circuits, **C)** Controlling translation with RBPs.

The synthetic circuit designs with non-coding RNAs, which are post-transcriptional regulators, stand out as remarkable approaches in tissue engineering and regenerative medicine applications. These synthetic platforms can be engineered with miRNAs, which are considered the key players of post-transcriptional regulators [138]. These behind-the-scenes players, miRNAs, in the regulation of developmental signals and determination of cell differentiation and fate, start from the embryonic stage [139]. Already, miRNAs behave as a natural circuit, with feed-back and feed-forward loops exposed through cross-interactions with transcription factors and other endogenous biomolecules [140, 141]. This allows miRNAs to be used as synthetic biological circuits and a vigorous tool in cell manipulation. Parr *et al.* suggested the miRNA-switch design by integrating the overexpression of the miR-302/367 cluster in human iPSCs with a modRNA design [142]. The mRNAs containing the gene for expression of resistance to puromycin (puroR), which enables specific cell selection from heterogeneous cell populations, were modified with sequences containing the miR-320 target sequence. Thus, in the presence of puromycin, overexpression of miR-320 successfully accomplished the elimination of the observed undifferentiated or partially differentiated cells and selection of neural cells. In particular, these well-designed methods are promising for overcoming the tumorigenesis challenge posed by undifferentiated cells.

Similarly, Elovic *et al.* eliminated undifferentiated cells through the lethal effect of expression of the mutant Bax (S184del) variant in human ESCs [143]. The translation inhibition activity of miR-499, which is overexpressed in differentiated cardiac cells, occurred, inhibiting the expression of the lethal Bax variant, when the mRNA containing the Bax variant was modified with miR-499 targeted sequences. It was reported that the undifferentiated human ESCs progressed to a rapid apoptotic process with lethal Bax expression due to the lack of miR499 expression [143]. Miki *et al.* developed another miRNA-switch approach for separating cells that differentiate from human iPSCs [144]. The reporter gene in mRNAs modified with different anti-miRNAs will not be translated if specific miRNAs are endogenous in the cell. In this case, the desired cell types can be selected from heterogeneous cell populations according to particular miRNAs specific to different cell types. The researchers suggested the use of miR-126-, miR-122-5p-, and miR-375 switches for endothelial cells, hepatocytes, and insulin-producing cells, respectively [144]. It is an undeniable fact that the elimination of undifferentiated cells and undesirable phenotype cells during cell differentiation will contribute to the success of platforms proposed for tissue engineering applications. The studies with complex logic operations were also corroborated in terms of enriching the tuning and control mechanisms of these miRNA switches, which are performed with basic logic operations (OFF/ON). Gong *et al.* improved the synthetic circuit designed with the hsa-miR-21 input-sensitive YES logic gate, with more complex multi-input logic gates (OR, AND, INHIBIT, XOR, XOR-AND, XOR-INHIBIT, and XOR-OR) [145]. Such systems, designed as sensors for endogenous miRNAs in the cell, could also enlighten tissue engineering applications of non-coding RNA-based biological circuits.

Orthogonality may be the main concern in miRNA-targeted genetic circuit designs. Cellular miRNA and mRNA transcripts do not interact in a one-to-one fashion. In other words, it may be that various miRNAs share a common target or that one miRNA interacts with different mRNA transcripts [146]. As mentioned previously, system designs sensitive to more than one miRNA may be the solution to overcome this challenge that can cause crosstalk. In addition, miRNAs can interact with other non-coding RNAs [147], making it necessary to define all transcript expression profiles of cells to be manipulated by a designed genetic circuit.

The other promising approach for constructing genetic circuits to control translational mechanisms is using RBPs, which have molecular functions to generate rapid and specific responses to environmental conditions in native cell functions [148, 149]. One of the most elegant examples of the use of RNA-binding proteins in mRNA-based genetic circuits is the development of mRNA devices designed with the human LIN28A protein binding site aptamer for the detection of cell differentiation. The reporter region of mRNA engineered to the LIN28A protein-sensitive in human iPSCs was not translated due to the LIN28A-aptamer interaction [150]. It was possible to sort iPSCs and differentiated cells. RNA may expose different structures in the translation of mRNAs. One of the best described in the literature is the k-turn structure formed by the binding of metal ions or proteins [151]. One of the most preferable proteins for designing programmable circuits in mRNA translation control mechanisms through functional RNA folding is the archaeal ribosomal protein L7Ae. After the modification of mRNAs with motifs to which L7Ae protein (*i.e.,* RNA binding protein, RBP) can bind, synthetic biological circuits for different purposes, such as determination of cell fate, repair or rewiring of intrinsic cellular defects, control of feedback mechanisms and protein expression, have been studied [152 - 154]. Endo *et al.* showed that it is possible to control the expression of more than one protein product by modifying two different mRNA transcripts using the L7Ae protein, thus controlling the expression of complex cellular functions or phenotype in cells [155]. Advances in mRNA device approaches have provided advanced tools for controlling and programming mammalian cells. Wagner *et al.* proposed an approach that converts mRNA-based translational control into a small molecule-sensitive manner. It was reported that mRNA modified with aptamer containing binding site to tetracycline-responsive repressor-dead box helicase 6 (TetR-DDX6) protein complex loses its translational inhibition mechanism with Dox molecule. Conversely, the mRNA device which activates translation inhibition mechanisms with small-molecule induction has also been addressed by researchers [156].

In fact, although researchers have proposed the phenomenon of mRNA devices with potency to enable many logic operations, these approaches are guiding tissue engineering approaches as they offer programming tools in mammalian cells.

RNA-based genetic circuit designs are superior and elegant tools in tissue engineering applications due to their advantages, such as eliminating the requirement of transcriptional control compared to DNA-based methods, using each modified synthetic RNA module as a serial circuit to form a cascade, and ease of logic gate modifications.

Multi-module Genetic Circuits

It is also reasonable to design tools as multi-layer modules to reduce the undesired effects of the circuits in cell programming on the cell function. The control mechanisms for using DNA-based and RNA-based genetic circuit designs together can be used as a potent tool (Fig. **3A**). One early research on that was conducted by Leisner *et al.* [157]. With activator and repressor transcription factor inputs in mammalian cells, the control of transcription of miRNAs was regulated as logic gates, and cell programming can be controlled through translational inhibition mechanisms on the targets of miRNAs. In addition, miRNAs can be directly used in the control of protein expression due to their translation inhibition mechanisms on their target mRNAs. The mechanism to control miRNA expression by Dox induction through a miRNA circuit was designed with cytomegalovirus containing Tet-responsive promoter [158].

Moreover, the researchers were able to adapt the miRNA repressor activity into MS2 aptamer-dependent dynamics. For this, the aptamer of the bacteriophage coat MS2 protein was inserted into the genetic circuit, and it was shown that the genetic circuit can also perform MS2 transcription and that the output can be controlled by acting competitively against the other target of the miRNA. Thus, it was shown that complex module designs and fine-tuning systems can be suggested to control repressor activities on miRNA targets.

On the other hand, multiple RNA genetic circuits can offer advanced techniques for constructing multilayer control circuits (Fig. **3B**). L7Ae protein to be expressed through one modified mRNA with miRNA can act as a translation regulator on another modRNA. Moreover, another miRNA may attend the L7Ae. Further, the L7Ae RBP can be used to control the expression of a mediator protein that regulates the expression of output-containing mRNA. Wroblewska *et al.* described an RNA-based circuit design that connects multiple modified mRNA circuits like a serial electronic circuit [159]. The first modRNA module, miR-21 target (anti-miR), and L7Ae modRNA were designed in two modRNA layered circuits controlled by the expression of multiple miRNAs. Another modRNA containing the L7Ae protein binding site and target sequences, miR-141, miR-142 and miR-146a, were implemented as the second element of the serial genetic circuit. Thus, with a single RNA (miRNA) input, the cell selectively produced RBP in a two-layer genetic circuit and was driven to apoptosis. As a result of designing these approaches as logic gates, the determination of the cell state or programming of the cell can also be achieved. Secondary modRNA translation was controlled by translation from L7Ae mRNA implemented with miR-21 and miR-302 AND gate [160]. The researchers amended the genetic circuits they designed to operate on AND, OR, NAND, NOR, and XOR gates. In addition, in order to increase the L7Ae sensitivity to the repressive activity of miRNA, the miRNA targeted sequences were integrated in large numbers at both ends of the modified mRNA, 5' and 3'. Of the gates tested, especially the AND gate, an advanced technique was employed for regulating the expression of the output protein and determining cell fate.

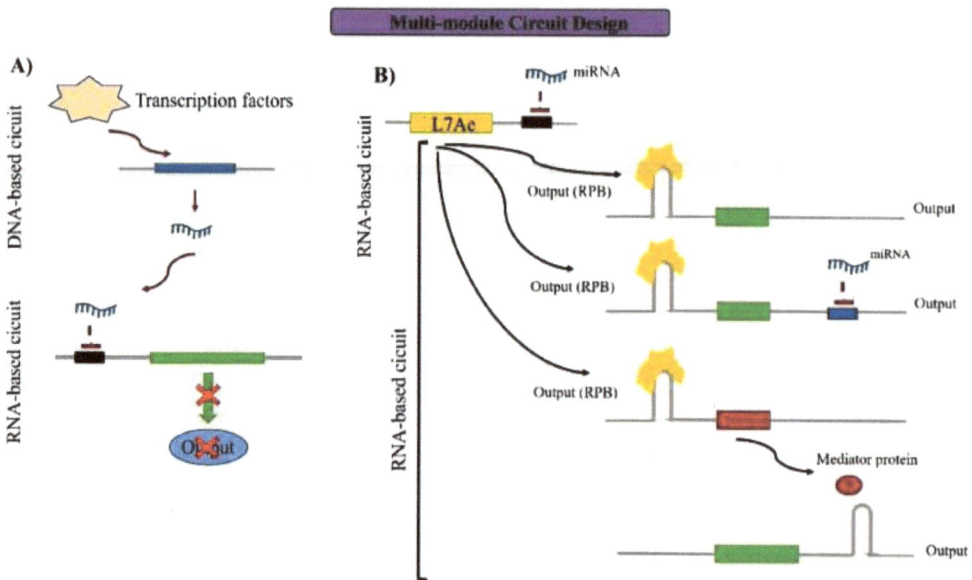

Fig. (3). Genetic circuits could be designed by constructing various types of circuits. **A**) DNA- and RNA-based module genetic circuit, **B**) L7Ae RBPs, produced through a genetic circuit, may induce another circuit to control gene expression and function.

The complex structure and specific functional properties of mammalian cells feed forward synthetic biology approaches in tissue engineering applications. Barros *et al.* used astrocyte cells as logic gates and designed them as molecular computing machinery [161]. Ca^{2+} signals and two different compounds and two different logic gate structures were used to induce the gene in human astrocytoma cells modified with pcDNA3.1 plasmid containing the hGPR17 gene. It was shown that if different cell groups in the cell population are modified with various gates and the signal transduction occurring throughout the population is induced by compounds, the cells can be manipulated depending on the concentration of the Ca^{2+} signals. Researchers emphasized that it is a superior approach that can form a basis for the development of advanced techniques of neural-molecular-based chips and even brain implants in the future.

In synthetic biology approaches, complex circuit designs that can implement multi-input, process and output contents can offer excellent approaches for high-order tissue engineering involving complex interactions. However, the highly complex nature of mammalian tissues is the major challenge in transforming these synthetic circuits into viable tissue-engineered parts. Ensuring orthogonality, which is one of the considerable parameters, especially in genetic circuit designs, is the most noteworthy challenge in mammalian cells. For this reason, it was also suggested to construct predictive models by simulating the molecular arrangements in mammalian cells with computational approaches [162]. Taking biological modules step-by-step, one-by-one, and even as integrated matters in the functions from transcription to post-translation by using circuits with synthetic biological tools has a driving likelihood in the development of real-life approaches from state-of-art.

These approaches can offer advanced techniques for the design of orthogonal modulating circuits by triggering specific regulations sequentially. These molecular mechanisms can also enable cells to be used in perfectly designed tissue engineering applications as a result of receiving signals induced by biomaterials used as scaffolds in tissue engineering applications and processing these inputs with molecular mechanisms.

BIOMATERIALS IN SYNTHETIC BIOLOGY

Early examples of synthetic biology focused mainly on using gene regulation systems to create new gene expression systems and transcription-based logic gates and circuits [17]. Although the term synthetic biology is based on the regulation and control of gene expression for genetic reprogramming of cells, this approach has also led up to the design of bioactive materials that can affect cellular processes in tissue engineering [19].

The ECM is an extensive molecular network that fills the extracellular spaces of all tissues and creates a 3D micro-environment for cells. In addition to providing physical support to cells, it also initiates biochemical and biomechanical cues necessary for cellular behaviors such as cell proliferation, differentiation, and migration [163, 164]. Therefore, tissue engineering, whose primary goal is to repair damaged or diseased tissues, has focused on biomaterials to mimic the properties of native tissue ECM [165].

Biomaterials are used to provide cells with a 3D micro-environment specific to *in vivo* environments. However, growing data has revealed that it is difficult to mimic and control the interaction between intrinsic (signal for gene expression) and extrinsic signals (signals from other cells and ECM) in these conventional scaffolds [166 - 168]. Within the framework of this approach, studies have been started on the production of various scaffolds, such as smart scaffolds that undergo reversible conformational changes due to the desired stimulus, and self-assembled peptide-based scaffolds. In addition, it was aimed to control cell behaviors by adding building blocks, such as genetic inducers, tissue-specific peptides, or virus-like particles to the scaffolds. In this part, the studies will be discussed on a material basis.

Hydrogels

The use of hydrogels as scaffolds is relevant because their mechanical properties can be adapted to mimic natural tissues. Moreover, these scaffolds provide a suitable environment for various molecules to move freely through the material. On the other hand, hydrogels lack various signals that induce cellular behaviors. Therefore, these scaffolds need to be modified to create 3D micro-environments in which the intrinsic and extrinsic signals can be controlled (Fig. **4**) [169, 170].

Deans *et al.* evaluated the applicability of combining synthetic biology with biomaterials using genetic inducers, first constructed polyethylene glycol (PEG) hydrogels containing the genetic inducer IPTG used these hydrogels for spatial control of gene expression (Fig. **4A**) [166]. IPTG was combined with PEG hydrogels *via* an ester bond, and as a result of the hydrolysis of the ester bond over time, this inducer was released into the medium of Chinese hamster ovary (CHO) cells transfected with LTRi_EGFP. LTRi_EGFP is a synthetic genetic switch able to regulate enhanced green fluorescent protein (EGFP) by the addition of IPTG [171]. After 4 days of culture, EGFP expression was detected in CHO cells. The results obtained showed an initial burst of IPTG release and a slower and sustained IPTG release over the subsequent days. Furthermore, studies were conducted to examine whether EGFP expression can be controlled by *in vivo* induction of genetic circuits. For this purpose, PEG hydrogels without IPTG but containing CHO cells transfected with LTRi_EGFP were first obtained, and then these hydrogels were implanted in the abdominal cavity of athymic mice. EGFP induction was started by adding IPTG to the drinking water of mice. After IPTG administration, EGFP expression was confirmed using fluorescent microscopy and PCR. The expression level was easily adjusted by adding different amounts of

IPTG to the water. In another study using a genetic circuit, IPTG was covalently attached to PEG hydrogels *via* 2-(2-azido-6-nitrophenyl)ethoxycarbonyl (ANPEOC) bond, which is biocompatible, easy to synthesize, and photodegradable. In addition, 1×10^6 CHO cells transfected with LTRi_EGFP were encapsulated in PEG-λIPTG hydrogels. The hydrogels were then exposed to 302 nm UV light. In the absence of light, the inducer molecules remained attached to the biomaterial, and EGFP expression was not observed. On the other hand, photolabile bonds were broken in light-exposed systems, releasing inducer molecules that activate the LTRi_EGFP genetic circuit in cells [168]. These studies showed that it is possible to construct highly defined engineering niches for the spatial and temporal control of gene expression.

The association or dissociation of the hydrogel due to a stimulus functioning under physiological conditions can be used to regulate cell behaviors (Fig. **4B**). Ehrbar *et al.* designed a hydrogel for the responsive release of human VEGF to aminocoumarin antibiotic levels [172]. The hydrogel was obtained by dimerizing the genetically engineered bacterial gyrase subunit B (GyrB) bound to polyacrylamide by adding coumermycin to the medium. The addition of novobiocin to the medium at increasing concentrations provided the separation of these subunits and accordingly the dissociation of the hydrogel. With the dissociation of the hydrogel, dose- and time-dependent release of VEGF entrapped in the structure occurred to trigger the proliferation of human umbilical vein endothelial cells (HUVECs). In another study, VEGF-containing hydrogels were obtained by binding FK-binding protein 12 to polyacrylamides instead of GyrB to avoid possible *in vivo* immunogenic effects of bacterial GyrB. The dose- and time-dependent release of VEGF was obtained by the administration of the FK506 inducer molecule, which monomerizes FM. In addition, these hydrogels were implanted subcutaneously in mice, and VEGF molecules were detected in mice serum by injection or oral administration of FK506 to mice [173]. In addition to hydrogels based on protein-protein interaction, hydrogels based on DNA-protein interaction were also developed. This hydrogel was obtained by modifying the TetO DNA motif and single-chain TetR, a protein that binds this motif and is separated from this motif by the presence of the antibiotic tetracycline, to polyacrylamide gels. Tetracycline-induced dissociation was used for the dose-dependent release of the cytokine interleukin 4 (IL-4) in the hydrogel structure [174]. The development of smart hydrogels, such as those mentioned above, has shown that dose- and time-dependent release of various molecules depending on the desired stimulus can be possible, thereby controlling cellular behaviors.

Another approach to mimic the complex nature of tissues and control cell behavior on the scaffold is to integrate tissue-specific peptide sequences into the design of biomaterials (Fig. **4C**). A comprehensive study in this area was done by Jansen *et al.* [175]. In this study, first, 20 bone marrow-specific peptide sequences were identified using proteomics-based bioinformatics and biomechanics. Subsequently, all these peptide sequences were attached to the PEG hydrogels. Thus, bone marrow hydrogels with bone marrow peptide sequences and elasticity were obtained. The effect of the hydrogels on the proliferation and differentiation of mesenchymal stem cells (MSCs) was analyzed and compared to RGD (Arg–Gly–Asp) functionalized PEG hydrogel and tissue culture polystyrene. The results showed that this bone marrow-inspired hydrogel improved cell proliferation and increased bone differentiation capacity. With this approach, which allows adding many tissue-specific components to hydrogels, it is possible to give many signals to cells together and to examine the effect of each signal on cell phenotype.

Today, various methods are being investigated for the synthesis of protein-based hydrogels with desired mechanical, physical, and functional properties and the cross-linking of tissue-specific protein building blocks to these hydrogels. One of these methods is the SpyTag-SpyCatcher system, a protein ligation tool (Fig. **4D**). This system was designed by cleaving and engineering the second immunoglobulin-like collagen adhesin domain (CnaB2) of *Streptococcus pyogenes* fibronectin-binding protein (FbaB), which forms a natural intramolecular isopeptide bond, into a pair of reactive protein domains (SpyTag and SpyCatcher) [176, 177]. SpyTag (a peptide tag with 13 amino acids) and SpyCatcher (a C-terminal fragment containing 138 amino acids) are able to self-assemble to reconstitute the intact CnaB2 domain [178]. This reaction takes place at temperatures between 4-37 °C without the need for an additional chemical reagent [179]. This system was used to design elastin-like polypeptide (ELP) hydrogels [180]. Moreover, this system also allows the insertion of various structures, such as cell adhesion ligands, matrix metalloproteinase cleavage sites, and full-length globular proteins into protein-based hydrogels. Sun *et al.* designed a self-crosslinking hydrogel by inserting the SpyTag and SpyCatcher into specific regions of the ELP sequences [181]. They also added an integrin-binding RGD sequence for cell adhesion and metalloproteinase-1 cleavage site for matrix remodeling to these hydrogels. Then, researchers encapsulated 3T3 fibroblasts into the hydrogels to see the potential of these hydrogels in tissue engineering applications. After 24 hours of incubation, cell viability was observed to be 95%, and viable cells exhibited 'long morphologies' indicative of cell adhesion and matrix remodeling. For the control of stem cell fate, leukemia inhibitory factor, a cytokine that inhibits the differentiation of ESCs, was added to the hydrogels. Mouse ESCs encapsulated in hydrogels continued to maintain their pluripotent properties in the presence of leukemia inhibitory factors. In the study by Gao *et al.*, instead of using ELP sequences, folded globular domains (GB1 and FnIII domain) were used as building blocks and multiple SpyTag or SpyCatcher sequences were designed to form protein hydrogels [182]. The analysis showed that these obtained hydrogels were soft, stable, and biocompatible. In addition, human lung fibroblasts were successfully encapsulated into the hydrogels and high cell viability (>90%) was observed after 96 hours of incubation. As an alternative to the SpyTag-SpyCatcher system, a new Tag-Catcher pair from *Streptococcus dysgalactiae* fibronectin binding protein was also designed and optimized under the name SdyTag-SdyCatcher. Moreover, methods for kinetically controllable directed protein ligation were developed by using these two systems together (dual tag (SpyTag-EGFP-SdyTag) and dual catcher (SpyCatcher-SdyCatcher) structures) [183]. Considering the designs obtained with these systems, it is thought that these systems will allow the production of new and dynamic protein architectures that can control the behavior of cells.

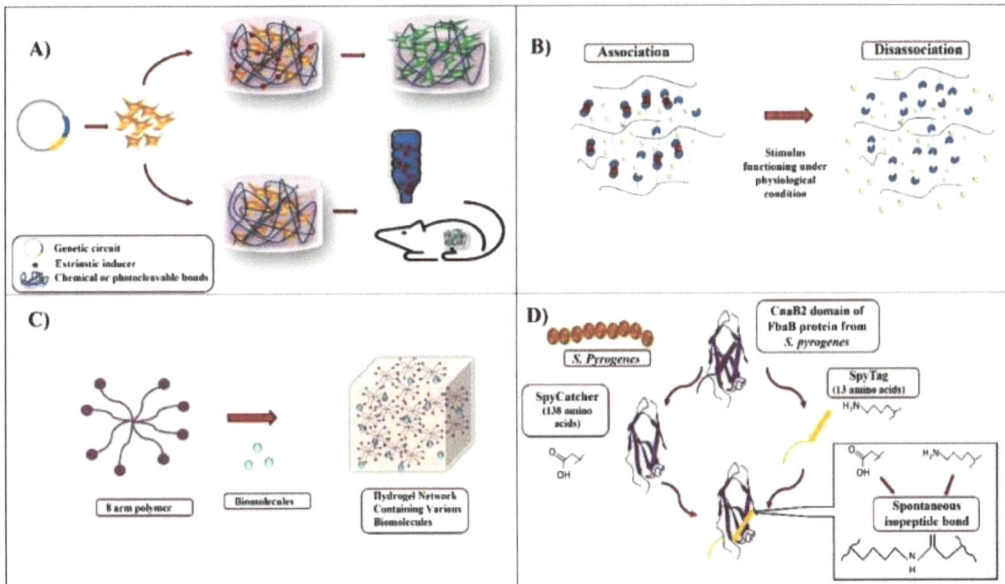

Fig. (4). Combining synthetic biology with biomaterials. **A)** Hydrogel-mediated induction and programming of cells transfected with synthetic genetic circuits, **B)** Controllable hydrogels integrated with genetically modified biomolecules, **C)** Multi-arm hydrogels that control cell behavior with many tissue-specific components, **D)** Development of new protein-based hydrogels using the SpyTag and SpyCatcher system.

Self-Assembling Peptide Hydrogels

Self-assembling peptides (SAPs) consist of periodic repeats of variable ionic hydrophilic and hydrophobic amino acids with a length of about 5-30 and have the ability to assemble into higher structures under different conditions or stimuli [184]. The first example of self-assembling peptides was reported by Zhang *et al.*, when it was discovered that the sequence AEAEAAKAKEAEAAKAK (EAK16-

II), derived from the Zuotin protein in yeast, readily induces hydrogelation of the solvent [185]. The EAK16-II peptide, which has charged and hydrophobic side groups, undergoes a conformational change in an aqueous environment, with the ionic groups oriented on the outside and the hydrophobic groups on the inside, and periodic repetition of this condition creates a stable structure with the effect of complementary forces and turns into typical β-sheet structures by self-assembly. The advanced aggregation of these β-sheets creates a nanofibrous hydrogel network [185, 186]. In subsequent studies, when the glutamic acid (Glu) and lysine (Lys) residues of the EAK16-II sequence were replaced by aspartic acid (Asp) and arginine (Arg), respectively, the resulting sequence named RADA16-II was found to show an even greater tendency for self-assembly and hydrogelation [187]. In the following years, many SAPs with different properties were developed from different amino acid types (Table **1**).

SAPs could be obtained by recombinant technologies or solid-phase synthesis methods, and it is possible to design SAPs with desired structures and functions by changing the number, type, and sequence of amino acids [215]. In addition, SAPs can be designed according to their secondary structure (β-sheets or α-helix) or chemistry (hydrogen bonding, hydrophobicity and/or electrostatic interaction), which can occur to produce self-assembled hydrogels under a given pH [214]. Cell cultures of neural cells, MSCs, chondrocytes, hepatocytes, cardiomyocytes, skin epithelial cells, and endothelial cells performed on synthetic peptide hydrogels have exhibited superior properties over conventional tissue cultures or polymer matrices [216, 217].

The importance of synthetic SAP hydrogels for tissue engineering and synthetic biology is because the peptide backbone can be modified by adding various bioactive functional motifs [218]. SAPs can be functionalized by modifications to include a variety of biological motifs to both the carbonyl and amino ends of the peptide backbone [215]. The approach to modification of the peptide backbone with functional motifs aims to directly control the tissue-directed biological activity of cells and better mimic the nanostructured architecture of the native ECM. To this end, a number of tailor-made peptide motifs have been developed, showing that biological cues that could actively participate in cellular signaling influence cell attachment, proliferation, migration, differentiation and gene expression [219]. Bioactive epitopes found in laminin, collagen and fibronectin proteins, such as RGD (Arg-Gly-Asp), IKVAV and YIGSR, are the most widely used bioactive motifs in tissue engineering [219].

Table 1. Some self-assembly peptides and tissue engineering applications.

Type of Gelling Stimuli	Peptide and Sequences	Applications	References
pH, ionic	**EAK16-II** (AEAEAKAK)$_2$	Nerve regeneration	[185, 188]
		Hydrophobic drug encapsulation	[189, 190]
	RADA16-1 (RADA)$_4$	Nerve regeneration	[187]
		Skin regeneration	[191]
		Hemostatic activity	[192]
		Controlled release	[193, 194]
	RAD16-II (RARADADARARADADA)	Hydrophobic drug carrier	[195]
		Nerve regeneration	[196]
		Myocardium regeneration	[197, 198]
		Wound healing	[199]
	FEFEFKFK	Nucleus pulposus regeneration	[200]
		Bone tissue engineering	[201]
		Cartilage tissue engineering	[202]
	MAX1 (VKVKVKVKVDPPTKVKVKVKV)	Tissue scaffold	[203]
		Antibacterial activity	[204]
	KFE-8 (FKFEFKFE)	Vascularization	[205]
	SPG-178 (RLDLRLALRLDLR)	Bone regeneration	[206]
		Hemostatic agent	[207]
	KLD-12 (KLDLKLDLKLDL)	MSC encapsulation	[208]
		Cartilage regeneration	[209]
Polar solvents	**P11-I** (QQRQQQQQEQQ) **P11-4** (QQRFEWEFEQQ) **P11-8** (QQRFOWOFEQQ)	Controlled release and bone repair Periodontal tissue regeneration	[210 - 212]
Temperature	**hSAFAAA p1** (IAALKAK-IAALKAE IAALEAENAALEA) **hSAFAAA p2** (IAALKAK-NAALKA- -IAALEAEIAALEA)	Neural tissue engineering	[213]
Ca ions	**E1Y9** (E-YEYKYEYKY)	Bone regeneration	[214]

From the synthetic biology perspective, SAPs functionalized with various motifs can affect the transcriptional activity of cells by acting on various cellular signaling pathways through their interaction with receptor domains or proteins [220]. For example, KLD-12 SAP hydrogel scaffolds functionalized with the N-cadherin-derived HAVDI motif, an important transmembrane protein in cell-cell interactions and intracellular signaling, have been shown to promote chondrogenic differentiation characterized by increased expression of chondrogenic marker genes in human MSCs [221]. Results have suggested that SAP hydrogel scaffolds suppress canonical Wnt signaling in human MSCs through decreasing β-catenin nuclear translocation and related transcriptional activity of the β-catenin/LEF-1/TCF complex, thereby increasing the chondrogenesis of human MSCs. In another study, Yang *et al.* reported that RAD16 peptide functionalized with laminin-derived motif IKVAV and brain-derived neurotrophic factor-mimetic peptide epitope RGI (RGIDKRHWNSQ) motifs have enhanced gene expression of nerve growth factor, brain-derived neurotrophic factor, ciliary neurotrophic factor, and peripheral myelin protein 22 in rat Schwann cells [222]. Some other studies of SAPs functionalized with various bioactive motifs are summarized in Table **2**.

Table 2. Some studies of SAPs functionalized with various bioactive motifs in tissue engineering applications.

Motif	Origin	SAP	Tissue Engineering Application	References
ALK n(ALKRQGRTLYGF)	Bone secretory signal peptides	E1Y9	High cell proliferation and differentiation activity for pre-osteoblast cell line MC3T3-E1. Increasing expression of RUNX2 and osteopontin. Promote the activity of alkaline phosphatase.	[223, 224]
PRG (PRGDSGYRGDS)	Two-unit RGD (integrin binding sequence) motif from collagen type IV	RADA16	Increased cell proliferation demonstrated in *in vivo* rat model of periodontal defects, enhancing the expression of VEGF.	[225]
			Increased proliferation of skin-derived precursors (SKPs) cells. Increased Akp2 and Bmp6 gene expressions. Endothelial cell activity for angiogenesis.	[226, 227]

(Table 2) cont.....

Motif	Origin	SAP	Tissue Engineering Application	References
PRG+ KLT (KLTWQELYQLKYKGI)	VEGF-mimicking peptide (17–25 helix region of VEGF)	RADA16	Improved proliferation rate and VEGF-secreting capability of human dental pulp stem cells (hDPSCs).	[228]
FPG (FPGERGVEGPGP)	Collagen type I	RADA16	Increased fibroblast migration.	[191]
HAVDIGG	N-cadherin mimetic peptide	KLD	Suppressed canonical Wnt signaling promoting chondrogenic differentiation in human MSCs.	[221]
IKVAV	α-laminin chain	RADA16	Increased expression of adhesion-associated genes. Enhanced neural progenitor cells (NPCs) proliferation and neuronal differentiation.	[229]
YIGSR	β-laminin chain	KLD	Enhanced vasculogenesis in human MSCs/HUVECs co-culture.	[230]
		RADA16	Decreasing beta-amyloi--mediated hippocampal apoptosis by attenuating downregulation of synapsin-1 in NSC.	[231]
KPSSAPTQLN	BMP-7 mimetic motif	RADA16	Induction of increase of collagen 2 and aggregate levels in bone marrow MSCs in nucleus pulposus regeneration, enhancing chemotactic migration.	[232]
SKP (SKPPGTSS) and PFS (PFSSTKT)	Bone marrow homing peptide (BMHP)	RADA16	VEGF-/HGF-secretion ability.	[233]
			Neural cell growth, migration, adhesion, and differentiation.	[234]
			Human adipose stem cells promoted cell survival, attachment, and proliferation.	[235]

Functionalized SAPs can maintain their ability to influence cellular events even when incorporated into polymer matrices. For example, Danesin *et al.* reported that functionalized SAPs maintain osteoblast differentiation of electrospun scaffolds enriched with polycaprolactone, stimulate osteoblast adhesion and expression of hALP, hOPN, and hBSP genes [236]. For peripheral nerve regeneration, Nune *et al.* developed RADA16-I-BMHP1/ poly(lactic-*co*-glycolic acid) electrospun hybrid scaffolds and evaluated their effectiveness in Schwann cells in terms of adhesion, proliferation and gene expression levels [237]. After 7 days of culture, the peptide-coated scaffolds exhibited significantly higher proliferation rates compared to the control poly(lactic-*co*-glycolic acid) nanofiber scaffolds. Gene expression analysis of PMP22, NCAM, and GFAP genes, showed that they significantly increase gene expression levels in peptide-coated scaffolds compared to poly(lactic-*co*-glycolic acid) scaffolds. All these studies show that SAP hydrogels functionalized with bioactive motifs can serve as adaptable and versatile 3D culture platforms to investigate the effect on cell behavior and to direct cellular processes [205, 215].

In addition, SAPs are attractive candidates for controlling the release of various gene transporters to increase residence time and transfection efficiency in a defined target region. For example, Rey-Rico *et al.* have used hyaluronic acid (RAD-HA)-conjugated peptide hydrogel RAD16-I in the controlled release of rAAV vectors to genetically modify primary human bone marrow-derived MSCs, reporting high cell viability and high transduction efficiencies [238]. In another study, SAP (DDIKVAVK) hydrogels derived from the laminin-derived IKVAV motif were investigated for localized viral vector delivery *in vivo* [239]. Researchers have reported that stereotaxic injection of mCherry lentivirus immobilized on SAP hydrogels into mouse brain produced a significant difference in transduction level compared to viral vector alone and affected transfection cell number and volume. In a recent study, Zhang *et al.* have used layer-by-layer (LbL) SAP coatings on nanofibers for the localized delivery of CRISPR/dCas9 systems to control specific gene expression [240]. In the study, efficient loading and sustained release of electrostatically coated peptide-DNA complexes with SAP+-RGD (FKFKFKFKGGRGDSP) on polycaprolactone nanofibers were achieved. Designed scaffolds to activate GDNF expression were reported to promote neurite outgrowth of rat neurons. The researchers suggest that SAP-coated nanofibers could be a novel way to create a bioactive interface that provides a simple and effective platform for the deployment of CRISPR/dCas9 systems for tissue engineering and regenerative medicine.

Virus-like Particles

Today, many studies are investigating the surface topographic properties of materials trying to establish connections between the adhesion, proliferation, and differentiation of cells to control them. Virus-like particles, which allow controlling their topology accurately thanks to their repetitive structures and self-assembly properties, are the leading building blocks of great interest for cell orientation [18, 241, 242].

Rong *et al.* produced unidirectional thin films from M13 bacteriophages conjugated with RGD to direct cell growth [243]. Two different cell lines (3T3 and CHO) were then cultured on these films to test the universality of cell alignment. As a result, the addition of RGD tripeptide increased cell adhesion and growth in one direction, responding to the oriented surface. In another study by Lin *et al.*, the self-assembly properties of rod-like tobacco mosaic virus (TMV) particles were used to coat a glass capillary tube [244]. In the study, different patterns were created in capillary tubes depending on parameters, such as TMV particle concentration, the salt concentration in the aqueous solution, and the properties of the inner surface of the capillary tube. Then, smooth muscle cells were cultured on these patterns, and it was observed that the growth and morphology of these cells changed depending on the patterns. Wang *et al.* designed an artificial ECM activated by M13 that contains biochemical cues RGD and PHSRN (Pro-His-Ser-Arg-Asn) to regulate morphology, proliferation, and osteoblastic differentiation of rat MSCs [245]. In this study, the researchers took advantage of the phage self-assembly in the ridge/groove structure. As a result of the study, it was shown that osteoblastic differentiation of MSCs was induced without any osteogenic supplements, thanks to ridge/groove nanotopography. Furthermore, it was observed in the study that the self-assembly of the phage was not affected by the added peptides, and the nanotopography remained stable. This result revealed that it is possible to form an ECM with different peptides using this method, and the effects of these peptides on cells can be systematically investigated.

CONCLUSION

In this chapter, we aimed to show how various synthetic biology tools can be applied using the intrinsic and extrinsic modules of cells in the context of tissue engineering. Although the classical tissue engineering paradigm, which involves culturing and transplanting cells differentiated to specific lineages with various biochemical molecules on 3D artificial matrices for new tissue formation, or regeneration, is effective for simple tissues, the formation and regeneration of more complex tissues/organs require more complex systems. Synthetic biological approaches are important candidates for constructing these complex systems. The genetic reprogramming of living systems, especially cells, with a variety of synthetic biology tools and the high control over this programming have revolutionized cell-based approaches in tissue engineering. Thanks to the ability to combine different genetic parts to serve new functional purposes, cells with specific characteristics can be endowed with transcriptional control with stimuli, such as small molecules and light, which have been designed and investigated for tissue engineering applications. In addition, with the success of genome editing tools, such as engineering ZFN, TALEN, and CRISPR in the field of stem cells in recent years, great strides have been made to produce reprogrammed ESCs and iPSCs to treat genetic diseases and understand developmental processes, such as morphogenesis. In addition to synthetic biology approaches that directly target intracellular mechanisms, biomaterials-mediated applications in tissue engineering are remarkable. Various studies have shown that cell fate decisions are also strongly influenced by extracellular factors. The design of synthetic materials that combine functional extracellular domains and modules that can guide the growth and behavior of cells has paved the way for the development of dynamic and adaptive systems in tissue engineering. The combined use of both intracellular and extracellular tools, such as biomaterials, in the process of controlled tissue formation and restoration, may enable the development of hybrid systems for temporal and spatial control of cells in tissue engineering.

On the other hand, although synthetic biology is one of the most promising approaches in tissue engineering, there are still some problems to be overcome. Integration difficulties of transgenes into mammalian cells due to effects, such as genomic localization, transcriptional disruption, chromatin rearrangement, or DNA methylation, are important research topics in synthetic biology. For example, *in vivo* monitoring of cells that have lost their differentiation ability or non-differentiated cells after iPSCs transplantations is very important to prevent conditions, such as tumorigenicity. The use of engineering genome editing tools, such as ZFN, TALEN, and CRISPR/Cas, for the integration, accuracy, stability, and reproducibility of transgenes in mammalian cells, is an important advance for cellular reprogramming and the stability of genetic networks. However, if applications of synthetic biology in tissue engineering are to be used in clinical applications, tools must be developed to repair damaged elements of synthetic networks or return systems to their original structures after unexpected changes.

Different tools regarding the synthetic biology approach make it possible to redesign the organism. Nonetheless, there are concerns, such as the risk of gene transfer from a synthetically created biological entity to the organism and a possible deterioration of natural genomes. In addition, sanitary issues are plausible if these engineered assets are thrown away into the environment; however, real consequences are yet unknown. Moreover, biotechnology may be misused for bioterrorism purposes. To date, no biosafety risks related to synthetic biology have been reported. However, it is thought that the potential advantages of these technologies should be evaluated as well as the risks that may arise.

CONSENT FOR PUBLICATION

Not applicable.

CONFLICT OF INTEREST

The author declares no conflict of interest, financial or otherwise.

ACKNOWLEDGEMENT

Declared none.

REFERENCES

[1]   Langer R, Vacanti JP. Tissue Engineering. Science 1993; 260(5110): 920-6.
[http://dx.doi.org/10.1126/science.8493529] [PMID: 8493529]

[2]   Khademhosseini A, Vacanti JP, Langer R. Progress in tissue engineering. Sci Am 2009; 300(5): 64-71.
[http://dx.doi.org/10.1038/scientificamerican0509-64] [PMID: 19438051]

[3]   Patrick CW, Mikos AG, McIntire LV. Frontiers in tissue engineering. Elsevier 1998; pp. 5-9.

[4]   Chen G, Ushida T, Tateishi T. Scaffold design for tissue engineering. Macromol Biosci 2002; 2(2): 67-77.
[http://dx.doi.org/10.1002/1616-5195(20020201)2:2<67::AID-MABI67>3.0.CO;2-F]

[5]   Chapekar MS. Tissue engineering: Challenges and opportunities. J Biomed Mater Res 2000; 53(6): 617-20.
[http://dx.doi.org/10.1002/1097-4636(2000)53:6<617::AID-JBM1>3.0.CO;2-C] [PMID: 11074418]

[6]   Saltzman WM, Kyriakides TR. Cell interactions with polymers. Lanza R, Langer R, Vacanti JP, Atala A, Eds Principles of tissue engineering. Academic press, 2014; pp. 385-406.
[http://dx.doi.org/10.1016/B978-0-12-398358-9.00020-3]

[7]   Atala A, Kasper FK, Mikos AG. Engineering complex tissues. Sci Transl Med 2012; 4(160): 160rv12.
[http://dx.doi.org/10.1126/scitranslmed.3004890] [PMID: 23152327]

[8]   Eltom A, Zhong G, Muhammad A. Scaffold techniques and designs in tissue engineering functions and purposes: a review. Adv Mater Sci Eng 2019; 2019: 1-13.
[http://dx.doi.org/10.1155/2019/3429527]

[9]   Vleminckx K, Kemler R. Cadherins and tissue formation: integrating adhesion and signaling. BioEssays 1999; 21(3): 211-20.
[http://dx.doi.org/10.1002/(SICI)1521-1878(199903)21:3<211::AID-BIES5>3.0.CO;2-P]   [PMID: 10333730]

[10]   Gao L, Kupfer ME, Jung JP, *et al.* Myocardial tissue engineering with cells derived from human-induced pluripotent stem cells and a native-like, high-resolution, 3-dimensionally printed scaffold. Circ Res 2017; 120(8): 1318-25.
[http://dx.doi.org/10.1161/CIRCRESAHA.116.310277] [PMID: 28069694]

[11]   Dee KC, Puleo DA, Bizios R. An introduction to tissue-biomaterial interactions. John Wiley & Sons 2003; pp. 149-72.
[http://dx.doi.org/10.1002/0471270598.ch8]

[12]   Dubey DK, Tomar V. Role of molecular level interfacial forces in hard biomaterial mechanics: a review. Ann Biomed Eng 2010; 38(6): 2040-55.
[http://dx.doi.org/10.1007/s10439-010-9988-3] [PMID: 20221805]

[13]   Stegemann JP, Dey NB, Lincoln TM, Nerem RM. Genetic modification of smooth muscle cells to control phenotype and function in vascular tissue engineering. Tissue Eng 2004; 10(1-2): 189-99.
[http://dx.doi.org/10.1089/107632704322791844] [PMID: 15009945]

[14]   Werner K, Weitz J, Stange DE. Organoids as model systems for gastrointestinal diseases: tissue engineering meets genetic engineering. Curr Pathobiol Rep 2016; 4(1): 1-9.

[http://dx.doi.org/10.1007/s40139-016-0100-z]

[15]   Benner SA, Sismour AM. Synthetic biology. Nat Rev Genet 2005; 6(7): 533-43.
        [http://dx.doi.org/10.1038/nrg1637] [PMID: 15995697]

[16]   Heinemann M, Panke S. Synthetic biology--putting engineering into biology. Bioinformatics 2006;
        22(22): 2790-9.
        [http://dx.doi.org/10.1093/bioinformatics/btl469] [PMID: 16954140]

[17]   Verbič A, Praznik A, Jerala R. A guide to the design of synthetic gene networks in mammalian cells.
        FEBS J 2021; 288(18): 5265-88.
        [PMID: 33289352]

[18]   Bryksin AV, Brown AC, Baksh MM, Finn MG, Barker TH. Learning from nature – Novel synthetic
        biology approaches for biomaterial design. Acta Biomater 2014; 10(4): 1761-9.
        [http://dx.doi.org/10.1016/j.actbio.2014.01.019] [PMID: 24463066]

[19]   Tang TC, An B, Huang Y, *et al.* Materials design by synthetic biology. Nat Rev Mater 2021; 6(4):
        332-50.
        [http://dx.doi.org/10.1038/s41578-020-00265-w]

[20]   Lambert SA, Jolma A, Campitelli LF, *et al.* The human transcription factors. Cell 2018; 172(4): 650-
        65.
        [http://dx.doi.org/10.1016/j.cell.2018.01.029] [PMID: 29425488]

[21]   Slattery M, Zhou T, Yang L, Dantas Machado AC, Gordân R, Rohs R. Absence of a simple code: how
        transcription factors read the genome. Trends Biochem Sci 2014; 39(9): 381-99.
        [http://dx.doi.org/10.1016/j.tibs.2014.07.002] [PMID: 25129887]

[22]   D'Alessio AC, Fan ZP, Wert KJ, *et al.* A systematic approach to identify candidate transcription
        factors that control cell identity. Stem Cell Reports 2015; 5(5): 763-75.
        [http://dx.doi.org/10.1016/j.stemcr.2015.09.016] [PMID: 26603904]

[23]   Kim S, Shendure J. Mechanisms of interplay between transcription factors and the 3D genome. Mol
        Cell 2019; 76(2): 306-19.
        [http://dx.doi.org/10.1016/j.molcel.2019.08.010] [PMID: 31521504]

[24]   Peng SL. Transcription factors in autoimmune diseases. Front Biosci 2008; 13: 4218-40.
        [http://dx.doi.org/10.2741/3001] [PMID: 18508507]

[25]   Zhu M, Liu CC, Cheng C. REACTIN: Regulatory activity inference of transcription factors underlying
        human diseases with application to breast cancer. BMC Genomics 2013; 14(1): 504.
        [http://dx.doi.org/10.1186/1471-2164-14-504] [PMID: 23885756]

[26]   Huilgol D, Venkataramani P, Nandi S, Bhattacharjee S. Transcription factors that govern development
        and disease: An achilles heel in cancer. Genes (Basel) 2019; 10(10): 794.
        [http://dx.doi.org/10.3390/genes10100794] [PMID: 31614829]

[27]   Khalil AS, Lu TK, Bashor CJ, *et al.* A synthetic biology framework for programming eukaryotic
        transcription functions. Cell 2012; 150(3): 647-58.
        [http://dx.doi.org/10.1016/j.cell.2012.05.045] [PMID: 22863014]

[28]   Gossen M, Bonin AL, Freundlieb S, Bujard H. Inducible gene expression systems for higher
        eukaryotic cells. Curr Opin Biotechnol 1994; 5(5): 516-20.
        [http://dx.doi.org/10.1016/0958-1669(94)90067-1] [PMID: 7765466]

[29]   Lutz R, Bujard H. Independent and tight regulation of transcriptional units in Escherichia coli *via* the
        LacR/O, the TetR/O and AraC/I1-I2 regulatory elements. Nucleic Acids Res 1997; 25(6): 1203-10.
        [http://dx.doi.org/10.1093/nar/25.6.1203] [PMID: 9092630]

[30]   Lewis M. The lac repressor. C R Biol 2005; 328(6): 521-48.
        [http://dx.doi.org/10.1016/j.crvi.2005.04.004] [PMID: 15950160]

[31]   Baron U, Bujard H. Tet repressor-based system for regulated gene expression in eukaryotic cells:

Principles and advances. Methods Enzymol 2000; 327: 401-21.
[http://dx.doi.org/10.1016/S0076-6879(00)27292-3] [PMID: 11044999]

[32]   Biard DSF, James MR, Cordier A, Sarasin A. Regulation of the Escherichia coli lac operon expressed in human cells. Biochim Biophys Acta Gene Struct Expr 1992; 1130(1): 68-74.
[http://dx.doi.org/10.1016/0167-4781(92)90463-A] [PMID: 1311956]

[33]   Furth PA, St Onge L, Böger H, *et al.* Temporal control of gene expression in transgenic mice by a tetracycline-responsive promoter. Proc Natl Acad Sci USA 1994; 91(20): 9302-6.
[http://dx.doi.org/10.1073/pnas.91.20.9302] [PMID: 7937760]

[34]   Caron L, Prot M, Rouleau M, Rolando M, Bost F, Binétruy B. The Lac repressor provides a reversible gene expression system in undifferentiated and differentiated embryonic stem cell. Cell Mol Life Sci 2005; 62(14): 1605-12.
[http://dx.doi.org/10.1007/s00018-005-5123-2] [PMID: 15968459]

[35]   Kuo CJ, Conley PB, Chen L, Sladek FM, Darnell JE Jr, Crabtree GR. A transcriptional hierarchy involved in mammalian cell-type specification. Nature 1992; 355(6359): 457-61.
[http://dx.doi.org/10.1038/355457a0] [PMID: 1734282]

[36]   Fussenegger M, Bailey JE. Control of mammalian cell proliferation as an important strategy in cell culture technology, cancer therapy and tissue engineering. In: Al-Rubeai M, Ed. Cell engineering. Dordrecht: Springer 1999; pp. 186-219.
[http://dx.doi.org/10.1007/978-0-585-37971-5_7]

[37]   Lee RJ, Springer ML, Blanco-Bose WE, Shaw R, Ursell PC, Blau HM. VEGF gene delivery to myocardium: deleterious effects of unregulated expression. Circulation 2000; 102(8): 898-901.
[http://dx.doi.org/10.1161/01.CIR.102.8.898] [PMID: 10952959]

[38]   Gidekel S, Pizov G, Bergman Y, Pikarsky E. Oct-3/4 is a dose-dependent oncogenic fate determinant. Cancer Cell 2003; 4(5): 361-70.
[http://dx.doi.org/10.1016/S1535-6108(03)00270-8] [PMID: 14667503]

[39]   Ueblacker P, Wagner B, Krüger A, *et al.* Inducible nonviral gene expression in the treatment of osteochondral defects. Osteoarthritis Cartilage 2004; 12(9): 711-9.
[http://dx.doi.org/10.1016/j.joca.2004.05.011] [PMID: 15325637]

[40]   Gersbach CA, Le Doux JM, Guldberg RE, García AJ. Inducible regulation of Runx2-stimulated osteogenesis. Gene Ther 2006; 13(11): 873-82.
[http://dx.doi.org/10.1038/sj.gt.3302725] [PMID: 16496016]

[41]   Tafuro S, Ayuso E, Zacchigna S, *et al.* Inducible adeno-associated virus vectors promote functional angiogenesis in adult organisms *via* regulated vascular endothelial growth factor expression. Cardiovasc Res 2009; 83(4): 663-71.
[http://dx.doi.org/10.1093/cvr/cvp152] [PMID: 19443424]

[42]   Tóth F, Gáll JM, Tőzsér J, Hegedűs C. Effect of inducible bone morphogenetic protein 2 expression on the osteogenic differentiation of dental pulp stem cells *in vitro*. Bone 2020; 132: 115214.
[http://dx.doi.org/10.1016/j.bone.2019.115214] [PMID: 31884130]

[43]   Marquardt LM, Ee X, Iyer N, *et al.* Finely tuned temporal and spatial delivery of GDNF promotes enhanced nerve regeneration in a long nerve defect model. Tissue Eng Part A 2015; 21(23-24): 2852-64.
[http://dx.doi.org/10.1089/ten.tea.2015.0311] [PMID: 26466815]

[44]   Chae J, Zimmerman LB, Grainger RM. Inducible control of tissue-specific transgene expression in Xenopus tropicalis transgenic lines. Mech Dev 2002; 117(1-2): 235-41.
[http://dx.doi.org/10.1016/S0925-4773(02)00219-8] [PMID: 12204263]

[45]   Saxena P, Bojar D, Zulewski H, Fussenegger M. Generation of glucose-sensitive insulin-secreting beta-like cells from human embryonic stem cells by incorporating a synthetic lineage-control network. J Biotechnol 2017; 259: 39-45.

[http://dx.doi.org/10.1016/j.jbiotec.2017.07.018] [PMID: 28739109]

[46] Bacchus W, Weber W, Fussenegger M. Increasing the dynamic control space of mammalian transcription devices by combinatorial assembly of homologous regulatory elements from different bacterial species. Metab Eng 2013; 15: 144-50.
[http://dx.doi.org/10.1016/j.ymben.2012.11.003] [PMID: 23178502]

[47] Hanna-Rose W, Hansen U. Active repression mechanisms of eukaryotic transcription repressors. Trends Genet 1996; 12(6): 229-34.
[http://dx.doi.org/10.1016/0168-9525(96)10022-6] [PMID: 8928228]

[48] Lodish H, Berk A, Zipursky SL, Matsudaira P, Baltimore D, Darnell J. Molecular mechanisms of eukaryotic transcriptional control. Molecular Cell Biology New York. CO: W.H. Freeman 2000.

[49] Zhong H, Voll RE, Ghosh S. Phosphorylation of NF-κ B p65 by PKA stimulates transcriptional activity by promoting a novel bivalent interaction with the coactivator CBP/p300. Mol Cell 1998; 1(5): 661-71.
[http://dx.doi.org/10.1016/S1097-2765(00)80066-0] [PMID: 9660950]

[50] Triezenberg SJ, Kingsbury RC, McKnight SL. Functional dissection of VP16, the trans-activator of herpes simplex virus immediate early gene expression. Genes Dev 1988; 2(6): 718-29.
[http://dx.doi.org/10.1101/gad.2.6.718] [PMID: 2843425]

[51] Ji H, Jiang Z, Lu P, *et al.* Specific reactivation of latent HIV-1 by dCas9-SunTag-VP64-mediated guide RNA targeting the HIV-1 promoter. Mol Ther 2016; 24(3): 508-21.
[http://dx.doi.org/10.1038/mt.2016.7] [PMID: 26775808]

[52] Hardwick JM, Tse L, Applegren N, Nicholas J, Veliuona MA. The Epstein-Barr virus R transactivator (Rta) contains a complex, potent activation domain with properties different from those of VP16. J Virol 1992; 66(9): 5500-8.
[http://dx.doi.org/10.1128/jvi.66.9.5500-5508.1992] [PMID: 1323708]

[53] Margolin JF, Friedman JR, Meyer WK, Vissing H, Thiesen HJ, Rauscher FJ III. Krüppel-associated boxes are potent transcriptional repression domains. Proc Natl Acad Sci USA 1994; 91(10): 4509-13.
[http://dx.doi.org/10.1073/pnas.91.10.4509] [PMID: 8183939]

[54] Nielsen AAK, Segall-Shapiro TH, Voigt CA. Advances in genetic circuit design: novel biochemistries, deep part mining, and precision gene expression. Curr Opin Chem Biol 2013; 17(6): 878-92.
[http://dx.doi.org/10.1016/j.cbpa.2013.10.003] [PMID: 24268307]

[55] Brophy JAN, Voigt CA. Principles of genetic circuit design. Nat Methods 2014; 11(5): 508-20.
[http://dx.doi.org/10.1038/nmeth.2926] [PMID: 24781324]

[56] Yin H, Kauffman KJ, Anderson DG. Delivery technologies for genome editing. Nat Rev Drug Discov 2017; 16(6): 387-99.
[http://dx.doi.org/10.1038/nrd.2016.280] [PMID: 28337020]

[57] Sharma T, Rawat N, Dua D, Singh MK, Alam A, Chauhan MS. Transfection methods affect cellular function and gene expression. Anim Sci Pap Rep 2018; 36: 431-51.

[58] Wu N, Ataai MM. Production of viral vectors for gene therapy applications. Curr Opin Biotechnol 2000; 11(2): 205-8.
[http://dx.doi.org/10.1016/S0958-1669(00)00080-X] [PMID: 10753765]

[59] Vannucci L, Lai M, Chiuppesi F, Ceccherini-Nelli L, Pistello M. Viral vectors: a look back and ahead on gene transfer technology. New Microbiol 2013; 36(1): 1-22.
[PMID: 23435812]

[60] Ren J, Lee J, Na D. Recent advances in genetic engineering tools based on synthetic biology. J Microbiol 2020; 58(1): 1-10.
[http://dx.doi.org/10.1007/s12275-020-9334-x] [PMID: 31898252]

[61] Heiderscheit EA, Eguchi A, Spurgat MC, Ansari AZ. Reprogramming cell fate with artificial

transcription factors. FEBS Lett 2018; 592(6): 888-900.
[http://dx.doi.org/10.1002/1873-3468.12993] [PMID: 29389011]

[62]   Miyaoka Y, Berman JR, Cooper SB, *et al.* Systematic quantification of HDR and NHEJ reveals effects of locus, nuclease, and cell type on genome-editing. Sci Rep 2016; 6(1): 23549.
[http://dx.doi.org/10.1038/srep23549] [PMID: 27030102]

[63]   Mani M, Kandavelou K, Dy FJ, Durai S, Chandrasegaran S. Design, engineering, and characterization of zinc finger nucleases. Biochem Biophys Res Commun 2005; 335(2): 447-57.
[http://dx.doi.org/10.1016/j.bbrc.2005.07.089] [PMID: 16084494]

[64]   Doyon Y, McCammon JM, Miller JC, *et al.* Heritable targeted gene disruption in zebrafish using designed zinc-finger nucleases. Nat Biotechnol 2008; 26(6): 702-8.
[http://dx.doi.org/10.1038/nbt1409] [PMID: 18500334]

[65]   Bedell VM, Wang Y, Campbell JM, *et al.* In vivo genome editing using a high-efficiency TALEN system. Nature 2012; 491(7422): 114-8.
[http://dx.doi.org/10.1038/nature11537] [PMID: 23000899]

[66]   Li L, Atef A, Piatek A, *et al.* Characterization and DNA-binding specificities of Ralstonia TAL-like effectors. Mol Plant 2013; 6(4): 1318-30.
[http://dx.doi.org/10.1093/mp/sst006] [PMID: 23300258]

[67]   Miller JC, Tan S, Qiao G, *et al.* A TALE nuclease architecture for efficient genome editing. Nat Biotechnol 2011; 29(2): 143-8.
[http://dx.doi.org/10.1038/nbt.1755] [PMID: 21179091]

[68]   Grau J, Boch J, Posch S. TALENoffer: genome-wide TALEN off-target prediction. Bioinformatics 2013; 29(22): 2931-2.
[http://dx.doi.org/10.1093/bioinformatics/btt501] [PMID: 23995255]

[69]   Gaj T, Gersbach CA, Barbas CF III. ZFN, TALEN, and CRISPR/Cas-based methods for genome engineering. Trends Biotechnol 2013; 31(7): 397-405.
[http://dx.doi.org/10.1016/j.tibtech.2013.04.004] [PMID: 23664777]

[70]   Lintner NG, Kerou M, Brumfield SK, *et al.* Structural and functional characterization of an archaeal clustered regularly interspaced short palindromic repeat (CRISPR)-associated complex for antiviral defense (CASCADE). J Biol Chem 2011; 286(24): 21643-56.
[http://dx.doi.org/10.1074/jbc.M111.238485] [PMID: 21507944]

[71]   Wiedenheft B, Sternberg SH, Doudna JA. RNA-guided genetic silencing systems in bacteria and archaea. Nature 2012; 482(7385): 331-8.
[http://dx.doi.org/10.1038/nature10886] [PMID: 22337052]

[72]   Ran FA, Hsu PD, Wright J, Agarwala V, Scott DA, Zhang F. Genome engineering using the CRISPR-Cas9 system. Nat Protoc 2013; 8(11): 2281-308.
[http://dx.doi.org/10.1038/nprot.2013.143] [PMID: 24157548]

[73]   Hoffman T, Antovski P, Tebon P, *et al.* Synthetic biology and tissue engineering: toward fabrication of complex and smart cellular constructs. Adv Funct Mater 2020; 30(26): 1909882.
[http://dx.doi.org/10.1002/adfm.201909882]

[74]   Takahashi K, Tanabe K, Ohnuki M, *et al.* Induction of pluripotent stem cells from adult human fibroblasts by defined factors. Cell 2007; 131(5): 861-72.
[http://dx.doi.org/10.1016/j.cell.2007.11.019] [PMID: 18035408]

[75]   Takahashi K, Yamanaka S. Induction of pluripotent stem cells from mouse embryonic and adult fibroblast cultures by defined factors. Cell 2006; 126(4): 663-76.
[http://dx.doi.org/10.1016/j.cell.2006.07.024] [PMID: 16904174]

[76]   Fletcher JC. The stem cell debate in historical context. In: Holland S, Lebacqz K, Zoloth L, Eds. The human embryonic stem cell debate: Science, ethics, and public policy. Cambridge: MIT Press 2001; pp. 27-34.

[77]   Ohara Y, Koganezawa N, Yamazaki H, *et al.* Early-stage development of human induced pluripotent stem cell-derived neurons. J Neurosci Res 2015; 93(12): 1804-13.
[http://dx.doi.org/10.1002/jnr.23666] [PMID: 26346430]

[78]   Chamberlain SJ, Chen PF, Ng KY, *et al.* Induced pluripotent stem cell models of the genomic imprinting disorders Angelman and Prader–Willi syndromes. Proc Natl Acad Sci USA 2010; 107(41): 17668-73.
[http://dx.doi.org/10.1073/pnas.1004487107] [PMID: 20876107]

[79]   Yang CS, Lopez CG, Rana TM. Discovery of nonsteroidal anti-inflammatory drug and anticancer drug enhancing reprogramming and induced pluripotent stem cell generation. Stem Cells 2011; 29(10): 1528-36.
[http://dx.doi.org/10.1002/stem.717] [PMID: 21898684]

[80]   Estève J, Blouin JM, Lalanne M, *et al.* Generation of induced pluripotent stem cells-derived hepatocyte-like cells for *ex vivo* gene therapy of primary hyperoxaluria type 1. Stem Cell Res (Amst) 2019; 38: 101467.
[http://dx.doi.org/10.1016/j.scr.2019.101467] [PMID: 31151050]

[81]   Shi Y, Inoue H, Wu JC, Yamanaka S. Induced pluripotent stem cell technology: a decade of progress. Nat Rev Drug Discov 2017; 16(2): 115-30.
[http://dx.doi.org/10.1038/nrd.2016.245] [PMID: 27980341]

[82]   Chang CJ, Bouhassira EE. Zinc-finger nuclease-mediated correction of α-thalassemia in iPS cells. Blood 2012; 120(19): 3906-14.
[http://dx.doi.org/10.1182/blood-2012-03-420703] [PMID: 23002118]

[83]   Ji Q, Fischer AL, Brown CR, *et al.* Engineered zinc-finger transcription factors activate OCT4 (POU5F1), SOX2, KLF4, c-MYC (MYC) and miR302/367. Nucleic Acids Res 2014; 42(10): 6158-67.
[http://dx.doi.org/10.1093/nar/gku243] [PMID: 24792165]

[84]   Gao X, Yang J, Tsang JCH, Ooi J, Wu D, Liu P. Reprogramming to pluripotency using designer TALE transcription factors targeting enhancers. Stem Cell Reports 2013; 1(2): 183-97.
[http://dx.doi.org/10.1016/j.stemcr.2013.06.002] [PMID: 24052952]

[85]   Höher T, Wallace L, Khan K, Cathomen T, Reichelt J. Highly efficient zinc-finger nuclease-mediated disruption of an eGFP transgene in keratinocyte stem cells without impairment of stem cell properties. Stem Cell Rev 2012; 8(2): 426-34.
[http://dx.doi.org/10.1007/s12015-011-9313-z] [PMID: 21874280]

[86]   Ma N, Liao B, Zhang H, *et al.* Transcription activator-like effector nuclease (TALEN)-mediated gene correction in integration-free β-thalassemia induced pluripotent stem cells. J Biol Chem 2013; 288(48): 34671-9.
[http://dx.doi.org/10.1074/jbc.M113.496174] [PMID: 24155235]

[87]   Fleischer A, Vallejo-Díez S, Martín-Fernández JM, *et al.* iPSC-derived intestinal organoids from cystic fibrosis patients acquire CFTR activity upon TALEN-mediated repair of the p. F508del mutation. Mol Ther Methods Clin Dev 2020; 17: 858-70.
[http://dx.doi.org/10.1016/j.omtm.2020.04.005] [PMID: 32373648]

[88]   Kwon YW, Ahn HS, Lee JW, *et al.* HLA DR Genome Editing with TALENs in Human iPSCs Produced Immune-Tolerant Dendritic Cells. Stem Cells Int 2021; 2021: 1-14.
[http://dx.doi.org/10.1155/2021/8873383] [PMID: 34093711]

[89]   Liu Y, Yang Y, Kang X, *et al.* One-step biallelic and scarless correction of a β-Thalassemia mutation in patient-specific iPSCs without drug selection. Mol Ther Nucleic Acids 2017; 6: 57-67.
[http://dx.doi.org/10.1016/j.omtn.2016.11.010] [PMID: 28325300]

[90]   Long C, McAnally JR, Shelton JM, Mireault AA, Bassel-Duby R, Olson EN. Prevention of muscular dystrophy in mice by CRISPR/Cas9–mediated editing of germline DNA. Science 2014; 345(6201):

1184-8.
[http://dx.doi.org/10.1126/science.1254445] [PMID: 25123483]

[91]   Chang CW, Lai YS, Westin E, *et al.* Modeling human severe combined immunodeficiency and correction by CRISPR/Cas9-enhanced gene targeting. Cell Rep 2015; 12(10): 1668-77.
[http://dx.doi.org/10.1016/j.celrep.2015.08.013] [PMID: 26321643]

[92]   Osborn MJ, Gabriel R, Webber BR, *et al.* Fanconi anemia gene editing by the CRISPR/Cas9 system. Hum Gene Ther 2015; 26(2): 114-26.
[http://dx.doi.org/10.1089/hum.2014.111] [PMID: 25545896]

[93]   Yin H, Xue W, Chen S, *et al.* Genome editing with Cas9 in adult mice corrects a disease mutation and phenotype. Nat Biotechnol 2014; 32(6): 551-3.
[http://dx.doi.org/10.1038/nbt.2884] [PMID: 24681508]

[94]   Wang L, Yi F, Fu L, *et al.* CRISPR/Cas9-mediated targeted gene correction in amyotrophic lateral sclerosis patient iPSCs. Protein Cell 2017; 8(5): 365-78.
[http://dx.doi.org/10.1007/s13238-017-0397-3] [PMID: 28401346]

[95]   Jacków J, Guo Z, Hansen C, *et al.* CRISPR/Cas9-based targeted genome editing for correction of recessive dystrophic epidermolysis bullosa using iPS cells. Proc Natl Acad Sci USA 2019; 116(52): 26846-52.
[http://dx.doi.org/10.1073/pnas.1907081116] [PMID: 31818947]

[96]   Moreno AM, Fu X, Zhu J, *et al. In situ* gene therapy *via* AAV-CRISPR-Cas9-mediated targeted gene regulation. Mol Ther 2018; 26(7): 1818-27.
[http://dx.doi.org/10.1016/j.ymthe.2018.04.017] [PMID: 29754775]

[97]   Chavez A, Scheiman J, Vora S, *et al.* Highly efficient Cas9-mediated transcriptional programming. Nat Methods 2015; 12(4): 326-8.
[http://dx.doi.org/10.1038/nmeth.3312] [PMID: 25730490]

[98]   Hartwell LH, Hopfield JJ, Leibler S, Murray AW. From molecular to modular cell biology. Nature 1999; 402(S6761) (Suppl.): C47-52.
[http://dx.doi.org/10.1038/35011540] [PMID: 10591225]

[99]   Costello A, Badran AH. Synthetic biological circuits within an orthogonal central dogma. Trends Biotechnol 2021; 39(1): 59-71.
[http://dx.doi.org/10.1016/j.tibtech.2020.05.013] [PMID: 32586633]

[100] MacDonald IC, Deans TL. Tools and applications in synthetic biology. Adv Drug Deliv Rev 2016; 105(Pt A): 20-34.
[http://dx.doi.org/10.1016/j.addr.2016.08.008] [PMID: 27568463]

[101] Black JB, Perez-Pinera P, Gersbach CA. Mammalian synthetic biology: engineering biological systems. Annu Rev Biomed Eng 2017; 19(1): 249-77.
[http://dx.doi.org/10.1146/annurev-bioeng-071516-044649] [PMID: 28633563]

[102] Szenk M, Yim T, Balázsi G. Multiplexed gene expression tuning with orthogonal synthetic gene circuits. ACS Synth Biol 2020; 9(4): 930-9.
[http://dx.doi.org/10.1021/acssynbio.9b00534] [PMID: 32167761]

[103] Bacchus W, Aubel D, Fussenegger M. Biomedically relevant circuit-design strategies in mammalian synthetic biology. Mol Syst Biol 2013; 9(1): 691.
[http://dx.doi.org/10.1038/msb.2013.48] [PMID: 24061539]

[104] Li B, You L. Division of logic labour. Nature 2011; 469(7329): 171-2.
[http://dx.doi.org/10.1038/469171a] [PMID: 21228867]

[105] Rao CV. Expanding the synthetic biology toolbox: engineering orthogonal regulators of gene expression. Curr Opin Biotechnol 2012; 23(5): 689-94.
[http://dx.doi.org/10.1016/j.copbio.2011.12.015] [PMID: 22237017]

[106] Liu CC, Jewett MC, Chin JW, Voigt CA. Toward an orthogonal central dogma. Nat Chem Biol 2018; 14(2): 103-6.
[http://dx.doi.org/10.1038/nchembio.2554] [PMID: 29337969]

[107] Bradley RW, Wang B. Designer cell signal processing circuits for biotechnology. N Biotechnol 2015; 32(6): 635-43.
[http://dx.doi.org/10.1016/j.nbt.2014.12.009] [PMID: 25579192]

[108] Bradley RW, Buck M, Wang B. Recognizing and engineering digital-like logic gates and switches in gene regulatory networks. Curr Opin Microbiol 2016; 33: 74-82.
[http://dx.doi.org/10.1016/j.mib.2016.07.004] [PMID: 27450541]

[109] Khalil AS, Collins JJ. Synthetic biology: applications come of age. Nat Rev Genet 2010; 11(5): 367-79.
[http://dx.doi.org/10.1038/nrg2775] [PMID: 20395970]

[110] Gaber R, Lebar T, Majerle A, *et al.* Designable DNA-binding domains enable construction of logic circuits in mammalian cells. Nat Chem Biol 2014; 10(3): 203-8.
[http://dx.doi.org/10.1038/nchembio.1433] [PMID: 24413461]

[111] Singh V. Recent advances and opportunities in synthetic logic gates engineering in living cells. Syst Synth Biol 2014; 8(4): 271-82.
[http://dx.doi.org/10.1007/s11693-014-9154-6] [PMID: 26396651]

[112] Anderson JC, Voigt CA, Arkin AP. Environmental signal integration by a modular AND gate. Mol Syst Biol 2007; 3(1): 133.
[http://dx.doi.org/10.1038/msb4100173] [PMID: 17700541]

[113] Tamsir A, Tabor JJ, Voigt CA. Robust multicellular computing using genetically encoded NOR gates and chemical 'wires'. Nature 2011; 469(7329): 212-5.
[http://dx.doi.org/10.1038/nature09565] [PMID: 21150903]

[114] Miyamoto T, Razavi S, DeRose R, Inoue T. Synthesizing biomolecule-based Boolean logic gates. ACS Synth Biol 2013; 2(2): 72-82.
[http://dx.doi.org/10.1021/sb3001112] [PMID: 23526588]

[115] Lonzarić J, Fink T, Jerala R. Design and applications of synthetic information processing circuits in mammalian cells. In: Ryadnov M, Brunsveld L, Suga H, Eds. Synthetic Biology. Royal Society of Chemistry 2017; Vol. 2: pp. 1-34.
[http://dx.doi.org/10.1039/9781782622789-00001]

[116] Smith LR, Cho S, Discher DE. Stem cell differentiation is regulated by extracellular matrix mechanics. Physiology (Bethesda) 2018; 33(1): 16-25.
[http://dx.doi.org/10.1152/physiol.00026.2017] [PMID: 29212889]

[117] Iismaa SE, Kaidonis X, Nicks AM, *et al.* Comparative regenerative mechanisms across different mammalian tissues. NPJ Regen Med 2018; 3(1): 6.
[http://dx.doi.org/10.1038/s41536-018-0044-5] [PMID: 29507774]

[118] Das D, Fletcher RB, Ngai J. Cellular mechanisms of epithelial stem cell self-renewal and differentiation during homeostasis and repair. Wiley Interdiscip Rev Dev Biol 2020; 9(1): e361.
[http://dx.doi.org/10.1002/wdev.361] [PMID: 31468728]

[119] Morrison SJ, Shah NM, Anderson DJ. Regulatory mechanisms in stem cell biology. Cell 1997; 88(3): 287-98.
[http://dx.doi.org/10.1016/S0092-8674(00)81867-X] [PMID: 9039255]

[120] Johnson MB, March AR, Morsut L. Engineering multicellular systems: Using synthetic biology to control tissue self-organization. Curr Opin Biomed Eng 2017; 4: 163-73.
[http://dx.doi.org/10.1016/j.cobme.2017.10.008] [PMID: 29308442]

[121] Brassard JA, Lutolf MP. Engineering stem cell self-organization to build better organoids. Cell Stem

Cell 2019; 24(6): 860-76.
[http://dx.doi.org/10.1016/j.stem.2019.05.005] [PMID: 31173716]

[122] Guye P, Ebrahimkhani MR, Kipniss N, *et al.* Genetically engineering self-organization of human pluripotent stem cells into a liver bud-like tissue using Gata6. Nat Commun 2016; 7(1): 10243.
[http://dx.doi.org/10.1038/ncomms10243] [PMID: 26732624]

[123] Cournil-Henrionnet C, Huselstein C, Wang Y, *et al.* Phenotypic analysis of cell surface markers and gene expression of human mesenchymal stem cells and chondrocytes during monolayer expansion. Biorheology 2008; 45(3-4): 513-26.
[http://dx.doi.org/10.3233/BIR-2008-0487] [PMID: 18836250]

[124] Diaz-Romero J, Nesic D, Grogan SP, Heini P, Mainil-Varlet P. Immunophenotypic changes of human articular chondrocytes during monolayer culture reflect bona fide dedifferentiation rather than amplification of progenitor cells. J Cell Physiol 2008; 214(1): 75-83.
[http://dx.doi.org/10.1002/jcp.21161] [PMID: 17559082]

[125] Yao Y, He Y, Guan Q, Wu Q. A tetracycline expression system in combination with Sox9 for cartilage tissue engineering. Biomaterials 2014; 35(6): 1898-906.
[http://dx.doi.org/10.1016/j.biomaterials.2013.11.043] [PMID: 24321708]

[126] Bacchus W, Lang M, El-Baba MD, Weber W, Stelling J, Fussenegger M. Synthetic two-way communication between mammalian cells. Nat Biotechnol 2012; 30(10): 991-6.
[http://dx.doi.org/10.1038/nbt.2351] [PMID: 22983089]

[127] Saxena P, Heng BC, Bai P, Folcher M, Zulewski H, Fussenegger M. A programmable synthetic lineage-control network that differentiates human IPSCs into glucose-sensitive insulin-secreting beta-like cells. Nat Commun 2016; 7(1): 11247.
[http://dx.doi.org/10.1038/ncomms11247] [PMID: 27063289]

[128] Ariyachet C, Tovaglieri A, Xiang G, *et al.* Reprogrammed stomach tissue as a renewable source of functional β cells for blood glucose regulation. Cell Stem Cell 2016; 18(3): 410-21.
[http://dx.doi.org/10.1016/j.stem.2016.01.003] [PMID: 26908146]

[129] Glass KA, Link JM, Brunger JM, Moutos FT, Gersbach CA, Guilak F. Tissue-engineered cartilage with inducible and tunable immunomodulatory properties. Biomaterials 2014; 35(22): 5921-31.
[http://dx.doi.org/10.1016/j.biomaterials.2014.03.073] [PMID: 24767790]

[130] Polstein LR, Juhas M, Hanna G, Bursac N, Gersbach CA. An engineered optogenetic switch for spatiotemporal control of gene expression, cell differentiation, and tissue morphogenesis. ACS Synth Biol 2017; 6(11): 2003-13.
[http://dx.doi.org/10.1021/acssynbio.7b00147] [PMID: 28793186]

[131] Yamada M, Suzuki Y, Nagasaki SC, Okuno H, Imayoshi I. Light control of the Tet gene expression system in mammalian cells. Cell Rep 2018; 25(2): 487-500.e6.
[http://dx.doi.org/10.1016/j.celrep.2018.09.026] [PMID: 30304687]

[132] Yamada M, Nagasaki SC, Suzuki Y, Hirano Y, Imayoshi I. Optimization of light-inducible Gal4/UAS gene expression system in mammalian cells. iScience 2020; 23(9): 101506.
[http://dx.doi.org/10.1016/j.isci.2020.101506] [PMID: 32919371]

[133] Bernal JA. RNA-based tools for nuclear reprogramming and lineage-conversion: towards clinical applications. J Cardiovasc Transl Res 2013; 6(6): 956-68.
[http://dx.doi.org/10.1007/s12265-013-9494-8] [PMID: 23852582]

[134] Rinaudo K, Bleris L, Maddamsetti R, Subramanian S, Weiss R, Benenson Y. A universal RNAi-based logic evaluator that operates in mammalian cells. Nat Biotechnol 2007; 25(7): 795-801.
[http://dx.doi.org/10.1038/nbt1307] [PMID: 17515909]

[135] Clancy S, Brown W. Translation: DNA to mRNA to protein. New Educator 2008; 1: 101.

[136] Ohno H, Akamine S, Saito H. Synthetic mRNA-based systems in mammalian cells. Adv Biosyst 2020; 4(5): 1900247.

[http://dx.doi.org/10.1002/adbi.201900247] [PMID: 32402126]

[137] Karagiannis P, Fujita Y, Saito H. RNA-based gene circuits for cell regulation. Proc Jpn Acad, Ser B, Phys Biol Sci 2016; 92(9): 412-22.
[http://dx.doi.org/10.2183/pjab.92.412] [PMID: 27840389]

[138] Filipowicz W, Bhattacharyya SN, Sonenberg N. Mechanisms of post-transcriptional regulation by microRNAs: are the answers in sight? Nat Rev Genet 2008; 9(2): 102-14.
[http://dx.doi.org/10.1038/nrg2290] [PMID: 18197166]

[139] Fazi F, Nervi C. MicroRNA: basic mechanisms and transcriptional regulatory networks for cell fate determination. Cardiovasc Res 2008; 79(4): 553-61.
[http://dx.doi.org/10.1093/cvr/cvn151] [PMID: 18539629]

[140] Ferro E, Bena CE, Grigolon S, Bosia C. From endogenous to synthetic microRNA-mediated regulatory circuits: An overview. Cells 2019; 8(12): 1540.
[http://dx.doi.org/10.3390/cells8121540] [PMID: 31795372]

[141] Cora' D, Re A, Caselle M, Bussolino F. MicroRNA-mediated regulatory circuits: outlook and perspectives. Phys Biol 2017; 14(4): 045001.
[http://dx.doi.org/10.1088/1478-3975/aa6f21] [PMID: 28586314]

[142] Parr CJC, Katayama S, Miki K, et al. MicroRNA-302 switch to identify and eliminate undifferentiated human pluripotent stem cells. Sci Rep 2016; 6(1): 32532.
[http://dx.doi.org/10.1038/srep32532] [PMID: 27608814]

[143] Elovic E, Etzion S, Cohen S. MiR-499 responsive lethal construct for removal of human embryonic stem cells after cardiac differentiation. Sci Rep 2019; 9(1): 14490.
[http://dx.doi.org/10.1038/s41598-019-50899-2] [PMID: 31601830]

[144] Miki K, Endo K, Takahashi S, et al. Efficient detection and purification of cell populations using synthetic microRNA switches. Cell Stem Cell 2015; 16(6): 699-711.
[http://dx.doi.org/10.1016/j.stem.2015.04.005] [PMID: 26004781]

[145] Gong X, Wei J, Liu J, Li R, Liu X, Wang F. Programmable intracellular DNA biocomputing circuits for reliable cell recognitions. Chem Sci (Camb) 2019; 10(10): 2989-97.
[http://dx.doi.org/10.1039/C8SC05217D] [PMID: 30996878]

[146] Xu J, Shao T, Ding N, Li Y, Li X. miRNA-miRNA crosstalk: from genomics to phenomics. Brief Bioinform 2017; 18(6): 1002-11.
[PMID: 27551063]

[147] Yamamura S, Imai-Sumida M, Tanaka Y, Dahiya R. Interaction and cross-talk between non-coding RNAs. Cell Mol Life Sci 2018; 75(3): 467-84.
[http://dx.doi.org/10.1007/s00018-017-2626-6] [PMID: 28840253]

[148] Harvey RF, Smith TS, Mulroney T, et al. Trans-acting translational regulatory RNA binding proteins. Wiley Interdiscip Rev RNA 2018; 9(3): e1465.
[http://dx.doi.org/10.1002/wrna.1465] [PMID: 29341429]

[149] Martínez-Salas E, Lozano G, Fernandez-Chamorro J, Francisco-Velilla R, Galan A, Diaz R. RNA-binding proteins impacting on internal initiation of translation. Int J Mol Sci 2013; 14(11): 21705-26.
[http://dx.doi.org/10.3390/ijms141121705] [PMID: 24189219]

[150] Kawasaki S, Fujita Y, Nagaike T, Tomita K, Saito H. Synthetic mRNA devices that detect endogenous proteins and distinguish mammalian cells. Nucleic Acids Res 2017; 45(12): e117-7.
[http://dx.doi.org/10.1093/nar/gkx298] [PMID: 28525643]

[151] Lilley DMJ. The K-turn motif in riboswitches and other RNA species. Biochim Biophys Acta Gene Regul Mech 2014; 1839(10): 995-1004.
[http://dx.doi.org/10.1016/j.bbagrm.2014.04.020] [PMID: 24798078]

[152] Saito H, Fujita Y, Kashida S, Hayashi K, Inoue T. Synthetic human cell fate regulation by protein-

driven RNA switches. Nat Commun 2011; 2(1): 160.
[http://dx.doi.org/10.1038/ncomms1157] [PMID: 21245841]

[153]   Saito H, Kobayashi T, Hara T, *et al.* Synthetic translational regulation by an L7Ae–kink-turn RNP switch. Nat Chem Biol 2010; 6(1): 71-8.
[http://dx.doi.org/10.1038/nchembio.273] [PMID: 20016495]

[154]   Stapleton JA, Endo K, Fujita Y, *et al.* Feedback control of protein expression in mammalian cells by tunable synthetic translational inhibition. ACS Synth Biol 2012; 1(3): 83-8.
[http://dx.doi.org/10.1021/sb200005w] [PMID: 23651072]

[155]   Endo K, Stapleton JA, Hayashi K, Saito H, Inoue T. Quantitative and simultaneous translational control of distinct mammalian mRNAs. Nucleic Acids Res 2013; 41(13): e135-5.
[http://dx.doi.org/10.1093/nar/gkt347] [PMID: 23685611]

[156]   Wagner TE, Becraft JR, Bodner K, *et al.* Small-molecule-based regulation of RNA-delivered circuits in mammalian cells. Nat Chem Biol 2018; 14(11): 1043-50.
[http://dx.doi.org/10.1038/s41589-018-0146-9] [PMID: 30327560]

[157]   Leisner M, Bleris L, Lohmueller J, Xie Z, Benenson Y. Rationally designed logic integration of regulatory signals in mammalian cells. Nat Nanotechnol 2010; 5(9): 666-70.
[http://dx.doi.org/10.1038/nnano.2010.135] [PMID: 20622866]

[158]   Bloom RJ, Winkler SM, Smolke CD. A quantitative framework for the forward design of synthetic miRNA circuits. Nat Methods 2014; 11(11): 1147-53.
[http://dx.doi.org/10.1038/nmeth.3100] [PMID: 25218181]

[159]   Wroblewska L, Kitada T, Endo K, *et al.* Mammalian synthetic circuits with RNA binding proteins for RNA-only delivery. Nat Biotechnol 2015; 33(8): 839-41.
[http://dx.doi.org/10.1038/nbt.3301] [PMID: 26237515]

[160]   Matsuura S, Ono H, Kawasaki S, Kuang Y, Fujita Y, Saito H. Synthetic RNA-based logic computation in mammalian cells. Nat Commun 2018; 9(1): 4847.
[http://dx.doi.org/10.1038/s41467-018-07181-2] [PMID: 30451868]

[161]   Barros MT, Doan P, Kandhavelu M, Jennings B, Balasubramaniam S. Engineering calcium signaling of astrocytes for neural–molecular computing logic gates. Sci Rep 2021; 11(1): 595.
[http://dx.doi.org/10.1038/s41598-020-79891-x] [PMID: 33436729]

[162]   Muldoon JJ, Kandula V, Hong M, *et al.* Model-guided design of mammalian genetic programs. Sci Adv 2021; 7(8): eabe9375.
[http://dx.doi.org/10.1126/sciadv.abe9375] [PMID: 33608279]

[163]   Kusindarta DL, Wihadmadyatami H. The role of extracellular matrix in tissue regeneration. Tissue Regen 2018; p. 65.
[http://dx.doi.org/10.5772/intechopen.75728]

[164]   Frantz C, Stewart KM, Weaver VM. The extracellular matrix at a glance. J Cell Sci 2010; 123(24): 4195-200.
[http://dx.doi.org/10.1242/jcs.023820] [PMID: 21123617]

[165]   Bambole V, Yakhmi JV. Tissue engineering: Use of electrospinning technique for recreating physiological functions. Grumezescu AM, Ed Nanobiomaterials in Soft Tissue Engineering. 387-455.Elsevier 2016; pp.

[166]   Deans TL, Singh A, Gibson M, Elisseeff JH. Regulating synthetic gene networks in 3D materials. Proc Natl Acad Sci USA 2012; 109(38): 15217-22.
[http://dx.doi.org/10.1073/pnas.1204705109] [PMID: 22927376]

[167]   Gübeli RJ, Burger K, Weber W. Synthetic biology for mammalian cell technology and materials sciences. Biotechnol Adv 2013; 31(1): 68-78.
[http://dx.doi.org/10.1016/j.biotechadv.2012.01.007] [PMID: 22286074]

[168] Singh A, Deans TL, Elisseeff JH. Photomodulation of cellular gene expression in hydrogels. ACS Macro Lett 2013; 2(3): 269-72.
[http://dx.doi.org/10.1021/mz300591m] [PMID: 35581895]

[169] El-Sherbiny IM, Yacoub MH. Hydrogel scaffolds for tissue engineering: Progress and challenges. Glob Cardiol Sci Pract 2013; 2013(3): 38.
[http://dx.doi.org/10.5339/gcsp.2013.38] [PMID: 24689032]

[170] Weisenberger MS, Deans TL. Bottom-up approaches in synthetic biology and biomaterials for tissue engineering applications. J Ind Microbiol Biotechnol 2018; 45(7): 599-614.
[http://dx.doi.org/10.1007/s10295-018-2027-3] [PMID: 29552703]

[171] Deans TL, Cantor CR, Collins JJ. A tunable genetic switch based on RNAi and repressor proteins for regulating gene expression in mammalian cells. Cell 2007; 130(2): 363-72.
[http://dx.doi.org/10.1016/j.cell.2007.05.045] [PMID: 17662949]

[172] Ehrbar M, Schoenmakers R, Christen EH, Fussenegger M, Weber W. Drug-sensing hydrogels for the inducible release of biopharmaceuticals. Nat Mater 2008; 7(10): 800-4.
[http://dx.doi.org/10.1038/nmat2250] [PMID: 18690239]

[173] Kämpf MM, Christen EH, Ehrbar M, *et al.* A gene therapy technology-based biomaterial for the trigger-inducible release of biopharmaceuticals in mice. Adv Funct Mater 2010; 20(15): 2534-8.
[http://dx.doi.org/10.1002/adfm.200902377]

[174] Christen EH, Karlsson M, Kämpf MM, *et al.* Conditional DNA-protein interactions confer stimulus-sensing properties to biohybrid materials. Adv Funct Mater 2011; 21(15): 2861-7.
[http://dx.doi.org/10.1002/adfm.201100731]

[175] Jansen LE, McCarthy TP, Lee MJ, Peyton SR. A synthetic, three-dimensional bone marrow hydrogel. Biorxiv 2018; 275842.

[176] Zakeri B, Howarth M. Spontaneous intermolecular amide bond formation between side chains for irreversible peptide targeting. J Am Chem Soc 2010; 132(13): 4526-7.
[http://dx.doi.org/10.1021/ja910795a] [PMID: 20235501]

[177] Zakeri B, Fierer JO, Celik E, *et al.* Peptide tag forming a rapid covalent bond to a protein, through engineering a bacterial adhesin. Proc Natl Acad Sci USA 2012; 109(12): E690-7.
[http://dx.doi.org/10.1073/pnas.1115485109] [PMID: 22366317]

[178] Li L, Fierer JO, Rapoport TA, Howarth M. Structural analysis and optimization of the covalent association between SpyCatcher and a peptide Tag. J Mol Biol 2014; 426(2): 309-17.
[http://dx.doi.org/10.1016/j.jmb.2013.10.021] [PMID: 24161952]

[179] Reddington SC, Howarth M. Secrets of a covalent interaction for biomaterials and biotechnology: SpyTag and SpyCatcher. Curr Opin Chem Biol 2015; 29: 94-9.
[http://dx.doi.org/10.1016/j.cbpa.2015.10.002] [PMID: 26517567]

[180] Zhang WB, Sun F, Tirrell DA, Arnold FH. Controlling macromolecular topology with genetically encoded SpyTag-SpyCatcher chemistry. J Am Chem Soc 2013; 135(37): 13988-97.
[http://dx.doi.org/10.1021/ja4076452] [PMID: 23964715]

[181] Sun F, Zhang WB, Mahdavi A, Arnold FH, Tirrell DA. Synthesis of bioactive protein hydrogels by genetically encoded SpyTag-SpyCatcher chemistry. Proc Natl Acad Sci USA 2014; 111(31): 11269-74.
[http://dx.doi.org/10.1073/pnas.1401291111] [PMID: 25049400]

[182] Gao X, Fang J, Xue B, Fu L, Li H. Engineering protein hydrogels using SpyCatcher-SpyTag chemistry. Biomacromolecules 2016; 17(9): 2812-9.
[http://dx.doi.org/10.1021/acs.biomac.6b00566] [PMID: 27477779]

[183] Tan LL, Hoon SS, Wong FT. Kinetic controlled tag-catcher interactions for directed covalent protein assembly. PLoS One 2016; 11(10): e0165074.

[http://dx.doi.org/10.1371/journal.pone.0165074] [PMID: 27783674]

[184]   Levin A, Hakala TA, Schnaider L, Bernardes GJL, Gazit E, Knowles TPJ. Biomimetic peptide self-assembly for functional materials. Nat Rev Chem 2020; 4(11): 615-34.
[http://dx.doi.org/10.1038/s41570-020-0215-y]

[185]   Zhang S, Holmes T, Lockshin C, Rich A. Spontaneous assembly of a self-complementary oligopeptide to form a stable macroscopic membrane. Proc Natl Acad Sci USA 1993; 90(8): 3334-8.
[http://dx.doi.org/10.1073/pnas.90.8.3334] [PMID: 7682699]

[186]   Xiao C, Pérez LM, Russell DH. Effects of charge states, charge sites and side chain interactions on conformational preferences of a series of model peptide ions. Analyst (Lond) 2015; 140(20): 6933-44.
[http://dx.doi.org/10.1039/C5AN00826C] [PMID: 26081298]

[187]   Zhang S, Lockshin C, Cook R, Rich A. Unusually stable? -sheet formation in an ionic self-complementary oligopeptide. Biopolymers 1994; 34(5): 663-72.
[http://dx.doi.org/10.1002/bip.360340508] [PMID: 8003624]

[188]   Zhang S, Holmes TC, DiPersio CM, Hynes RO, Su X, Rich A. Self-complementary oligopeptide matrices support mammalian cell attachment. Biomaterials 1995; 16(18): 1385-93.
[http://dx.doi.org/10.1016/0142-9612(95)96874-Y] [PMID: 8590765]

[189]   Keyes C, Duhamel J, Fung SY, Bezaire J, Chen P. Self-assembling peptide as a potential carrier of hydrophobic compounds. J Am Chem Soc 2004; 126(24): 7522-32.
[http://dx.doi.org/10.1021/ja0381297] [PMID: 15198599]

[190]   Bawa R, Fung SY, Shiozaki A, et al. Self-assembling peptide-based nanoparticles enhance cellular delivery of the hydrophobic anticancer drug ellipticine through caveolae-dependent endocytosis. Nanomedicine 2012; 8(5): 647-54.
[http://dx.doi.org/10.1016/j.nano.2011.08.007] [PMID: 21889478]

[191]   Bradshaw M, Ho D, Fear MW, Gelain F, Wood FM, Iyer KS. Designer self-assembling hydrogel scaffolds can impact skin cell proliferation and migration. Sci Rep 2015; 4(1): 6903.
[http://dx.doi.org/10.1038/srep06903] [PMID: 25384420]

[192]   Wang T, Zhong X, Wang S, Lv F, Zhao X. Molecular mechanisms of RADA16-1 peptide on fast stop bleeding in rat models. Int J Mol Sci 2012; 13(12): 15279-90.
[http://dx.doi.org/10.3390/ijms131115279] [PMID: 23203125]

[193]   Zhou A, Chen S, He B, Zhao W, Chen X, Jiang D. Controlled release of TGF-beta 1 from RADA self-assembling peptide hydrogel scaffolds. Drug Des Devel Ther 2016; 10: 3043-51.
[http://dx.doi.org/10.2147/DDDT.S109545] [PMID: 27703332]

[194]   Liu J, Zhang L, Yang Z, Zhao X. Controlled release of paclitaxel from a self-assembling peptide hydrogel formed *in situ* and antitumor study *in vitro*. Int J Nanomed 2011; 6: 2143-53.
[http://dx.doi.org/10.2147/IJN.S24038] [PMID: 22114478]

[195]   Li F, Wang J, Tang F, et al. Fluorescence studies on a designed self-assembling peptide of RAD16-II as a potential carrier for hydrophobic drug. J Nanosci Nanotechnol 2009; 9(2): 1611-4.
[http://dx.doi.org/10.1166/jnn.2009.C214] [PMID: 19441582]

[196]   Holmes TC, de Lacalle S, Su X, Liu G, Rich A, Zhang S. Extensive neurite outgrowth and active synapse formation on self-assembling peptide scaffolds. Proc Natl Acad Sci USA 2000; 97(12): 6728-33.
[http://dx.doi.org/10.1073/pnas.97.12.6728] [PMID: 10841570]

[197]   Davis ME, Motion JPM, Narmoneva DA, et al. Injectable self-assembling peptide nanofibers create intramyocardial microenvironments for endothelial cells. Circulation 2005; 111(4): 442-50.
[http://dx.doi.org/10.1161/01.CIR.0000153847.47301.80] [PMID: 15687132]

[198]   Dubois G, Segers VFM, Bellamy V, et al. Self-assembling peptide nanofibers and skeletal myoblast transplantation in infarcted myocardium. J Biomed Mater Res B Appl Biomater 2008; 87B(1): 222-8.
[http://dx.doi.org/10.1002/jbm.b.31099] [PMID: 18386833]

[199]   Cho H, Balaji S, Sheikh AQ, *et al.* Regulation of endothelial cell activation and angiogenesis by injectable peptide nanofibers. Acta Biomater 2012; 8(1): 154-64.
[http://dx.doi.org/10.1016/j.actbio.2011.08.029] [PMID: 21925628]

[200]   Wan S, Borland S, Richardson SM, Merry CLR, Saiani A, Gough JE. Self-assembling peptide hydrogel for intervertebral disc tissue engineering. Acta Biomater 2016; 46: 29-40.
[http://dx.doi.org/10.1016/j.actbio.2016.09.033] [PMID: 27677593]

[201]   Castillo Diaz LA, Saiani A, Gough JE, Miller AF. Human osteoblasts within soft peptide hydrogels promote mineralisation *in vitro.* J Tissue Eng 2014; 5.
[http://dx.doi.org/10.1177/2041731414539344] [PMID: 25383164]

[202]   Mujeeb A, Miller AF, Saiani A, Gough JE. Self-assembled octapeptide scaffolds for *in vitro* chondrocyte culture. Acta Biomater 2013; 9(1): 4609-17.
[http://dx.doi.org/10.1016/j.actbio.2012.08.044] [PMID: 22963851]

[203]   Kretsinger JK, Haines LA, Ozbas B, Pochan DJ, Schneider JP. Cytocompatibility of self-assembled β-hairpin peptide hydrogel surfaces. Biomaterials 2005; 26(25): 5177-86.
[http://dx.doi.org/10.1016/j.biomaterials.2005.01.029] [PMID: 15792545]

[204]   Salick DA, Kretsinger JK, Pochan DJ, Schneider JP. Inherent antibacterial activity of a peptide-based β-hairpin hydrogel. J Am Chem Soc 2007; 129(47): 14793-9.
[http://dx.doi.org/10.1021/ja076300z] [PMID: 17985907]

[205]   Sieminski AL, Semino CE, Gong H, Kamm RD. Primary sequence of ionic self-assembling peptide gels affects endothelial cell adhesion and capillary morphogenesis. J Biomed Mater Res A 2008; 87A(2): 494-504.
[http://dx.doi.org/10.1002/jbm.a.31785] [PMID: 18186067]

[206]   Tsukamoto J, Naruse K, Nagai Y, *et al.* Efficacy of a self-assembling peptide hydrogel, SPG-178-gel, for bone regeneration and three-dimensional osteogenic induction of dental pulp stem cells. Tissue Eng Part A 2017; 23(23-24): 1394-402.
[http://dx.doi.org/10.1089/ten.tea.2017.0025] [PMID: 28530133]

[207]   Komatsu S, Nagai Y, Naruse K, Kimata Y. The neutral self-assembling peptide hydrogel SPG-178 as a topical hemostatic agent. PLoS One 2014; 9(7): e102778.
[http://dx.doi.org/10.1371/journal.pone.0102778] [PMID: 25047639]

[208]   Sun J, Zheng Q. Experimental study on self-assembly of KLD-12 peptide hydrogel and 3-D culture of MSC encapsulated within hydrogel *in vitro.* J Huazhong Univ Sci Technolog Med Sci 2009; 29(4): 512-6.
[http://dx.doi.org/10.1007/s11596-009-0424-6] [PMID: 19662373]

[209]   Kisiday J, Jin M, Kurz B, *et al.* Self-assembling peptide hydrogel fosters chondrocyte extracellular matrix production and cell division: Implications for cartilage tissue repair. Proc Natl Acad Sci USA 2002; 99(15): 9996-10001.
[http://dx.doi.org/10.1073/pnas.142309999] [PMID: 12119393]

[210]   Fishwick CWG, Beevers AJ, Carrick LM, Whitehouse CD, Aggeli A, Boden N. Structures of helical β-tapes and twisted ribbons: the role of side-chain interactions on twist and bend behavior. Nano Lett 2003; 3(11): 1475-9.
[http://dx.doi.org/10.1021/nl034095p]

[211]   Gharaei R, Tronci G, Goswami P, Davies RPW, Kirkham J, Russell SJ. Biomimetic peptide enriched nonwoven scaffolds promote calcium phosphate mineralisation. RSC Advances 2020; 10(47): 28332-42.
[http://dx.doi.org/10.1039/D0RA02446E] [PMID: 35519117]

[212]   Koch F, Wolff A, Mathes S, *et al.* Amino acid composition of nanofibrillar self-assembling peptide hydrogels affects responses of periodontal tissue cells *in vitro.* Int J Nanomed 2018; 13: 6717-33.
[http://dx.doi.org/10.2147/IJN.S173702] [PMID: 30425485]

[213] Banwell EF, Abelardo ES, Adams DJ, *et al.* Rational design and application of responsive α-helical peptide hydrogels. Nat Mater 2009; 8(7): 596-600.
[http://dx.doi.org/10.1038/nmat2479] [PMID: 19543314]

[214] Fukunaga K, Tsutsumi H, Mihara H. Self-assembling peptides as building blocks of functional materials for biomedical applications. Bull Chem Soc Jpn 2019; 92(2): 391-9.
[http://dx.doi.org/10.1246/bcsj.20180293]

[215] Millar-Haskell CS, Dang AM, Gleghorn JP. Coupling synthetic biology and programmable materials to construct complex tissue ecosystems. MRS Commun 2019; 9(2): 421-32.
[http://dx.doi.org/10.1557/mrc.2019.69] [PMID: 31485382]

[216] Acar H, Srivastava S, Chung EJ, *et al.* Self-assembling peptide-based building blocks in medical applications. Adv Drug Deliv Rev 2017; 110-111: 65-79.
[http://dx.doi.org/10.1016/j.addr.2016.08.006] [PMID: 27535485]

[217] Matson JB, Stupp SI. Self-assembling peptide scaffolds for regenerative medicine. Chem Commun (Camb) 2012; 48(1): 26-33.
[http://dx.doi.org/10.1039/C1CC15551B] [PMID: 22080255]

[218] Genové E, Shen C, Zhang S, Semino CE. The effect of functionalized self-assembling peptide scaffolds on human aortic endothelial cell function. Biomaterials 2005; 26(16): 3341-51.
[http://dx.doi.org/10.1016/j.biomaterials.2004.08.012] [PMID: 15603830]

[219] Matson JB, Zha RH, Stupp SI. Peptide self-assembly for crafting functional biological materials. Curr Opin Solid State Mater Sci 2011; 15(6): 225-35.
[http://dx.doi.org/10.1016/j.cossms.2011.08.001] [PMID: 22125413]

[220] Gelain F, Bottai D, Vescovi A, Zhang S. Designer self-assembling peptide nanofiber scaffolds for adult mouse neural stem cell 3-dimensional cultures. PLoS One 2006; 1(1): e119.
[http://dx.doi.org/10.1371/journal.pone.0000119] [PMID: 17205123]

[221] Li R, Xu J, Wong DSH, Li J, Zhao P, Bian L. Self-assembled N-cadherin mimetic peptide hydrogels promote the chondrogenesis of mesenchymal stem cells through inhibition of canonical Wnt/β-catenin signaling. Biomaterials 2017; 145: 33-43.
[http://dx.doi.org/10.1016/j.biomaterials.2017.08.031] [PMID: 28843065]

[222] Yang S, Wang C, Zhu J, *et al.* Self-assembling peptide hydrogels functionalized with LN- and BDNF-mimicking epitopes synergistically enhance peripheral nerve regeneration. Theranostics 2020; 10(18): 8227-49.
[http://dx.doi.org/10.7150/thno.44276] [PMID: 32724468]

[223] Horii A, Wang X, Gelain F, Zhang S. Biological designer self-assembling peptide nanofiber scaffolds significantly enhance osteoblast proliferation, differentiation and 3-D migration. PLoS One 2007; 2(2): e190.
[http://dx.doi.org/10.1371/journal.pone.0000190] [PMID: 17285144]

[224] Tsutsumi H, Kawamura M, Mihara H. Osteoblastic differentiation on hydrogels fabricated from Ca^{2+}-responsive self-assembling peptides functionalized with bioactive peptides. Bioorg Med Chem 2018; 26(12): 3126-32.
[http://dx.doi.org/10.1016/j.bmc.2018.04.039] [PMID: 29699909]

[225] Matsugami D, Murakami T, Yoshida W, *et al.* Treatment with functionalized designer self-assembling peptide hydrogels promotes healing of experimental periodontal defects. J Periodontal Res 2021; 56(1): 162-72.
[http://dx.doi.org/10.1111/jre.12807] [PMID: 33022075]

[226] Wang X, Wang J, Guo L, *et al.* Self-assembling peptide hydrogel scaffolds support stem cell-based hair follicle regeneration. Nanomedicine 2016; 12(7): 2115-25.
[http://dx.doi.org/10.1016/j.nano.2016.05.021] [PMID: 27288668]

[227] Wang X, Horii A, Zhang S. Designer functionalized self-assembling peptide nanofiber scaffolds for

growth, migration, and tubulogenesis of human umbilical vein endothelial cells. Soft Matter 2008; 4(12): 2388-95.
[http://dx.doi.org/10.1039/b807155a]

[228]   Xia K, Chen Z, Chen J, *et al.* RGD-and VEGF-mimetic peptide epitope-functionalized self-assembling peptide hydrogels promote dentin-pulp complex regeneration. Int J Nanomedicine 2020; 15: 6631-47.
[http://dx.doi.org/10.2147/IJN.S253576] [PMID: 32982223]

[229]   Cheng TY, Chen MH, Chang WH, Huang MY, Wang TW. Neural stem cells encapsulated in a functionalized self-assembling peptide hydrogel for brain tissue engineering. Biomaterials 2013; 34(8): 2005-16.
[http://dx.doi.org/10.1016/j.biomaterials.2012.11.043] [PMID: 23237515]

[230]   Onak Pulat G, Gökmen O, Çevik ZBY, Karaman O. Role of functionalized self-assembled peptide hydrogels in *in vitro* vasculogenesis. Soft Matter 2021; 17(27): 6616-26.
[http://dx.doi.org/10.1039/D1SM00680K] [PMID: 34143171]

[231]   Cui G, Shao S, Yang J, Liu J, Guo H. Designer self-assemble peptides maximize the therapeutic benefits of neural stem cell transplantation for Alzheimer's disease *via* enhancing neuron differentiation and paracrine action. Mol Neurobiol 2016; 53(2): 1108-23.
[http://dx.doi.org/10.1007/s12035-014-9069-y] [PMID: 25586060]

[232]   Wu Y, Jia Z, Liu L, *et al.* Functional self-assembled peptide nanofibers for bone marrow mesenchymal stem cell encapsulation and regeneration in nucleus pulposus. Artif Organs 2016; 40(6): E112-9.
[http://dx.doi.org/10.1111/aor.12694] [PMID: 27153338]

[233]   Liu X, Wang X, Wang X, *et al.* Functionalized self-assembling peptide nanofiber hydrogels mimic stem cell niche to control human adipose stem cell behavior *in vitro*. Acta Biomater 2013; 9(6): 6798-805.
[http://dx.doi.org/10.1016/j.actbio.2013.01.027] [PMID: 23380207]

[234]   Koutsopoulos S, Zhang S. Long-term three-dimensional neural tissue cultures in functionalized self-assembling peptide hydrogels, Matrigel and Collagen I. Acta Biomater 2013; 9(2): 5162-9.
[http://dx.doi.org/10.1016/j.actbio.2012.09.010] [PMID: 22995405]

[235]   Chen Y, Lu J, Chen B, *et al.* PFS-functionalized self-assembling peptide hydrogel for the maintenance of human adipose stem cell *in vitro*. J Biomater Tissue Eng 2017; 7(10): 943-51.
[http://dx.doi.org/10.1166/jbt.2017.1663]

[236]   Danesin R, Brun P, Roso M, *et al.* Self-assembling peptide-enriched electrospun polycaprolactone scaffolds promote the h-osteoblast adhesion and modulate differentiation-associated gene expression. Bone 2012; 51(5): 851-9.
[http://dx.doi.org/10.1016/j.bone.2012.08.119] [PMID: 22926428]

[237]   Nune M, Krishnan UM, Sethuraman S. Decoration of PLGA electrospun nanofibers with designer self-assembling peptides: a "Nano-on-Nano" concept. RSC Advances 2015; 5(108): 88748-57.
[http://dx.doi.org/10.1039/C5RA13576A]

[238]   Rey-Rico A, Venkatesan JK, Frisch J, *et al.* Effective and durable genetic modification of human mesenchymal stem cells *via* controlled release of rAAV vectors from self-assembling peptide hydrogels with a maintained differentiation potency. Acta Biomater 2015; 18: 118-27.
[http://dx.doi.org/10.1016/j.actbio.2015.02.013] [PMID: 25712390]

[239]   Rodriguez AL, Wang TY, Bruggeman KF, *et al.* Tailoring minimalist self-assembling peptides for localized viral vector gene delivery. Nano Res 2016; 9(3): 674-84.
[http://dx.doi.org/10.1007/s12274-015-0946-0]

[240]   Zhang K, Chooi WH, Liu S, *et al.* Localized delivery of CRISPR/dCas9 *via* layer-by-layer self-assembling peptide coating on nanofibers for neural tissue engineering. Biomaterials 2020; 256: 120225.
[http://dx.doi.org/10.1016/j.biomaterials.2020.120225] [PMID: 32738650]

[241] Lee SY, Lim JS, Harris MT. Synthesis and application of virus-based hybrid nanomaterials. Biotechnol Bioeng 2012; 109(1): 16-30.
[http://dx.doi.org/10.1002/bit.23328] [PMID: 21915854]

[242] Liu Z, Qiao J, Niu Z, Wang Q. Natural supramolecular building blocks: from virus coat proteins to viral nanoparticles. Chem Soc Rev 2012; 41(18): 6178-94.
[http://dx.doi.org/10.1039/c2cs35108k] [PMID: 22880206]

[243] Rong J, Lee LA, Li K, *et al.* Oriented cell growth on self-assembled bacteriophage M13 thin films. Chem Commun (Camb) 2008; 41(41): 5185-7.
[http://dx.doi.org/10.1039/b811039e] [PMID: 18956063]

[244] Lin Y, Balizan E, Lee LA, Niu Z, Wang Q. Self-assembly of rodlike bio-nanoparticles in capillary tubes. Angew Chem Int Ed 2010; 49(5): 868-72.
[http://dx.doi.org/10.1002/anie.200904993] [PMID: 20013831]

[245] Wang J, Wang L, Li X, Mao C. Virus activated artificial ECM induces the osteoblastic differentiation of mesenchymal stem cells without osteogenic supplements. Sci Rep 2013; 3(1): 1242.
[http://dx.doi.org/10.1038/srep01242] [PMID: 23393624]

CHAPTER 3

Innovative Approaches to Prosthetics and Implants

Sıtkı Kocaoğlu[1,*] and **Erhan Akdoğan**[2,3]

[1] *Department of Biomedical Engineering, Faculty of Engineering and Natural Sciences, Ankara Yıldırım Beyazıt University, Ankara, Turkey*

[2] *Department of Mechatronics Engineering, Faculty of Mechanical Engineering, Yildiz Technical University, İstanbul, Turkey*

[3] *Health Institutes of Türkiye, İstanbul, Turkey*

Abstract: The use of prosthesis plays an important role in rehabilitation in the case of congenital absence or loss of an extremity. Apart from lower and upper extremity prostheses, there is a wide variety of prostheses used in different parts of the body. Unlike limb prostheses, these are permanently placed in the body by surgical intervention and are also called implants. New studies emerge every day in the development of innovative prostheses and implants. These innovations include material selection, new material development, control strategies, feedback system development, sensor and actuator development, power supply methods, and power equipment development work. Besides, many studies aim to increase user comfort as well as acceptance rate and the useful life of prostheses. Some researchers are working to develop prostheses exclusively for the use of children. Innovative developments in prostheses and implants are examined in this section. Developments are presented from various aspects, and information is given about the research that has made significant contributions to the field. As an example of technological development in prosthetics, an autonomic tumor prosthesis developed for children with bone cancer is introduced at the end of the section as a case study.

Keywords: Active prosthesis, Implant, Innovation, Material selection, Orthopedic implant, Pediatric prosthetics, Prosthesis classification, Prosthetics, Tumor prosthesis.

INTRODUCTION

In the event of a loss of any limb in the human body, the artificial components produced in order to fulfill the function of this limb partially or completely are called prostheses. The artificial body attachments used in cases that require support, protection, or correction in any limb without loss of the limb are called orthoses. Implant, on the other hand, is the name given to all solid substances that are surgically placed in the body for treatment.

---

* **Corresponding author Sıtkı Kocaoğlu:** Department of Biomedical Engineering, Faculty of Engineering and Natural Sciences, Ankara Yıldırım Beyazıt University, Ankara, Turkey; E-mail: sitkikocaoglu@hotmail.com

Felipe López-Saucedo (Ed.)

Prosthetics, orthoses, and implants can be in a very simple structure formed by shaping a raw material with classical methods, or they can be in a structure that includes high technology and active works. For this reason, prostheses, orthoses, and implants can be examined and separated from each other not in terms of their structure or technology but in terms of their functions and methods of placing them on the body. Implant is a more general expression and can be reclassified as prosthesis, orthosis and other (for example, it may be used for some medical measurements), depending on its function, provided that the implants are placed inside the body. In this section, prostheses, including implant prostheses, are examined in detail.

For most amputated, an artificial limb increases mobility and the ability to perform daily activities, reducing the rate of dependency on other people. Limb amputations are one of the oldest surgeries in history that affect patients both psychologically and physically. Today, when the causes of amputation are examined, accident-induced trauma takes the first place due to the developing technologies and the increase in motor vehicles. Apart from this, non-traumatic reasons, such as diabetes, atherosclerosis, hypertension, and peripheral vascular diseases, may also require amputation. There is a wide variety of limb prostheses that are designed to function as a natural arm, leg, hand, or foot and are, in most cases, visible from the outside. Although there are many products available on the market and used clinically, most of them consist of similar parts. For conventional prostheses, these consist of a socket surrounding the residual limb, the suspension holding the prosthesis on top of the residual limb, the shaft, the limb-like body it replaces, and a cosmetic coating. The socket part is usually covered with silicone or similar soft material to protect the contact area and is worn on the residual limb after wearing socks for adaptation. In determining the prosthesis to be used, the body area where the prosthesis will be used, the amputation level, the condition of the residual limb, the activity level and the expectations of the person play significant roles.

Prostheses have been developed since the first ages of history. The first examples of prostheses are simple structures, such as legs made of wood or hooks to replace the forearm and hand limb. The earliest recorded use of a limb prosthesis was reported by Herodotus. According to Herodotus's account, the Persian soldier Hegesistratus cut off one of his feet in 484 BC and fled using a wooden foot instead [1]. The oldest known artificial limb is a leg, unearthed in Capri, Italy, in 1858, made of copper and wood around 300 BC. In the 15th century, artificial hands made of iron were used by knights. The Alt-Ruppin hand, which is exhibited with other hands belonging to the 15th century in the Stibbert Museum in

Florence, Italy, is one of these examples.

Amputation cases have increased in post-war periods as a result of the injuries of many people and diseases of the musculoskeletal system and neuromuscular diseases, calling for an increase in prostheses development. The United States government has started to support research projects on orthoses and prostheses to improve the quality and performance of assistive devices, especially for amputated veterans [2]. Research and education committees have been established and developed several projects in this regard. With advances in modular components and bioengineering, the use of myoelectric prostheses began in 1950. Special attention has been paid to improving the biomechanical design of prostheses during this period.

The development of new materials that could be used in prostheses after the Second World War led to further advances in the field. The newly emerging transparent plastics were then used in prosthesis production. As a result of the expectation of higher performance from prostheses by special groups, such as athletes, many specially designed prosthetics feet were developed. Innovative designs for prostheses have thus been made possible thanks to carbon composite technology.

The development of computer-aided design/computer-aided manufacturing (CAD/CAM) systems for prostheses in the 1970s was a major technological advance. In the late 1980s and early 1990s, as computers became more economical, prosthetic manufacturers began to integrate CAD/CAM systems into their applications [3].

The use of hand and foot prostheses is possible by performing a surgical intervention called amputation. Ligature in surgery was first described by Hippocrates but was lost in the dark ages [4]. The ligature was reintroduced in 1529 by the French military surgeon Ambroise Pare. Thus, the risk of death due to excessive blood loss in amputation surgeries was reduced [5]. Morel introduced the tourniquet in 1674, which showed a similar effect, reducing the risk of death and accelerating amputation surgeries. In 1536, Pare performed the first elbow disarticulation procedure. Sir James Syme reported the ankle amputation procedure in 1843 [6]. The antiseptic technique was introduced by Lord Lister in 1867 [7]. Ever since, the introduction of antiseptic, chloroform and ether has contributed greatly to the overall success of amputation surgery. Kineplasty, the act of shaping the residual limb to allow comfortable use of the prosthesis in amputation and aimed at strengthening upper limb prostheses by direct muscle contraction, was discovered by Vanghetti in 1898 while trying to improve the prosthetic function of Italian soldiers whose hands were amputated [8]. Ceci performed the first kineplasty operation on humans in 1900 [9]. In 1916, Sauerbruch and Horn developed the skin-covered muscle tunnel in Germany, and Bosch Arana first performed clinical studies of this procedure in Argentina in the 1920s [10, 11]. In 1900, Bier proposed the osteoplastic procedure in which the incised end was covered with a cortical bone flap connected by a periosteal hinge to allow the remaining distal end of the incised bone to bear weight [12]. This procedure was never widespread, but in the late 1940s, Ertl went one step further and developed a new procedure to create a bone bridge between the cut ends of the fibula and the tibia [13]. A few years later, Mondry combined the bone bridge technique with myodesis (the interconnection of cut muscles over the distal end of the residual limb). Researchers, such as Dederich and Weiss, have adopted and popularized these procedures in various fields [14].

Today, developments in medicine, materials science, software, engineering, and technology have created a multidisciplinary field of science called biomechatronics, which adapts mechanical and electronic design principles onto biological systems. The field of biomechatronics covers the topics of supporting human limb movements, strengthening limbs, improving problematic body functions, or performing the functions of the limbs [15]. New studies emerge every day in the development of innovative prostheses. Unlike ordinary prostheses, these innovative prostheses not only assume the static function of the limb they replace but also imitate the function dynamically. After processing and interpreting the signals received from the user's body, thanks to intelligent decision-making algorithms, the motor elements in their structures are activated and attempt to imitate the original limb in the best way. Later in this section, the innovative aspects of biomechatronic prostheses will be discussed in detail.

INNOVATIONS IN MATERIALS USED IN PROSTHETICS

Material selection for prosthetics depends on the individual needs of the prosthesis wearer. In the past, prostheses were generally made of wood and leather. In recent years, however, with significant technological advances in the field of material sciences, the materials used in prostheses have diversified. The demand for strong and lightweight components in the aviation and marine industries has led to the production of a variety of new materials with mechanical properties suitable for prosthetic construction. The new plastics have brought significant cosmetic improvements with greater durability and strength. Until recently, prostheses were only used outside of the human body and implant prostheses (orthopedic implants) were not invented. Although there is a wide variety of materials that can be used in limb prostheses today, traditional materials are still widely used. When deciding which material is suitable for limb prostheses, five important properties are taken into consideration: strength, stiffness, durability, density, and corrosion resistance [3].

The strength of a material is determined by the maximum external load that the material can support or carry. Strength is more important for lower extremity prostheses that for which loading forces can be very high during walking due to body weight.

Stiffness is a measure of the material's resistance to relative atomic splitting. This measurement, called Young's modulus, is greater when the materials are harder and is related to the amount of force required to disrupt the atomic structure of the material [16]. The hardness of the material is inversely proportional to its flexibility; when stability is important, sturdy materials are usually chosen. On the other hand, a more flexible material is used when the prosthesis needs to act in harmony with body segments.

The durability of a material is the material's ability to withstand repeated loading-unloading cycles. Repeated loading reduces the strength of the material and increases the breakage risk. The durability is more problematic on the junction surfaces of materials.

Density is the mass of the unit volume of the material and is the main determinant of the amount of energy spent by the patient during functional activities while using the prosthesis. While the prosthesis is intended to be as light as possible, strength, durability, and fatigue resistance may inversely require a denser material [3].

Corrosion resistance is the degree to which the material is resistant to chemical deterioration. Corrosion can occur in various ways when liquid material and solid material react. Generally, the surrounding fluid interacts with the bondings of the solid, which weakens the solid matter. Most of the materials used in prostheses retain heat, leading to abnormal perspiration. Waterproof materials are easier to clean and have a lower corrosion risk than porous materials [16].

Ease of manufacture is another important parameter for material selection. While some materials can be easily shaped, others require special equipment or techniques. Leather, different metals, wood, plastics, and various composite materials are used most commonly. Leather is generally used in prostheses as a contact surface material to protect the skin from irritation. Sometimes it is used as supportive material when strength and flexibility are required.

Steels are strong, hard, and durable materials, but their high density and corrosion susceptibility are the main problems. Various steel types have been developed to meet engineering needs. It is possible to achieve a high strength-to-weight ratio with steel, which is an important factor in repetitive loading situations in prostheses. Stainless steels are widely used in prostheses for durability and corrosion protection. Martensitic steel is broadly used in prostheses as it allows hardening by heat treatment.

Aluminum alloys are very suitable for prostheses thanks to their high strength-to-weight ratio and corrosion resistance. Processed Al alloys are suitable in prostheses for structural needs. The high compression bending stress needs of lower limb prostheses are well-suited to the use of processed Al alloys. Although Al alloys are very resistant to atmospheric and chemical corrosion of some substances, acids and alkalis in urine, sweat and other body fluids disrupt the natural protective oxides on the surface of Al, making it more susceptible to corrosion.

Components made of titanium alloys are becoming more common in prostheses. Although Ti alloys are stronger than Al and have strength comparable to some steels, their density is 60% lower than that of steel. Ti alloys are also more resistant to corrosion than Al and steel. However, Ti alloys are generally more difficult to process and fabricate, not to mention that it is more expensive than Al and steel.

Wood is frequently used in prostheses due to its durability, lightness, and ability to be easily shaped. Plastics are also widely used materials for prostheses for the same reason. Plastics used in prosthetics can be classified into two groups: thermoplastics and thermosetting plastics [3]. Thermoplastic materials can be shaped when heated and harden after cooling. Thermoplastics are the preferred materials for prosthetic components where structural strength is required. A kind of thermoplastics, namely polypropylene, is a hard plastic material that is relatively inexpensive, lightweight, and easy to thermoform. Polypropylene can withstand impact and several million cycles of repetitive stretching. However, the material is sensitive to ultraviolet light and extreme cold, and thus is susceptible to scratches. In prostheses, polypropylene is frequently used in components, such as sockets, pelvic bands, hip joints, or knee joints. Polyethylene, which has a long fatigue life during repeated loading, is one of the plastics used in prostheses.

Fiber-reinforced plastics, also called composites, are often used in dentures, mainly because they can be designed to have strength properties optimized for specific loadings. While these materials have high strength and hardness values, they are also resistant to compression and bending stresses. Foamed plastics can be used in areas vulnerable to pressure, such as bone protrusions. The microcellular structure of foamed plastic materials allows displacement in several planes, which is an ideal physical property for reducing shear forces. Because closed-cell foams are liquid-proof, they absorb fewer body fluids, such as sweat or urine, and act as insulators.

Unlike extremity prostheses, the material used in body prostheses (orthopedic implants) must be biocompatible. The materials used in removable joint prostheses are subject to wear [17], which causes the patient to feel pain over time and may result in aseptic loosening of the prosthesis. Therefore, the selection of materials used in implant prostheses is an important issue. In general, the most common materials used in orthopedic implants are metals and polyethylene. These two types of materials are combined in most joint implants, meaning one component is made of metal and one of polyethylene. When properly designed and implanted, friction-induced wear between the two components is minimized. The materials commonly used in implant prostheses are stainless steel, Co-Cr alloys, Ti alloys, Ti, Zr, Mo, and polyethylene (ultra-high molecular weight polyethylene (UHMWPE) and cross-linked polyethylene) [17].

In most orthopedic implants, stems that connect the prosthesis to biological tissues are manufactured using Ti-6Al-4V alloy. This material is preferred thanks to its high mechanical strength, corrosion resistance and excellent biocompatibility [18]. But Ti-6Al-4V-based stems have two main disadvantages: the relatively large mechanical stress protection and the relatively large micro movement. Both disadvantages lead to implant loosening and bone resorption in the surrounding bone, reducing the life of the prosthesis.

Various innovative studies have been carried out to provide more suitable designs and materials to overcome the problem of excessive mechanical stress as well as to improve implant life. One approach is to use Ti-based alloys with a low modulus of elasticity [19]. Among these materials, Ti-13Nb-13Zr and Ti-29N--13Ta-4.6Zr alloys provide sufficient strength, corrosion resistance and high biocompatibility [20]. Compared to conventional biomedical Ti-6Al-4V, low modulus alloys are effective in shielding against excessive stress and allow the bone to be remodeled although their modulus of elasticity is four to five times higher than that of adjacent bone. The disadvantages of these materials are their high cost, wear problems and low corrosion resistance.

Various studies have focused on the material optimization of orthopedic implants. For this purpose, a numerical approach to three-dimensional material optimization of prostheses using fiber-reinforced composite materials has been developed [21]. This optimization procedure can be used to minimize the stress concentration at the bone-implant interface in order to reduce the risk of mechanical failure. Kuiper and Huiskes developed a numerical approach to minimize the stress effect and interface micro-motion. Their results show that the use of a variable module prosthesis produces the desired load transfer distribution [22].

For the same purpose, long fiber composite materials have been developed as an alternative to hard metal alloys in an attempt to enhance stress protection and load transfer to the bone. These composite biomaterials appear to be the solution for orthopedic implants due to their potential benefits, such as special mechanical properties, anisotropy, mechanical reliability, environmental stability, and improved biocompatibility [23]. Currently, laminated fiber-reinforced composites are used in most composite material prostheses [24]. The findings of a study on orthopedic implants produced using vinyl ester resin reinforced with a braided carbon fiber preform showed that direct knitting of fibers over the joint could improve the quality of the prosthesis and ultimately its mechanical performance [25]. Akay and Aslan highlight the effect of fiber orientation on the mechanical performance of the prosthesis with a comparative stress analysis using carbon fiber-reinforced polyetherketone produced by injection molding [24].

New studies continue to appear on material selection for prostheses used both inside and outside the body. Numerous projects prepared with the aim of developing long-lasting prostheses and increasing user comfort receive intensive support around the world. The most recent studies include the use of soft robotic and mechanical devices that avoid traditional rigid components in favor of flexible technology that is more adaptable and capable of moving with multiple degrees of freedom.

CLASSIFICATION OF PROSTHESES

Prostheses are generally classified according to where they are used in the body. They can also be classified by control technique, size, activity, the feedback method (if any), and the material from which they are produced. After classifying the prostheses as internal and external prostheses, it is possible to divide the prostheses into two groups as lower limb prostheses and upper limb prostheses based on where in the body they are used. Structures, such as knee prostheses, tumor prostheses and skull bone prostheses, fall under other replacements, as shown in Fig. (**1**).

Upper limb amputations account for less than 20% of total amputations in developed countries [26]. The main purpose of the upper limb prosthesis is to fulfill the function of the lost limb and to give as good an aesthetic appearance as possible [27]. Upper extremity prostheses are divided into three subgroups: below-elbow prostheses (transradial), above-elbow prostheses (transhumeral) and partial hand prostheses. While transradial prostheses meet the function and appearance of the forearm and hand, transhumeral prostheses are intended to perform the function and appearance of the upper arm, elbow, forearm, and hand. Partial hand prostheses are intended for patients who have lost all or a certain part of the hand limb. Here, the structure completely changes according to the amputation level and position.

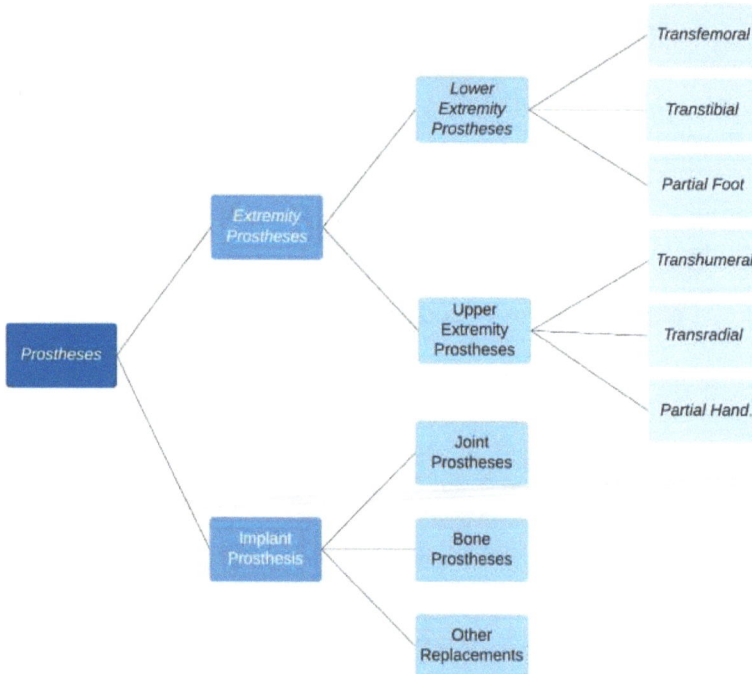

Fig. (1). Classification of prostheses.

Upper limb prostheses come in four groups: passive, mechanical (receiving force from the body), motorized (receiving energy from external source) and hybrid. Passive prostheses are only used to provide an appearance close to the natural limb and are rarely used as a push-pull arm for simple tasks in daily life. Mechanical prostheses use the patient's muscles as an energy source. Generally, as a result of pulling a strap with the forward movement of the shoulder joint, the hand fingers of the prosthesis are closed, and the holding function is fulfilled. This holding function is of low precision and low strength. Therefore, mechanical prostheses are used only for grabbing and picking up light and relatively large objects and cannot be used for holding and transporting small or heavy objects. Motorized prostheses often use myoelectric technology, in which electromyogram (EMG) signals are received over the muscles in the residual limb [28]. Generally, the state of contraction in relatively large muscles, such as elbow or wrist extensors, is measured by EMG, and a motor is driven accordingly. Thus, the holding function is fulfilled by using the simplest griper.

Transhumeral prostheses can offer up to 15 degrees of freedom, seven active and eight passive. These prostheses support the active movement of the first three fingers and have touch sensors, tension sensors and independent motors for each active finger. The wrist joint is usually passively designed with 2 degrees of freedom. These devices with multi-channel EMG have integrated batteries and central control units that allow the adjustment of prosthetic components by transferring data from the control system to the PC via Bluetooth®. This allows the user to choose different programs according to the physical activity to be performed at different times of the day. This configuration software also facilitates the evaluation of muscle signals and the optimal adjustment of the electrodes. Regarding aesthetics, it allows the patient to choose from several shades to obtain the tone most similar to the patient's skin [28]. There are several commercially available sophisticated myoelectric prostheses using this technology. However, since the costs of these prostheses are too high, they cannot become widespread enough. The most important factors affecting the costs are the high prices of actuators and sensors. With the addition of pressure sensors to myoelectric systems, there are studies that try to more accurately classify the motion the patient is trying to perform [29].

Lower extremity prostheses are much more critical than upper extremity prostheses, as they directly determine the patient's mobility and carry the patient's body weight during the day. These prostheses are divided into three groups as below-knee prostheses (transtibial), above-knee prostheses (transfemoral) and partial foot prostheses. Transtibial prostheses are intended to provide movements and appropriate external appearance in the ankle and foot. Transfemoral prostheses aim to simulate the appearance of the whole limb while meeting all degrees of freedom in the leg, excluding those in the toes, including the knee joint and upper leg. Partial foot prostheses are generally designed for cosmetic purposes and as a force transfer support. The reason for not creating a degree of freedom in the foot area is that the benefit-to-cost and benefit-to-complexity ratios are relatively low. However, in developed lower limb prostheses, it is seen that the foot part is also actively presented.

The number of studies on lower extremity prostheses is increasing day by day. Detailed studies are carried out on all components of the prosthesis with the aim of a lighter and more durable structure, high performance, and comfort. Material selection studies, sensor and motor improvements and control software development studies are carried out for this purpose. The spreading and improvement of 3D scanners and printers have also accelerated the work in this field. 3D printers have also contributed to the manufacture of specific prostheses

[30]. People with limb loss often complain of back pain due to improper fitting of the prosthesis, postural changes, leg length mismatch, amputation level, and general deconditioning [31]. Improving the prosthetic design can eliminate these problems.

The most important part of transfemoral prostheses is the knee joint. This joint can be designed as active or passive. The expected function of the knee joint is walking when active, controlled rolling when passive, and stability in all situations. The knee joint is determined in a way that provides the most safety for the patient and best suits their functional needs. Passive knees are divided into mechanical knees and microprocessor-controlled variable damping knees. Mechanical knees have the advantage of being lightweight and low maintenance. However, they are usually limited to a single-speed and straight walking function. There are also those produced specifically for sports, such as knees for running, cycling, or skiing [32]; these are often preferred thanks to their lower cost than advanced prostheses. Also called variable damping knees, gyroscope and accelerometer signals are processed in microprocessor-controlled knees to provide variable resistance to motion, and these knees adjust themselves according to the walking style and speed desired by the amputee. Microprocessors can continuously change friction during the swing and support phases by interpreting the position of the knee and foot and the loading forces. Some models also provide the user with a mobile device application, allowing the patient to control articulation and enable or disable functions.

When choosing the foot part in lower limb prostheses, the patient's conditions are prioritized. The intensity of physical activity and the environment, in which the prosthesis is meant to be used, are the most determining factors. As with other prosthetic components, there have been improvements in the design and functionality of this component in recent years [33]. There are quite different designs from each other in terms of both degrees of freedom and control approaches, such as the transfer of kinetic energy during walking.

Apart from lower and upper extremity prostheses, there is a wide variety of prostheses used in different parts of the body. Unlike limb prostheses, these are permanently placed in the body by surgical intervention and are also called implants. Skull, jaw, and facial bone implants are among the most widely used orthopedic implants. Further, tumor prostheses used for bone cancer patients and replacing cancerous bone fragments, penile prostheses used in urology, breast prostheses, other aesthetic prostheses, eye prostheses and dental prostheses are typical examples of special prostheses.

RECENT DEVELOPMENTS IN THE CONTROL OF ACTIVE PROSTHESIS

Limb prostheses that work with body muscles are still widely used today. However, since these prostheses cannot be controlled intuitively, over time, active prostheses have started to take their place. In active prostheses, the intuitive control provides neuroprosthetic interfaces that enable communication with the user's nervous system [34]. Recording the electrical activity on the nerves due to electrical stimulation of nerves and muscles in the relevant body region is the basis of prosthesis control. In this method, neuromuscular electrical stimulation (NMES) is applied to the relevant region to provide depolarization of the membranes of excitable cells. NMES produces contractions in certain muscles using electric current. Thus, neural activity is recorded as action by electrodes placed near neurons [35]. The stimulation process is limited to just enough time to provide the necessary clinical response in order to minimize the nerve damage they may incur.

Myoelectric systems or nerve electrodes are used to interface with peripheral nerves. Invasive and non-invasive muscle electrodes are used in myoelectric systems. Nerve and muscle electrodes used invasively are classified by their insertion depth [36]. While the nerve electrodes form a direct interface to the peripheral nervous system, the myoelectric systems are primarily connected to the muscles and form an indirect interface with the peripheral nerves. The main way of interfacing is to use the innervation of the muscle groups in the residual limb after amputation. However, since this muscle group usually fulfills a different purpose, users have serious difficulties getting used to their new task. Thankfully, with months of exercise repetitions, the muscle group can be adapted to its new tasks. Myoelectric systems can be created in a simple non-invasive structure or in a highly developed and invasive structure [34, 37].

The simplest myoelectric method is to use surface electrodes. In this method, control of the prosthetic limb depends on the activation of the remaining muscles after amputation [38]. This method has some disadvantages, such as the need to place and calibrate the electrodes every day, the need to maintain skin condition, motion artifacts, recording from unintended muscles, and a low signal-to-noise ratio [39]. In addition, this method is not used if the residual muscle mass of the limb is insufficient. An alternative method for recording impulses and stimulating muscles uses implantable muscle electrodes. This method provides greater resolution and signal-to-noise ratio. It also eliminates the need for daily insertion and problems with skin problems [40].

An alternative approach is to direct the severed nerves to a new muscle group. In this method, the entire nerve connection of the target muscle must be disconnected during surgery to avoid unwanted EMG signals that would complicate prosthesis control [41]. This control of the prosthesis is more natural than traditional myoelectric methods, and patients are more successful in performing desired tasks [42].

Another option is to use regenerative peripheral nerve interfaces surgically created from muscle grafts obtained from skeletal muscle that can be harvested from that limb or from another part of the body [43]. The biggest advantage of this method over the technique of guiding severed nerves to the new muscle group is that the operation is not limited to a specific anatomical area. In this method, muscle electrodes need to be implanted within 1–2 weeks after surgery to ensure biocompatible interfacing [44].

In interface design, several points need to be considered point-by-point. First, any substance placed in the body should be evaluated in terms of biocompatibility, as electrode interfaces can cause undesirable effects or tissue response. Active medical devices implanted in the body should also be tested for the susceptibility of the device to electromagnetic interference from any source [34, 45]. In addition, an interface should communicate with the target at the maximum level while causing a minimal disturbance in the surrounding tissues [46]. Another point to consider is the communication between the implant and the external control unit. Wired or wireless connections can be used for implantable electrical interfaces [47]. The use of energy harvested from the body's internal resources or wireless power transfer strategies can be used to overcome battery problems [48].

Unlike other prostheses, lower extremity prostheses must be a component of the walking cycle. Therefore, more advanced strategies are needed for the self-control of lower limb prostheses. Controllers of lower limb prosthetics use finite state controllers and divide gait into phases. Here, a generalized control frame consisting of four main sub-blocks as controller, device, user and environment can be mentioned [49].

In upper limb prostheses, prosthetic hand control is the compelling part that will take on the task of the natural hand and perform the function of holding and lifting. Upper limb prosthesis sensors are used to detect external stimuli, such as force and pressure on the fingertips of the prosthesis. Electronic components are used here that sense force and convert information directly into electrical signals [50].

Resistive sensors measure changes in resistance caused by an externally applied force. Resistance changes can be measured through Micro-Electro-Mechanical Systems (MEMS), strain gauges or liquid-based designs. Conductive liquid resistance sensors are custom-made sensors that measure changes in resistance through fluid displacement caused by mechanical load. In this structure, a core with outward-facing electrodes is surrounded by a compressive sheath with conductive particles, and the impedance changes created by the altered fluid path are measured. Capacitive sensors that measure changes in capacitive coupling are used in designs that allow pressure sensing, shear detection and tissue recognition. Piezoelectric material, which naturally converts mechanical voltage into electrical potential, is another common sensing element [51]. Optical sensors used for tactile signal transmission are employed to measure optical changes occurring in the translucent environment because of physical deformation resulting from contact pressure [52].

The design of the prosthetic sensors is inspired by biological components. Prosthetic sensors also mimic the biological properties of the skin, including signal mechanisms, structure, and mechanical properties. Biomimetic sensors try to replicate the skin's tightness as well as its ability to stretch. Most tactile sensors use stretchable materials such as elastomers to mimic the essential viscoelastic properties of the skin surface. However, the sensing elements themselves are under the skin layer. The sensor can be produced in a flexible structure to adapt to skin-like mechanical properties [53].

MEETING THE POWER REQUIREMENTS OF ACTIVE PROSTHESES

One of the important factors that reduce the widespread use of prostheses is the need for power. Active prostheses require a significant amount of electrical energy to operate sensors, actuators and microprocessors during motion and use. The required electrical energy is usually provided by rechargeable lithium polymer and lithium-ion batteries. The low energy density and discharge current capacity of rechargeable batteries make it very difficult to use them in today's prostheses, which are increasing in weight for biomimetic limb design. The use of supercapacitors with fast charging and discharging features instead of batteries has also been among the designs of recent years. We also notice a growing tendency to use a combination of batteries and supercapacitors.

Especially the batteries of the lower limb prostheses need to be recharged very frequently. For example, Össur Rheo Knee and Ottobock C-Leg need to be recharged approximately every 36 hours [54, 55]. Therefore, patients prefer semi-powered or passive prostheses to fully battery-powered active prostheses [56].

One of the methods that can be used to solve the prosthesis power problem is biomechanical energy harvesting. Energy harvesting means converting any environmental energy into electrical energy. In the technique of biomechanical energy harvesting, mechanical energy lost during the movement of human limbs is converted into electrical energy to recharge prosthetic batteries. Thus, a lighter battery can be used than the existing battery or a longer operating time can be obtained by using the same battery [57].

Human joints do mechanical work in opposite directions during acceleration and deceleration of movement. Especially during walking, the human body spends a large amount of energy. It is possible to use negative power at a certain stage of the gait and increase the energy efficiency of the gait.

The knee prostheses generally use energy dissipation systems to achieve biomimetic oscillation and stance control. Instead of dissipating mechanical energy as heat, the mentioned energy can be converted into electrical energy to charge internal batteries. Energy harvesting can be done by methods, such as biofuel cells, magnetic induction, thermoelectricity, and vibration. The system needs to be carefully optimized, as each new component to be added to the prosthetic structure or a different limb for energy harvesting will add extra weight.

One of the parameters to be considered in optimization is the movement speed of human joints. A natural walk consists of steps equivalent to approximately 20 complete rotations per minute (20rpm). Electromagnetic machines that will convert mechanical energy into electrical energy generally operate most efficiently at higher rotation speeds [54, 55]. Mechanical power transmission organs like gears can convert lower limb joint rotational speeds to speeds suitable for electric machines [55, 58].

Minimizing the device mass and moment of inertia will reduce the torque requirements of the electromagnetic machine used. So, lightweight composite materials can be employed for spares and chassis weight reduction. Structures, such as elastic elements, hydraulic accumulators and rotating flywheels, are used to store mechanical energy in active lower extremity prostheses [55]. The use of energy collected in the early stages of walking using springs in other stages is one of the frequently preferred designs [59, 60]. Walking support systems designed with this method can reduce the metabolic cost of walking under natural conditions by approximately 7.2% for healthy people [61].

AMPUTATION, IMPLANTATION AND REHABILITATION

Amputation is one of the most common surgical operations. With 30 million people living with limb loss worldwide, amputation is one of the most contributing factors to disability [62]. Most amputations are done to treat complications of peripheral vascular disease. Approximately 40% of amputations are performed in diabetics. Other indications for amputation include a disabled limb due to trauma, aggressive infection, malignant tumors, congenital deformity, chronic pain, or neurological injury. Preoperative evaluation of the patient is made with the opinions of the surgical and anesthesia team, prosthesis specialist, nurse, physiotherapist, occupational therapist, diabetes specialist and psychologists. A careful examination is required to detect the presence of previous orthopedic prostheses or vascular bypass grafts that may be encountered during surgery. Starting at this stage, it is important to manage patients' expectations and familiarize them with limb loss and the challenges of prosthesis use [63, 64].

Amputation surgery and rehabilitation can be divided into 9 stages. In the preoperative planning stage, the patient's body condition is examined, the patient is informed about the surgery, the amputation level is discussed by the surgical team, and postoperative prosthesis plans are evaluated. Amputations are defined by the anatomical level of the bone removed during surgery. 39% of amputations are transtibial, 31% transfemoral, 15% transradial and 8% transhumeral level, and the remaining 7% include hip and shoulder disarticulation, knee, elbow, and wrist level amputation. Right arm amputation is more common due to upper extremity work injuries. 60% of those with amputated arms are between the ages of 21 and 64, and 10% are under the age of 21 [65]. Congenital upper limb insufficiency has an incidence of approximately 4.1 per 10,000 live births [66]. While the goal of most lower limb amputees is walking, 20% of transtibial and 60% of transfemoral amputees cannot walk after amputation [64]. The second stage is performing the amputation operation and reshaping the limb. At this stage, myoplastic closure, soft tissue covering, nerve-related procedures and hard dressing applications are in question. The furthest possible amputation level is determined at the transtibial and transradial level according to the vitality of the soft tissues and the condition of the skin covering with sufficient sensation. In order to improve the prosthetic suspension, an attempt is made to maintain the length of the residual limb and to ensure that force transmission from the residual limb to the socket is possible. The residual limb must be carefully structured to allow it to relieve the necessary pressure to maintain muscle balance and fulfill its new function [67]. Bone spurs, skin scars, soft tissue shrinkage, cutting, and sweating can complicate this function [68]. The next stage is called the acute postoperative stage. The first day after the operation is the first evaluation day by a physiotherapist who is an expert in amputated rehabilitation. At this stage, wound healing is followed, pain control is provided, proximal body movement is observed, and emotional support is given to the patient. After that, the pre-prosthesis stage, which includes increasing the patient's muscle strength and regaining regional muscle control, is started, when the patient is adapted to using the prosthesis. The next stage is the prosthesis production stage, in which the prosthesis is adapted and prepared according to the patient. The next stage is the prosthetic exercise phase, during which the patient is expected to adapt slowly to the prosthesis. Subsequently, the social adaptation phase is started. At this stage, the patient tries to take part in family and community activities as before. Ensuring the emotional balance of the patient and coping with amputation may prolong the process. After that, the process of returning the patient to their pre-amputation professional activities and if the new body condition is not suitable for continuing the same job, the process of rescheduling or changing his job should begin. Observation of the long-term consequences of prosthesis use and lifetime follow-up of the case are the last steps [66].

With the provision of appropriate prostheses after amputation, rehabilitation helps people build their lives, regain independence and dignity, and continue their former activities. Following post-amputation healing, a prosthesis should be inserted immediately. Quickly applying a postoperative stiff dressing can speed up wound healing and maturation. In upper limb amputees, this is particularly important because there is a direct correlation between time of application and long-term prosthesis use. There is a 6-month opportunity period for unilateral upper limb amputees with a much greater acceptance rate of the artificial arm and functional integration [68].

Pain perceived by the patient as a result of amputation can be divided into four possible categories. These are postoperative pain, remaining limb pain, prosthetic pain (most often caused by standing and walking with the prosthesis), and phantom pain (perceived pain coming from the cut body part) [69]. Pain can originate from other areas of the body that are not associated with the amputation or are related to the limb being amputated. Once the nature of the pain is clarified, appropriate interventions can be initiated to allow the patient to work comfortably [65]. Prevention of falls that could cause damage to the residual limb is crucial for further rehabilitation [70].

When a patient is referred to a rehabilitation center after surgery, an amputee rehabilitation specialist together with a multidisciplinary team should perform the evaluation. For example, lower limb amputees with significant cardiovascular risks may not be able to walk with the weight of a lower limb prosthesis and may gain better mobility using a wheelchair [71].

Osseointegration for limb loss began in the 1990s, based on successful experience in dental osseointegration. Osseointegration is the direct structural and functional connection between living bone and the surface of the load-bearing artificial implant. Osseointegration is defined as functional ankylosis where the residual bone is placed directly on the implant surface, and the implant has mechanical stability [72]. This technique allows the user to have more precise and reliable motion control in the prosthetic limb.

Numerous research groups are operating around the world to develop these innovations. For example, new implanted electrodes are being developed to directly sense information from the surface of the residual limb to the cut peripheral nerves. Thus, structures that are easier to use can be included in more functional tasks.

PATIENT'S ACCEPTANCE OF THE PROSTHESIS

Prostheses are medical devices used by people to meet the function and appearance of lost limbs and that people should use for a lifetime. Therefore, prostheses must be easy to attach and remove, aesthetically pleasing, light, durable, working properly and easy to maintain. The socket part that allows the prosthesis to be attached to the patient's body is the main component that affects the comfort of the patient and affects the acceptance of the prosthesis by the patient.

The socket is the top part of the prosthesis that contacts the residual limb and transfers walking forces in lower limb prostheses. Therefore, socket design is very important in lower limb prostheses. According to many amputees, the most important features of a prosthesis are durability and comfort. It is seen that amputees often reject their prostheses because of discomfort or because they are not necessary for daily life [73].

Patients using limb prostheses experience problems in the region of the residual limb due to the stresses created by the prosthetic socket, especially when the socket is not fully seated or when the sling mechanism does not adequately limit the relative movement between the socket and the limb [74]. Conditions caused by improper prosthetic sockets are the most common symptoms reported in prosthetic clinics. Although advances in technology and socio-economic factors have led to significant advances in prosthetic design, discomfort in prosthetic fitting remains a major problem among many amputees [75]. Basically, comfort is about the pressure between the socket and the residual limb.

Especially in lower limb prostheses, the socket is designed to carry the body weight. Therefore, it should be snug and tight in the remaining limb. Poor socket design can have various consequences, such as skin irritation and deterioration, decreased prosthesis usage with heat and sweating, dermatitis and infections that can lead to unwanted mobility, limb pain, decreased blood flow, neuroma formation, and discomfort from a poorly fitted prosthesis [76]. Therefore, it causes more disability for amputees. While designing the socket, regions of pain in the remaining limb should be examined. The socket should fit comfortably and be free of problems.

There is no standardized scale for determining the comfort level, and only descriptive terms are used. Many amputees report feeling pain at the distal end of the limb when walking with the prosthesis. Jia *et al.* and Pirouzi *et al.* found that human skin is not capable of remaining undamaged under prosthetic pressure [75, 77]. Skin suffers from external and internal forces. At the same time, mechanical changes can occur, causing irritation and damage to the skin and the underlying area, which has a long-lasting negative effect. Therefore, the design of the socket must meet the comfort and functionality requirements of amputees. The current clinical management of socket fit and tension problems to reduce residual limb problems is primarily based on prosthetic user feedback and visual inspection of the residual limb.

Periodic adjustment and replacement of the prosthetic limbs are required due to increased or decreased body weight associated with the tightness and looseness of the socket [78]. Therefore, there is a constant need for new prosthetic sockets for amputees, even if materials are expensive. Typically, in this situation, patients often wear multi-layer stockings to accommodate fluctuations in remaining limb volume [79]. In addition, silicone liner is widely used among amputees. The purpose of this is to protect residual limb skin, better suspension and improve cosmetic appearance. However, itching and excessive sweating during the use of silicone liners can have adverse effects [73].

Detection and tracking technologies can be used to continuously monitor the condition, progress, and outcomes of prosthetic wearers, as well as to improve rehabilitation by obtaining meaningful objective data. Today, sensors, wires and cables are used to make physical measurements inside the socket. Such tools are not suitable for daily clinical use. The use of sensing and monitoring tools should be cost-effective for clinical use and lead to an increase in the quality and effectiveness of care. Flexible 3D printed sensors, liners and removable inserts, and epidermal sensors for skin-based monitoring have been developed. Integrating such a system into clinical practice would allow for multidimensional, bidirectional information exchange between the prosthesis wearer and the prosthesis specialist [80].

Amputees are apprehensive about tracking technologies. Problems raised in relation to the monitoring of the elderly include loss of privacy and autonomy, reduced human contact, discomfort with technology, personalization, affordability, and security [81, 82].

In addition to physical change, amputation of a limb can have unpleasant psychological consequences. Prosthetic appearance has a major role in the mental recovery of amputees. It has been noted that younger amputees are more interested in modern-type devices with a futuristic appearance. Many prefer to add some graphics to the prosthetic socket to customize it to reflect their identity and are attracted to unrealistic artificial-looking prostheses that do not resemble a human limb. In contrast, elderly amputees prefer prostheses that are as human limb-like as possible and are not interested in the visual design of the socket.

Older individuals prefer not to be obvious and not to attract attention in the community [83].

For amputees, the appearance of a prosthesis is often important, but some patients must accept the prosthesis without considering the appearance of their artificial limb. Patients usually buy their prostheses through social security institution or insurance companies. However, the company needs to consider the fact that, in addition to functionality, the appearance of the prosthesis is also very important to amputees, as it can improve their mental health and body image [73].

Further development is required to improve existing prosthetic sockets, which must be realized with special attention to aspects, such as design, material, and mechanics. Developments in these areas will improve the quality of life of patients and provide them with the opportunity to improve their ability to work in daily life.

LIFE CYCLE AND COST OF THE PROSTHESES

The rapid emergence of innovative and often expensive technologies in the field of prosthetics has increased the need for economic evaluation studies to guide decision-makers. The application of a prosthesis is a delicate stage in a patient's life because the selection of the prosthesis is an important step in the patient's general rehabilitation [83].

Although the laws vary from country to country, social security institutions and insurance companies generally cover the cost of procurement and maintenance of prostheses for those who suffer from limb loss. When an insured individual is injured leading to amputation, the support provided by funding agencies includes medical care, financial compensation, rehabilitation, education, workplace and home renovation, and prosthetic costs. Determining the technology level and function of the prosthesis to be presented to patients is controversial [84]. Life cycle cost analysis, maintenance requirements and reliability of prostheses are issues that need to be investigated. In 2011, Biddiss reported that the average cost of prosthetic components and their annual maintenance was $ 9,574 and $ 1,936, respectively [85].

Chan *et al.* made a retro respective cost analysis of upper limb prostheses. In that study, the average number of prosthetic repairs per amputee was approximately 1.64 times a year. According to the results of the study, there are more problems with transradial prostheses than transhumeral prostheses, and while a trans-radial prosthesis has an average of 1.96 problems per year, transhumeral prostheses require maintenance 1.26 times a year. When grouped as prosthesis types, it has been reported that myoelectric prostheses experience an average of 0.98 times a year, while prostheses that are powered by the body fail an average of 0.9 times a year. There is no serious difference between body-powered prostheses and myoelectric prosthesis maintenance requirements. Apart from repair work, prostheses require occasional adjustments (*e.g.,* adjustment of cable and harness for body-powered prosthesis) to maintain their functional effectiveness. On average, an upper limb prosthesis user needs to adjust their prosthesis every 2 years. Body-powered prostheses need more frequent adjustments than myoelectric prostheses. In addition, wearing components such as gloves and liners need to be replaced. Transradial prosthesis wearers show a higher frequency of component replacement, indicating that transradial amputees use their prostheses more than transhumeral amputees [86].

Prosthetic technology has improved significantly in recent years. An important innovation in the field is the development of the microprocessor based electronically controlled prosthetic knee mechanism (C-leg) [87]. A case study conducted in the United States concluded that walking with the C-leg is the most effective method known for bilateral amputated ambulation [88]. The relatively high price of C-leg (€ 18.616) prevents it from becoming widespread.

The cost of a mechanical knee prosthesis varies according to the model. The average acquisition cost of a multicenter mechanical joint prosthesis is € 3,000. Pharmaceutical treatment costs, diagnostic and laboratory tests, specialist visits, primary health center visits and caregiver fees should also be considered when calculating the costs of prostheses.

When compared in terms of physical mobility, daily use times and quality of life, it is seen those amputees using C-leg experience a higher level of satisfaction than amputees using mechanical knee joint prosthesis. In terms of healthcare costs, there appears to be no difference between patients using C-leg and those using conventional knee joint prostheses [87, 88].

The comfort offered by lower extremity prostheses to the user is of particular importance. Individuals with lower limb transfemoral amputation equipped with traditional prosthetic limbs often experience socket-related problems leading to a decrease in quality of life. Most of these problems can be remedied by replacing the socket with a surgically implanted osseointegrated fixation with a prosthesis that attaches directly to the residual bone. These structures are based on screw-type or push-fit designs that are commercially available today [89]. Surgical implantation of the fixation probably requires single or two-stage operations followed by a rehabilitation program lasting 6 to 12 months [90].

Both screw-type and push-fit bone-anchored prostheses have been used clinically for more than ten years. Bone-integrated prostheses can potentially reduce medical and financial burdens by reducing the treatment of skin socket interface problems that a user may encounter during their lifetime. Few studies have made an economic analysis of bone-anchored prostheses. According to these studies, the number of malfunctions seen in bone-anchored prostheses is quite low compared to socket prostheses [91 - 93].

Typical prosthetic care for individuals with transfemoral amputation begins with initial socket insertion and fixed costs for a temporary prosthesis, followed by the cost of fitting the permanent prosthesis that the individual will use throughout his life. For bone-anchored prostheses, there is a cost of fitting the light prosthesis to be used before permanent prosthesis installation. The preliminary costs mentioned are around $2000 for the socket-type prosthesis and around $3300 for bone-anchored prostheses [89]. The cost that can be covered by insurance companies to provide bone-anchored prostheses to patients pays back over time when evaluated from different frameworks. For example, these prostheses will provide better opportunities for amputees to increase their mobility and will play a socioeconomic role by facilitating their employment.

PROSTHESES FOR CHILDREN

Frequently used in individuals under the age of 18 are limb prostheses due to amputation or congenital limb deficiency, and tumor prostheses replacing a tumor bone fragment due to bone cancer. The main causes of amputation in children are congenital limb deficiency, trauma, and cancer. McLarney *et al.* found the prevalence of major lower limb loss to be 38.5 per 100,000 in individuals under 18 years of age. Congenital limb deficiency accounts for 84% of cases, followed by trauma at 13.5% [94].

Congenital partial absence of the forearm is the most common upper limb deficiency. Longitudinal deficiencies of the upper limb are less common. Conversely, longitudinal deficiencies of the lower limb occur more often than transverse ones. The most common is the longitudinal partial reduction of the toes, followed by the longitudinal defects of the femur, fibula, and tibia [95].

Children are among the most active prosthetic wearers and often require pediatric special prosthetic components to engage in a variety of high-level activities. Prosthetic options for pediatric prosthetic limb users have increased in recent years. The improved prosthetic design and technology allow not only daily use but also sports and activity-specific uses [96]. Some of the prosthetic technologies developed for adults, which we examined in previous chapters, are also useful for adolescents and young people.

Traumatic amputations in children occur twice as often as amputations caused by a tumor or infection [95]. Finger amputations in toddlers are the most common traumatic amputation due to door injury [96]. Adolescents are more likely to have injuries from tools or lawnmowers [97]. Traumatic injuries typically result in transverse or complete amputation of a limb and may be complicated by severe soft tissue loss. Amputations associated with tumors are the second most common amputations [95].

The general purpose of prosthetic treatment in children is to facilitate the development process and to prevent the onset of secondary disorders and functional limitations such as contracture, fatigue, and addiction in self-care. Although limb prostheses used in children are similar in many respects to prostheses used in adults, some special exceptions are limited body size and low level of development of motor functions. In addition, children grow and develop throughout the rehabilitation process. Therefore, the need for frequent replacement of prostheses arises [98].

Children are among the most active prosthetic wearers and often require pediatric special prosthetic components to engage in a variety of high-level activities. Prosthetic options for pediatric prosthetic limb users have increased in recent years. The improved prosthesis design and technology allow not only daily use but also sports and specific activities [99]. Some of the prosthetic technologies developed for adults, which we examined in the previous sections, are also useful for adolescents and young people.

The first prosthesis is usually attached to babies when they are about 6-9 months old. Some babies undergo surgery as a result of trauma or as part of tumor treatment to convert an abnormal limb from birth to a more suitable limb for prosthesis. Another intervention applicable to children is bone lengthening using an Ilizarov apparatus, particularly in the lower limbs [100]. For children born with proximal focal femoral insufficiency, knee rotationplasty is usually performed. This surgical procedure involves segmenting the limb and rotating the distal portion posteriorly [101].

Prostheses in babies help to gain motor skills in normal time. A baby with a lower extremity anomaly is ready for prosthesis at about 6 months when it has sufficient body control to sit and is ready to stand. The prosthesis improves symmetrical sitting balance and helps the baby in attempts to stand up, equalizes leg length, adds weight to the abnormal side, and prevents the development of a one-legged standing pattern. The use of a lower limb prosthesis before 6 months may prevent the baby from attempting to turn from the prone to the supine position and back. Motor skills in children develop at a predictable time. In the absence of cerebral development or malformation, a baby born with a limb abnormality or a small child with an amputation exhibits physical control at approximately the same time as an unaffected child. However, limb deficiency often alters how developmental tasks and activities are performed. While a baby with limb loss or abnormality will take a similar time crawling to normal babies, the crawling pattern will not coincide [102].

Physical conditioning programs are beneficial in improving overall health and endurance for children wearing prostheses. Games develop coordination and muscle strength. While planning the prosthesis, attention should be paid to comfortable socket placement and maintaining equal limb length. Preschool children need a new prosthesis approximately every year. Children of primary school age require a new prosthesis every 12 to 18 months. Prostheses should be replaced with a larger one every 18 to 24 months in young people whose growth period is not completed. There are also prosthetic knee designs that can be used in infancy in children [103]. However, this knee is not suitable for use during the transition from crawling to walking. Hydraulic knees are suitable for activities, such as walking and running. There are several pediatric hydraulic or pneumatic knees with improved stability provided in a multicenter design. These are generally suitable for 7–14-year-old children.

Children are involved in much more diverse activities in their daily lives than adults. The design of a prosthesis is usually modifiable to help the child overcome a certain limitation rather than a general design suitable for all activities. These activities often require the provision of a separate (secondary) prosthesis with activity-specific components that meet the needs of the child when it is unable to perform that activity in any other way. Activity-specific terminal devices are available for children with upper limb differences that can be replaced with a conventional terminal device or integrated into a secondary prosthesis. Examples include designs for grip bicycle handlebars and flexible terminal devices for weightlifting and sports. For children interested in music, there are terminal devices designed to hold a guitar pick, drumstick, or violin bow. Examples of activity-specific components for lower limb prosthesis patients include legs with carbon blades for running, feet on ski boots, and knees designed for cycling. Leg prostheses are not generally used for swimming, but many children use waterproof prostheses in pools, water parks, showers, or beaches [104]. In addition, the external appearance of the prosthesis may be an important concern for children, especially when they enter adolescence. There are also studies on prosthetic parts that can be extended with the growth of children. This practice is usually accomplished by replacing modular parts [104].

Upper limb prostheses are traditionally made with an exoskeleton design that simulates the shape of the limb and provides strength. In addition, lighter and adjustable components that allow endoskeletal designs in upper extremity pediatric prostheses have also been developed. Nevertheless, these structures with lower durability are only suitable for use in later childhood.

Appearance plays an important role in upper limb prosthesis design. Children who do not mind that their two limbs look different from each other in primary school age may feel uncomfortable when answering questions about their limbs in middle school or high school age. These adolescents may prefer a prosthesis that closely resembles the skin tone and nail length of their arms. These prostheses are more fragile and require more maintenance than a typical prosthesis but they can increase the child's confidence.

Today, unlike in the past, pediatric-sized endoskeleton pylons, adapters, feet, knees, and liners are produced by manufacturers and are easily accessible. Modular components found in a child's endoskeletal prosthesis can be replaced, realigned, and adjusted for growth. The availability of modern modular pediatric components makes it easy to customize the prosthesis for children.

Reconstructive surgery, especially the Ilizarov fixator, can be used in children with a low limb length discrepancy (LLD). Amputation becomes necessary in cases of larger limb loss. The LLD causes serious problems, especially in patients using lower limb prostheses. The LLD can cause erroneous walking and extra energy consumption and may cause new damage to healthy parts of the body. In upper extremity prostheses, the LLD that cannot be noticed by the eye does not cause any problem. For this reason, the lower extremity prostheses must be extensible.

Flexible feet with energy storage and re-rotation technology are used as prosthetic feet for children, so a structure suitable for the child's highly active lifestyle is selected. The feet, which are multi-axis and dynamic functionally, are designed in smaller sizes and lighter and are produced specifically for children.

A toddler with transfemoral amputation may use a knee joint-locking prosthesis

[105]. Teenagers often prefer hydraulic, pneumatic, or microprocessor-controlled knees. The C-leg can be attached to adolescents tall enough to fit the size of the microprocessor-controlled knee unit [106]. Microprocessor knees are often too big and heavy for adolescents and children. Still, patients with bilateral amputation can benefit from the enhanced stability and control that microprocessor knees provide on stairs, slopes, and uneven terrain.

Options, such as the Nintendo Wii game system, for children using lower limb prostheses can contribute to rehabilitation. In this way, the game system is used as a fun, interactive and motivating tool and serves to develop the child's awareness of weight distribution through prosthetics.

Prostheses have been observed to wear out more quickly in children and adolescents compared to adults. The reason for this is higher mobility and speed of movement. The socket, which starts to feel small due to the growth, tends to shift from the limb. In addition, pain or skin redness caused by socket tension and flesh rolling around the edge of the socket can be seen. As a precaution, larger sockets with socket liners are to be preferred. Thus, the socket liners can be removed as the child grows by adapting to the environmental growth.

A CASE STUDY: PEDIATRIC AUTONOMOUS EXPANDABLE TUMOR PROSTHESIS

Around 3000 new cases of osteosarcoma and Ewing's sarcoma are seen worldwide every year, which kills approximately 1500 people annually. Bone cancer is usually seen in the femur and tibia. Bone cancer that commonly occurs in adolescence usually affects children and adolescents between the ages of 15-25. Osteosarcoma often occurs in teens, while Ewing's sarcoma occurs in teens and young adulthood. 30% of the patients are less than 10 years old [107, 108].

Surgical intervention is used in the treatment of bone cancer, depending on the progression stage of the cancer. Usually, surgical intervention results in the removal of the cancerous part by protecting the limb or the removal of the limb completely if a piece of bone is removed, and then bone grafting is performed or a prosthesis is inserted [109, 110]. The prosthesis placed in the leg should be extended with the growth of the individual and the elongation of his healthy leg. For this reason, either periodic surgical interventions are performed or an extendable tumor prosthesis is used [111].

Extendable tumor prostheses have been used since the late 1970s and have shown great improvements over time. Almost all the designs are seen to have been made for the distal part of the femur bone. The two main reasons for this are that bone tumor is most common in this area and most of the total elongation in the leg occurs here (60-70% of total leg extension). Extensible tumor prostheses can be examined under 3 basic categories according to the lengthening technique. The first type of prosthesis is extended by adding modular parts to the prosthesis with surgical intervention while the patient is under general anesthesia. The second type of prosthesis, developed later, is extended from the outside of the patient's body using a screw mechanism. Here, the procedure is performed by turning the screw placed under the skin under general anesthesia in a minimally invasive manner. The third type of prosthesis is usually extended by the electromagnetic field delivered from outside. In this method, a rotatable magnet or a wound rotor is used in the prosthesis [112].

Surgical intervention is not required for the application of the lengthening procedure after implantation in the 3rd type prosthesis used today. Still, though, it has many disadvantages. The patient must be in the clinical environment during the lengthening of the existing extendable prostheses. The amount of elongation during the procedure is monitored using medical imaging techniques. The main problems here are: the patient has to go to the clinic frequently for measurements and lengthening procedures, the physician's workload increases, the errors caused by the person that may occur during the elongation measurement, the patient is exposed to radiation at each measurement and the extension is performed in relatively large sizes. In these systems, the patient must go to the clinic at least once a month to measure the LLD. If this period is extended, the LLD value of the patient may increase too much, reducing the patient's quality of life. If the implantation age of the tumor prosthesis in the patient is accepted as 10, this situation continues for about 8 years. This results in a large number of clinic visits and a large number of radiation exposures. Considering the growth suppressing aspect of radiation, it is seen that LLD detection methods of existing systems are problematic. This method of detection significantly increases the workload of the clinician.

Kocaoğlu and Akdoğan proposed a type 3 tumor prosthesis that can automatically elongate with the lengthening of the patient in order to reduce or eliminate the effects of all of the above-mentioned problems [113]. As an innovative application, the prosthesis developed by Kocaoğlu and Akdoğan is detailed in this section. The basic block diagram of their system is given in Fig. (**2**).

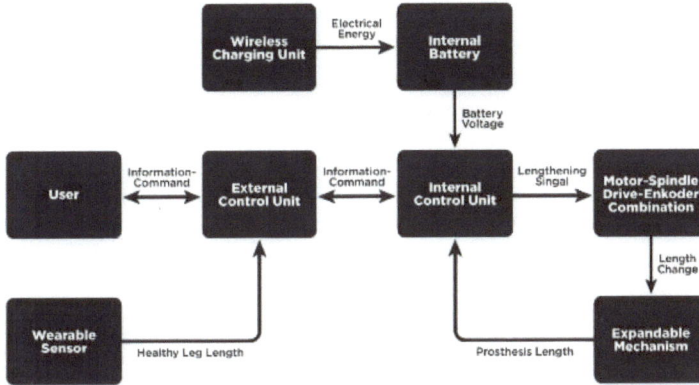

Fig. (2). Basic block diagram of an expandable tumor prosthesis system.

The elongation mechanism of the developed prosthesis consists of two intertwined tubes. The inner tube is connected to the femur bone, and the lower part of the artificial knee joint is connected to the tibia bone. The prosthetic mechanism is designed using Ti in the implant norm. In knee prostheses, abrasion and deformation are created due to friction, particles, chemical contact of the surface, and periodic loading. Therefore, UHMWPE material was preferred as a tibial component in the design due to its low friction force and abrasion resistance compared to other materials. While designing the knee joint to reduce wear, the contact surface has been kept as wide as possible.

The internal control unit, which provides control of the internal components of the prosthesis, is placed in the artificial knee joint. With the driving of the motor, the spindle drive enables the two tubes to move away from each other (Fig. 3).

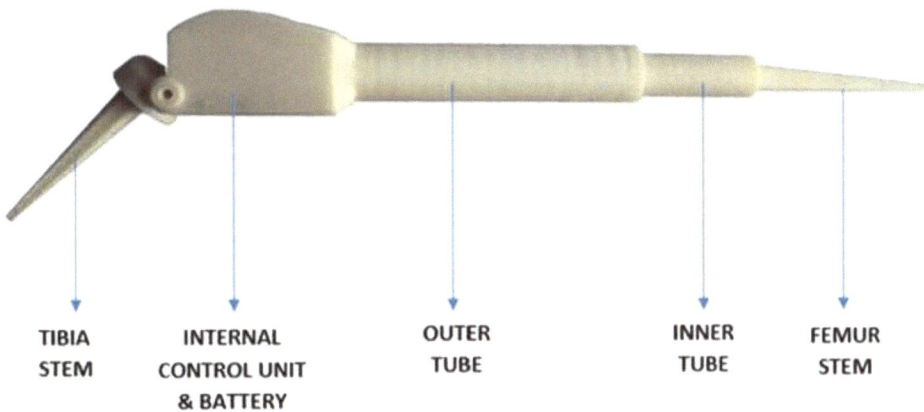

Fig. (3). Expandable tumor prosthesis section view.

Elongation amount, temperature, internal battery charge level and patient posture status are measured by the internal control unit sensors located in the prosthetic knee joint. The internal control unit wirelessly shares the information received from the sensors with the external control unit when necessary (increase in internal temperature, decrease in internal battery charge, *etc.*). When the elongation process is to be performed, it receives a command from the external control unit and checks through sensors whether the conditions are suitable for the elongation. It is the unit that controls the motor driver and monitors the elongation process. The internal temperature of the prosthesis is continuously monitored by the temperature sensor connected to the internal control unit. If the temperature rises excessively, the internal control unit does not allow elongation by pulling the components to the passive position until the desired temperature is achieved [114].

The external control unit is a control unit with a human-machine interface (HMI) panel that allows the patient to communicate with the system. It evaluates the data coming from the internal control unit and informs the patient when necessary. By comparing the length information obtained from the healthy limb with the length value of the implanted limb, it determines the need for extension, shows it to the patient and transmits it to the internal control unit. It also generates the extension command when necessary. Before starting the lengthening procedure, there should be no load on the prosthesis, that is, the patient should not be standing. The external control unit instructs the patient to go supine before the lengthening procedure. It controls the patient's position through the sensor inside the prosthesis, and if there is no extra load on the prosthesis, it sends the extension command to the internal control unit [114].

The wearable femur length measurement sensor is attached to the healthy limb from 3 points externally by the patient when necessary (Fig. **4**). The healthy limb length is determined and transferred to the internal control unit through the external control unit as a result of the communication between the RFID readers in the wearable sensor unit and the implantable RFID chips placed on the lower and upper ends of the patient's healthy limb femur bone. LLDs of 1mm and above, formed by the elongation of the healthy limb, can be detected by the wearable sensor. When the length difference occurs, the system initiates the lengthening process by guiding the patient.

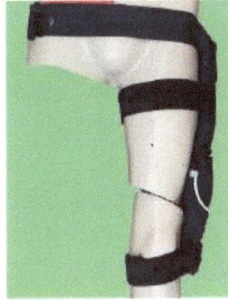

Fig. (4). Wearable femur length measuring sensor.

The energy of the prosthesis is provided by the battery placed in the artificial knee joint. Battery charge information is continuously measured by the internal control unit. When it falls below the specified value for charging, a warning is sent to the external control unit. In this case, the battery is charged by Radio Frequency (RF) method using the wireless charging unit (Fig. **5**) that the patient will use by wrapping it around the knee joint.

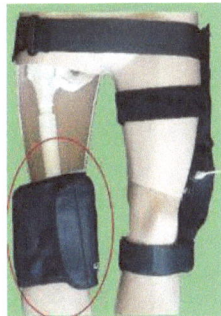

Fig. (5). Wireless charging unit.

A battery charge control unit is used to ensure security in the system and to protect the battery. Thus, if the battery voltage falls below a threshold, the discharge process is terminated, and if it exceeds the threshold, the charging process is terminated. There is a transmitter charging coil in the wireless charger. The receiver charging coil is placed at the back of the artificial knee joint within the internal control unit.

The developed system can be used in two different modes, autonomous and user-controlled. In the autonomous mode, after the necessary sensor information is collected, the system decides to elongate itself. The amount of elongation, temperature, internal battery charge level and patient posture status through the prosthesis are measured by the internal control unit sensors located in the knee joint. The internal control unit can communicate wirelessly with the unit that provides communication between the patient and the physician, which is called the external control unit. In this mode, the patient's healthy limb length information is measured by the developed wearable sensor. If a difference is detected between the length of the prosthetic limb and the healthy limb, the physical conditions are checked by the sensors in the prosthesis to perform the lengthening procedure. The lengthening process is limited to a maximum of 1mm per day. Thus, limb lengthening can be performed safely and without the need to go to the clinic. In the user mode, the system works as a decision support system. The physician/specialist can access the Manual Extension screen, which can be accessed with their password in the external control unit, request the sensor information from the system, determine the amount of extension and start the system.

Since the system is meant to work within the human body, the developers have taken measures to ensure that:

i. The prosthesis's internal temperature is continuously controlled by the temperature sensor. When the temperature exceeds the acceptable upper limit value, the prosthesis is deactivated.

ii. The internal battery is controlled by the charge protection and control module against over-discharge and overcharging. Thus, although the user continues to keep the wireless charging module connected to his body after the battery is fully charged, charging does not continue.

iii. In both autonomous and user mode, the patient or physician can use an emergency button to terminate the prosthesis extension due to reasons, such as feeling pain or elongation beyond the desired value.

iv. System communication and sending the extension command are carried out wirelessly. As a wireless communication module, two XBee modules that communicate using the IEEE 802.15.4 network protocol, one in the external control unit and one in the internal control unit, are used. With 128-bit encryption, it is only possible to exchange information between previously paired modules.

v. Important risks that may arise during the use of extendable tumor prostheses are involuntary and uncontrolled elongation. It is seen that these errors are encountered especially with electromagnetically extended prostheses [115]. One of the reasons why the elongation process is performed 1 mm per day is the mechanical resistance of the soft tissue against the greater elongation. This mechanical resistance will cause excessive current to be drawn even if the patient is lying down at the time of failure. Digital EC Controller (DEC) module, which is used as a motor driver in the developed system, has an adjustable current protection feature [116]. Thus, even if an error occurs in the internal control unit, the undesired elongation amount will not exceed a few mm.

vi. By performing Failure Mode and Effects Analysis (FMEA), problems that may arise during use have been determined, and measures have been taken regarding these [117].

The main advantage of the developed system over existing systems is that it can perform the extension process autonomously. The prerequisite for this situation is that it can be determined when there is a need for an extension. Within the scope of the study, a wearable sensor unit that can detect the length of the patient's healthy limb femur bone was developed. The external control unit communicates with the wearable sensor unit via the serial port and receives the healthy limb length information.

In clinical practice, tumor prosthesis lengthening procedures are performed when the patient is in the supine position. Thus, the extra load on the prosthesis due to the patient's body weight is eliminated. In an autonomous system, the posture status of the patient should be determined with the help of sensors, and if there is no load on the prosthesis, the extension process should be started. In this study, patient posture was determined by machine learning using the inertial measurement unit [118]. In this study, the data collected were classified in MATLAB using popular classification methods. By using the created model on the implanted microcontroller, the classification of patient lying/patient standing was made with high accuracy.

The most important difference between the developed system from existing studies in the literature is that it is autonomous. Within the scope of the study, the mechanical parts of the prosthesis were designed, strength tests were carried out in a simulation environment, and a prototype was created using a 3D printer. An autonomously operable tumor prosthesis control system has been developed by creating units, such as the internal control unit, external control unit, wearable sensor unit and wireless charging unit. On the experimental set-up, machine learning and AHRS sensor data were processed, and patient posture and elongation against maximum soft tissue resistance were tested. Unlike previous studies, an internal battery with wireless charging capability was used in a tumor prosthesis for the first time.

It is seen that the developed system has the following advantages over existing systems:

• It decreases the number of clinic visits.

• It reduces physician workload.

• It prevents patient exposure to radiation for measurement and extension purposes.

• It increases the accuracy in the measurement of the need for an extension.

• It allows elongation to be done in small pieces to increase patient comfort.

CONCLUSION

New developments in the fields of implants and prosthetics emerge every day. Medical and engineering researchers working in these fields try to follow these new developments and contribute to these fields with their ideas. The development of material technology for prostheses and implants is an important subject of study. Innovative materials can improve certain crucial parameters of prostheses and implants, such as durability, lightweight, low cost, and longevity. Another research topic is using prostheses and implants in areas that have not been used up to now. Until recently, cases of bone cancer often resulted in amputation, and after the development of tumor prostheses, the cancerous part of the bone is replaced with a prosthesis by preserving the limb. This procedure is a good example of the use of prostheses in contemporary prosthetic applications. Studies on the control of motorized prostheses are thriving, especially in detecting the prosthesis users' intentions and performing the desired movements correctly. Meeting the energy needs of active prostheses still seems to be a significant challenge. This issue will be overcome by developing light, high capacity, and batteries with a high discharge rate or by making energy harvesting techniques more efficient.

When examined in terms of surgery, it is seen that significant developments have been achieved in both extremity prostheses and implant prostheses recently. In order for the prosthesis to be used efficiently, the body must be well prepared for the prosthesis. Rehabilitation, on the other hand, is a highly critical process in terms of the user's adaptation to the prosthesis and should be managed in a multi-faceted manner by a multidisciplinary team. The rehabilitation process determines whether the patient accepts the prosthesis and starts using it or rejects it. Apart from these issues, studies will continue to prolong the prostheses' lifespan and make them usable throughout society by reducing the price if possible.

In this section, the developments in the mentioned fields are examined by referring to the related studies. The details of the studies show that improvements can be made in the mentioned fields only by way of multidisciplinary teams. It is a common issue that one-way studies in engineering, control, medicine, biomedical, or materials cannot be implemented due to conflicts in other fields.

CONSENT FOR PUBLICATION

Not applicable.

CONFLICT OF INTEREST

The author declares no conflict of interest, financial or otherwise.

ACKNOWLEDGEMENT

Declared none.

REFERENCES

[1]   Hansen A. The Prosthetic Hinge: Saints, Kings and Knights in Late Medieval England (Doctoral dissertation) 2014; p. 5.

[2]   Krajbich JI, Pinzur MS, Potter BK, Stevens PM. Atlas of Amputations & Limb Deficiencies. Canada: American Academy of Orthopaedic Surgeons 2018; p. 15.

[3]   Lusardi MM, Nielsen CC. Orthotics and Prosthetics in Rehabilitation. Boston: Butterworth-Heinemann 2000.

[4]   Lumley J, Robinson KP. Amputations. Vascular Surgery. Berlin, Heidelberg: Springer 2009; pp. 385-415.
        [http://dx.doi.org/10.1007/978-3-540-68816-7_33]

[5]   Mioton LM, Dumanian GA. Targeted muscle reinnervation and prosthetic rehabilitation after limb loss. J Surg Oncol 2018; 118(5): 807-14.
        [http://dx.doi.org/10.1002/jso.25256] [PMID: 30261116]

[6]   Grady JF, Winters CL. The Boyd amputation as a treatment for osteomyelitis of the foot. J Am Podiatr Med Assoc 2000; 90(5): 234-9.
        [http://dx.doi.org/10.7547/87507315-90-5-234] [PMID: 10833871]

[7]   Jessney B. Joseph Lister (1827–1912): a pioneer of antiseptic surgery remembered a century after his death. J Med Biogr 2012; 20(3): 107-10.
        [http://dx.doi.org/10.1258/jmb.2011.011074] [PMID: 22892302]

[8]   Debè A, Polenghi S. Assistance and education of mutilated soldiers of World War I. The Italian case History of Education & Children's Literature 2016; 11(2).

[9]   Smith LL. Prosthetic Limb Development: A Historical Review. Discussions 2006; p. 9.

[10]   Smit G. Mechanical evaluation of the "Hüfner hand" prosthesis. Prosthet Orthot Int 2020; 1-9.
        [PMID: 33834745]

[11]   Vallejo G. Zero hour of eugenics in Argentina: disputes and ideologies surrounding the emergence of a scientific field, 1916-1932. Hist Cienc Saude Manguinhos 2018; 25 (Suppl. 1): 15-32.
        [http://dx.doi.org/10.1590/s0104-59702018000300002] [PMID: 30133580]

[12]   Goerig M, Agarwal K, Schulte am Esch J. The versatile August Bier (1861–1949), father of spinal anesthesia. J Clin Anesth 2000; 12(7): 561-9.
        [http://dx.doi.org/10.1016/S0952-8180(00)00202-6] [PMID: 11137420]

[13]   Grannis KA, Cox JT, Graham MA, Laughlin RT. Retained internal fixation as a cause for late prosthetic fitting problems in ertl transtibial amputation. J Prosthet Orthot 2013; 25(4): 184-7.
        [http://dx.doi.org/10.1097/JPO.0000000000000004]

[14]   Pinto MAGS, Harris WW. Fibular segment bone bridging in trans-tibial amputation. Prosthet Orthot

Int 2004; 28(3): 220-4.
[http://dx.doi.org/10.3109/03093640409167753] [PMID: 15658634]

[15]   Akgün G, Ülkır O and Kaplanoğlu E. Introduction to Biomechatronics Systems, Istanbul: Papatya Publishing, Istanbul, 2017.

[16]   Callister WJ, *et al.* Materials Science and Engineering An Introduction. 9th ed., Hoboken, NJ: John Wiley & Sons, Inc 2014.

[17]   Hung J, Wu JSS. A comparative study on wear behavior of hip prosthesis by finite element simulation. Biomedical Engineering: Applications, Basis and Communications 2002; 14(04): 139-48.

[18]   Bougherara H, Bureau M, Campbell M, Vadean A, Yahia LH. Design of a biomimetic polymer-composite hip prosthesis. J Biomed Mater Res A 2007; 82A(1): 27-40.
[http://dx.doi.org/10.1002/jbm.a.31146] [PMID: 17265439]

[19]   Betekhtin VI, Kolobov YR, Golosova OA, *et al.* Elastoplastic properties of a low-modulus titanium-based β alloy. Tech Phys 2013; 58(10): 1432-6.
[http://dx.doi.org/10.1134/S1063784213100046]

[20]   Kamachimudali U, Sridhar TM, Raj B. Corrosion of bio implants. Sadhana 2003; 28(3-4): 601-37.
[http://dx.doi.org/10.1007/BF02706450]

[21]   Katoozian H, Davy DT, Arshi A, Saadati U. Material optimization of femoral component of total hip prosthesis using fiber reinforced polymeric composites. Med Eng Phys 2001; 23(7): 505-11.
[http://dx.doi.org/10.1016/S1350-4533(01)00079-0] [PMID: 11574257]

[22]   Kuiper J, Huiskes R. Numerical optimization of hip-prosthetic stem material. In: Middleton J, Jones ML, Shrive NG, Pande GN, Eds. Computer Methods in Biomechanics and Biomedecine. New York: Gordon and Breach 1993; pp. 76-84.

[23]   Ramakrishna S, Mayer J, Wintermantel E, Leong KW. Biomedical applications of polymer-composite materials: a review. Compos Sci Technol 2001; 61(9): 1189-224.
[http://dx.doi.org/10.1016/S0266-3538(00)00241-4]

[24]   Akay M, Aslan N. Numerical and experimental stress analysis of a polymeric composite hip joint prosthesis. J Biomed Mater Res 1996; 31(2): 167-82.
[http://dx.doi.org/10.1002/(SICI)1097-4636(199606)31:2<167::AID-JBM3>3.0.CO;2-L]   [PMID: 8731205]

[25]   Reinhardt A, Advani SG, Santare MH, Miller F. Preliminary study on composite hip prosthesis made by resin transfer molding. J Compos Mater 1999; 33(9): 852-70.
[http://dx.doi.org/10.1177/002199839903300904]

[26]   Narres M, Kvitkina T, Claessen H, *et al.* Incidence of lower extremity amputations in the diabetic compared with the non-diabetic population: A systematic review. PLoS One 2017; 12(8): e0182081.
[http://dx.doi.org/10.1371/journal.pone.0182081] [PMID: 28846690]

[27]   Lunsford C, Grindle G, Salatin B, Dicianno BE. Innovations with 3-D dimensional printing in physical medicine and rehabilitation: a review of the literature. PM R 2016; 8(12): 1201-12.
[http://dx.doi.org/10.1016/j.pmrj.2016.07.003] [PMID: 27424769]

[28]   Román-Casares AM, García-Gómez O, Guerado E. Prosthetic limb design and function: latest innovations and functional results. Curr Trauma Rep 2018; 4(4): 256-62.
[http://dx.doi.org/10.1007/s40719-018-0150-2]

[29]   Pasquina PF, Perry BN, Miller ME, Ling GSF, Tsao JW. Recent advances in bioelectric prostheses. Neurol Clin Pract 2015; 5(2): 164-70.
[http://dx.doi.org/10.1212/CPJ.0000000000000132] [PMID: 29443190]

[30]   ten Kate J, Smit G, Breedveld P. 3D-printed upper limb prostheses: a review. Disabil Rehabil Assist Technol 2017; 12(3): 300-14.
[http://dx.doi.org/10.1080/17483107.2016.1253117] [PMID: 28152642]

[31]   Seminati E, Canepa Talamas D, Young M, Twiste M, Dhokia V, Bilzon JLJ. Validity and reliability of a novel 3D scanner for assessment of the shape and volume of amputees' residual limb models. PLoS One 2017; 12(9): e0184498.
[http://dx.doi.org/10.1371/journal.pone.0184498] [PMID: 28886154]

[32]   Amici C, Borboni A, Taveggia G, Legnani G. Bioelectric prostheses: review of classifications and control strategies]. G Ital Med Lav Ergon 2015; 37(3) (Suppl.): 39-44.
[PMID: 26731956]

[33]   Hagberg K, Brånemark R. One hundred patients treated with osseointegrated transfemoral amputation prostheses–rehabilitation perspective. J Rehabil Res Dev 2009; 46: 331–44.
[http://dx.doi.org/10.1682/JRRD.2008.06.0080] [PMID: 19675986]

[34]   Yildiz KA, Shin AY, Kaufman KR. Interfaces with the peripheral nervous system for the control of a neuroprosthetic limb: a review. J Neuroeng Rehabil 2020; 17(1): 43.
[http://dx.doi.org/10.1186/s12984-020-00667-5] [PMID: 32151268]

[35]   Cogan SF. Neural stimulation and recording electrodes. Annu Rev Biomed Eng 2008; 10(1): 275-309.
[http://dx.doi.org/10.1146/annurev.bioeng.10.061807.160518] [PMID: 18429704]

[36]   Navarro X, Krueger TB, Lago N, Micera S, Stieglitz T, Dario P. A critical review of interfaces with the peripheral nervous system for the control of neuroprostheses and hybrid bionic systems. J Peripher Nerv Syst 2005; 10(3): 229-58.
[http://dx.doi.org/10.1111/j.1085-9489.2005.10303.x] [PMID: 16221284]

[37]   Svensson P, Wijk U, Björkman A, Antfolk C. A review of invasive and non-invasive sensory feedback in upper limb prostheses. Expert Rev Med Devices 2017; 14(6): 439-47.
[http://dx.doi.org/10.1080/17434440.2017.1332989] [PMID: 28532184]

[38]   Castellini C, van der Smagt P. Surface EMG in advanced hand prosthetics. Biol Cybern 2009; 100(1): 35-47.
[http://dx.doi.org/10.1007/s00422-008-0278-1] [PMID: 19015872]

[39]   Hewson DJ, Hogrel JY, Langeron Y, Duchêne J. Evolution in impedance at the electrode-skin interface of two types of surface EMG electrodes during long-term recordings. J Electromyogr Kinesiol 2003; 13(3): 273-9.
[http://dx.doi.org/10.1016/S1050-6411(02)00097-4] [PMID: 12706606]

[40]   Kilgore KL, Hoyen HA, Bryden AM, Hart RL, Keith MW, Peckham PH. An implanted upper-extremity neuroprosthesis using myoelectric control. J Hand Surg Am 2008; 33(4): 539-50.
[http://dx.doi.org/10.1016/j.jhsa.2008.01.007] [PMID: 18406958]

[41]   Kuiken TA, Barlow AK, Hargrove LJ, Dumanian GA. Targeted Muscle Reinnervation for the Upper and Lower Extremity. Tech Orthop 2017; 32(2): 109-16.
[http://dx.doi.org/10.1097/BTO.0000000000000194] [PMID: 28579692]

[42]   Dumanian GA, Potter BK, Mioton LM, et al. Targeted muscle reinnervation treats neuroma and phantom pain in major limb amputees. Ann Surg 2019; 270(2): 238-46.
[http://dx.doi.org/10.1097/SLA.0000000000003088] [PMID: 30371518]

[43]   Frost CM, Ursu DC, Flattery SM, et al. Regenerative peripheral nerve interfaces for real-time, proportional control of a Neuroprosthetic hand. J Neuroeng Rehabil 2018; 15(1): 108.
[http://dx.doi.org/10.1186/s12984-018-0452-1] [PMID: 30458876]

[44]   Kung TA, Langhals NB, Martin DC, Johnson PJ, Cederna PS, Urbanchek MG. Regenerative peripheral nerve interface viability and signal transduction with an implanted electrode. Plast Reconstr Surg 2014; 133(6): 1380-94.
[http://dx.doi.org/10.1097/PRS.0000000000000168] [PMID: 24867721]

[45]   Anderson JM, Rodriguez A, Chang DT. Foreign body reaction to biomaterials. Semin Immunol 2008; 20(2): 86-100.
[http://dx.doi.org/10.1016/j.smim.2007.11.004] [PMID: 18162407]

[46] Christensen MB, Pearce SM, Ledbetter NM, Warren DJ, Clark GA, Tresco PA. The foreign body response to the Utah Slant Electrode Array in the cat sciatic nerve. Acta Biomater 2014; 10(11): 4650-60.
[http://dx.doi.org/10.1016/j.actbio.2014.07.010] [PMID: 25042798]

[47] Joung YH. Development of implantable medical devices: from an engineering perspective. Int Neurourol J 2013; 17(3): 98-106.
[http://dx.doi.org/10.5213/inj.2013.17.3.98] [PMID: 24143287]

[48] Agarwal K, Jegadeesan R, Guo YX, Thakor NV. Wireless power transfer strategies for implantable bioelectronics. IEEE Rev Biomed Eng 2017; 10: 136-61.
[http://dx.doi.org/10.1109/RBME.2017.2683520] [PMID: 28328511]

[49] Tschiedel M, Russold MF, Kaniusas E. Relying on more sense for enhancing lower limb prostheses control: a review. J Neuroeng Rehabil 2020; 17(1): 99.
[http://dx.doi.org/10.1186/s12984-020-00726-x] [PMID: 32680530]

[50] Masteller A, *et al.* Recent developments in prosthesis sensors, texture recognition, and sensory stimulation for upper limb prostheses. Ann Biomed Eng 2020; 1-18.
[PMID: 33140242]

[51] Lucarotti C, Oddo C, Vitiello N, Carrozza M. Synthetic and bio-artificial tactile sensing: a review. Sensors (Basel) 2013; 13(2): 1435-66.
[http://dx.doi.org/10.3390/s130201435] [PMID: 23348032]

[52] Zou L, Ge C, Wang Z, Cretu E, Li X. Novel tactile sensor technology and smart tactile sensing systems: a review. Sensors (Basel) 2017; 17(11): 2653.
[http://dx.doi.org/10.3390/s17112653] [PMID: 29149080]

[53] Kim J, Lee M, Shim HJ, *et al.* Stretchable silicon nanoribbon electronics for skin prosthesis. Nat Commun 2014; 5(1): 5747.
[http://dx.doi.org/10.1038/ncomms6747] [PMID: 25490072]

[54] Li Q, Naing V, Donelan JM. Development of a biomechanical energy harvester. J Neuroeng Rehabil 2009; 6(1): 22.
[http://dx.doi.org/10.1186/1743-0003-6-22] [PMID: 19549313]

[55] Riemer R, Shapiro A. Biomechanical energy harvesting from human motion: theory, state of the art, design guidelines, and future directions. J Neuroeng Rehabil 2011; 8(1): 22.
[http://dx.doi.org/10.1186/1743-0003-8-22] [PMID: 21521509]

[56] Laschowski B, McPhee J, Andrysek J. Lower-limb prostheses and exoskeletons with energy regeneration: Mechatronic design and optimization review. J Mech Robot 2019; 11(4): 040801.
[http://dx.doi.org/10.1115/1.4043460]

[57] Selinger JC, Donelan JM. Myoelectric control for adaptable biomechanical energy harvesting. IEEE Trans Neural Syst Rehabil Eng 2016; 24(3): 364-73.
[http://dx.doi.org/10.1109/TNSRE.2015.2510546] [PMID: 26841402]

[58] Elery T, Rezazadeh S, Nesler C, Doan J, Zhu H, Gregg RD. Design and Benchtop Validation of a Powered Knee-Ankle Prosthesis with High-Torque, Low-Impedance Actuators. Proceedings of the IEEE International Conference on Robotics and Automation. Brisbane, Australia. May 21-25, 2018; 2788-95.
[http://dx.doi.org/10.1109/ICRA.2018.8461259]

[59] Au SK, Weber J, Herr H. Powered ankle-foot prosthesis improves walking metabolic economy. IEEE Trans Robot 2009; 25(1): 51-66.
[http://dx.doi.org/10.1109/TRO.2008.2008747]

[60] Shultz AH, Lawson BE, Goldfarb M. Variable cadence walking and ground adaptive standing with a powered ankle prosthesis. IEEE Trans Neural Syst Rehabil Eng 2016; 24(4): 495-505.
[http://dx.doi.org/10.1109/TNSRE.2015.2428196] [PMID: 25955789]

[61]   Feng Y, Mai J, Agrawal SK, Wang Q. Energy regeneration from electromagnetic induction by human dynamics for lower extremity robotic prostheses. IEEE Trans Robot 2020; 36(5): 1442-51.
[http://dx.doi.org/10.1109/TRO.2020.2991969]

[62]   Eide AH, Oderud T. Assistive technology in low-income countries. In: MacLachlan M, Swartz L, Eds. Disability and international development: towards inclusive global health. New York: Springer 2009.
[http://dx.doi.org/10.1007/978-0-387-93840-0_10]

[63]   Guest F, Marshall C, Stansby G. Amputation and rehabilitation. Surgery 2019; 37(2): 102-5.
[http://dx.doi.org/10.1016/j.mpsur.2018.12.008]

[64]   Devinuwara K, Dworak-Kula A, O'Connor RJ. Rehabilitation and prosthetics post-amputation. Orthop Trauma 2018; 32(4): 234-40.
[http://dx.doi.org/10.1016/j.mporth.2018.05.007]

[65]   Esquenazi A. Amputation rehabilitation and prosthetic restoration. From surgery to community reintegration. Disabil Rehabil 2004; 26(14-15): 831-6.
[http://dx.doi.org/10.1080/09638280410001708850] [PMID: 15497912]

[66]   Esquenazi A, Meier RH. Rehabilitation in limb deficiency. 4. Limb amputation. Arch Phys Med Rehabil 1996; 77(3) (Suppl.): S18-28.
[http://dx.doi.org/10.1016/S0003-9993(96)90239-7] [PMID: 8599542]

[67]   Gottschalk F. Transfemoral Amputation. Clin Orthop Relat Res 1999; 361(361): 15-22.
[http://dx.doi.org/10.1097/00003086-199904000-00003] [PMID: 10212591]

[68]   Esquenazi A. Upper limb amputee rehabilitation and prosthetic restoration. In: Braddom RL, Ed. Physical Medicine and Rehabilitation. 2nd ed. Philadelphia, PA: W.B. Saunders Co. 2000; pp. 263-78.

[69]   Esquenazi A. Pain management post amputation. In: Monga TN, Grabois M, Eds. Pain Management in Rehabilitation. New York Demos Medical Publishing 2002; pp. 191-202.

[70]   Sansam K, Neumann V, O'Connor R, Bhakta B. Predicting walking ability following lower limb amputation: A systematic review of the literature. J Rehabil Med 2009; 41(8): 593-603.
[http://dx.doi.org/10.2340/16501977-0393] [PMID: 19565152]

[71]   Smith S, Pursey H, Jones A, *et al.* Clinical guidelines for the pre and post operative physiotherapy management of adults with lower limb amputations. British Association of Chartered Physiotherapists in Amputee Rehabilitation 2016.

[72]   Frölke JPM, Leijendekkers RA, van de Meent H. Osseointegrated prosthesis for patients with an amputation. Unfallchirurg 2017; 120(4): 293-9.
[http://dx.doi.org/10.1007/s00113-016-0302-1] [PMID: 28097370]

[73]   Mohd Hawari N, Jawaid M, Md Tahir P, Azmeer RA. Case study: survey of patient satisfaction with prosthesis quality and design among below-knee prosthetic leg socket users. Disabil Rehabil Assist Technol 2017; 12(8): 868-74.
[http://dx.doi.org/10.1080/17483107.2016.1269209] [PMID: 28068847]

[74]   Highsmith MJ, Kahle JT, Klenow TD, *et al.* Interventions to manage residual limb ulceration due to prosthetic use in individuals with lower extremity amputation: a systemtatic review of the literature. Technol Innov 2016; 18(2): 115-23.
[http://dx.doi.org/10.21300/18.2-3.2016.115] [PMID: 28066521]

[75]   Pirouzi G, Abu Osman NA, Eshraghi A, Ali S, Gholizadeh H, Wan Abas WAB. Review of the socket design and interface pressure measurement for transtibial prosthesis. ScientificWorldJournal 2014; 2014: 1-9.
[http://dx.doi.org/10.1155/2014/849073] [PMID: 25197716]

[76]   Meulenbelt HEJ, Dijkstra PU, Jonkman MF, Geertzen JHB. Skin problems in lower limb amputees: A systematic review. Disabil Rehabil 2006; 28(10): 603-8.
[http://dx.doi.org/10.1080/09638280500277032] [PMID: 16690571]

[77]   Jia X, Zhang M, Lee WCC. Load transfer mechanics between trans-tibial prosthetic socket and residual limb—dynamic effects. J Biomech 2004; 37(9): 1371-7.
[http://dx.doi.org/10.1016/j.jbiomech.2003.12.024] [PMID: 15275844]

[78]   Rosalam Che M, Rahinah Ibrahim PMT. Natural based biocomposite material for prosthetic socket fabrication. Alam Cipta 2012; 5: 27-34.

[79]   Silver-Thorn MB. Design of artificial limbs for lower extremity amputees Standard handbook of biomedical engineering and design. New York: McGraw-Hill 2004; pp. 1-30.

[80]   Hafner BJ, Sanders JE. Considerations for development of sensing and monitoring tools to facilitate treatment and care of persons with lower-limb loss: A review J Rehabil Res Dev 2014; 51(1): 1-14.
[http://dx.doi.org/10.1682/JRRD.2013.01.0024] [PMID: 24805889]

[81]   Zwijsen SA, Niemeijer AR, Hertogh CMPM. Ethics of using assistive technology in the care for community-dwelling elderly people: An overview of the literature. Aging Ment Health 2011; 15(4): 419-27.
[http://dx.doi.org/10.1080/13607863.2010.543662] [PMID: 21500008]

[82]   Rigby M. Applying emergent ubiquitous technologies in health: The need to respond to new challenges of opportunity, expectation, and responsibility. Int J Med Inform 2007; 76 (Suppl. 3): S349-52.
[http://dx.doi.org/10.1016/j.ijmedinf.2007.03.002] [PMID: 17434338]

[83]   Brauer CA, Rosen AB, Olchanski NV, Neumann PJ. Cost-utility analyses in orthopaedic surgery. J Bone Joint Surg Am 2005; 87(6): 1253-9.
[PMID: 15930533]

[84]   Brenner CD, Brenner JK. The use of preparatory/evaluation/training prostheses in developing evidenced-based practice in upper limb prosthetics. J Prosthet Orthot 2008; 20(3): 70-82.
[http://dx.doi.org/10.1097/JPO.0b013e31817c59fb]

[85]   Biddiss E, McKeever P, Lindsay S, Chau T. Implications of prosthesis funding structures on the use of prostheses. Prosthet Orthot Int 2011; 35(2): 215-24.
[http://dx.doi.org/10.1177/0309364611401776] [PMID: 21515898]

[86]   Chan A, Ezra K, Petcharatana B. Cost of Ownership of Upper Limb Prostheses: A Retrospective Analysis. CMBES Proceedings 2013.

[87]   Gerzeli S, Torbica A, Fattore G. Cost utility analysis of knee prosthesis with complete microprocessor control (C-leg) compared with mechanical technology in trans-femoral amputees. Eur J Health Econ 2009; 10(1): 47-55.
[http://dx.doi.org/10.1007/s10198-008-0102-9] [PMID: 18379831]

[88]   Perry J, Burnfield JM, Newsam CJ, Conley P. Energy expenditure and gait characteristics of a bilateral amputee walking with C-leg prostheses compared with stubby and conventional articulating prostheses. Arch Phys Med Rehabil 2004; 85(10): 1711-7.
[http://dx.doi.org/10.1016/j.apmr.2004.02.028] [PMID: 15468036]

[89]   Frossard L, Berg D, Merlo G, Quincey T, Burkett B. Cost comparison of socket-suspended and bone-anchored transfemoral prostheses. J Prosthet Orthot 2017; 29(4): 150-60.
[http://dx.doi.org/10.1097/JPO.0000000000000142]

[90]   Muderis MA, Tetsworth K, Khemka A, *et al.* The Osseointegration Group of Australia Accelerated Protocol (OGAAP-1) for two-stage osseointegrated reconstruction of amputated limbs. Bone Joint J 2016; 98-B(7): 952-60.
[http://dx.doi.org/10.1302/0301-620X.98B7.37547] [PMID: 27365474]

[91]   Fish D. The development of coverage policy for lower extremity prosthetics: the influence of the payer on prosthetic prescription. J Prosthet Orthot 2006; 18(Proceedings): P125-9.
[http://dx.doi.org/10.1097/00008526-200601001-00017]

[92]   Heinemann AW, Fisher WP Jr, Gershon R. Improving health care quality with outcomes management. J Prosthet Orthot 2006; 18(Proceedings): P46-50.
[http://dx.doi.org/10.1097/00008526-200601001-00005]

[93]   Frossard L, Merlo G, Quincey T, Burkett B, Berg D. Development of a procedure for the government provision of bone-anchored prosthesis using osseointegration in Australia. PharmacoEconom Open 2017; 1(4): 301-14.
[http://dx.doi.org/10.1007/s41669-017-0032-5] [PMID: 29441506]

[94]   McLarney M, Pezzin LE, McGinley EL, Prosser L, Dillingham TR. The prevalence of lower limb loss in children and associated costs of prosthetic devices: A national study of commercial insurance claims. Prosthet Orthot Int 2020; 0309364620968645.
[PMID: 33158398]

[95]   Le JT, Scott-Wyard PR. Pediatric limb differences and amputations. Phys Med Rehabil Clin N Am 2015; 26(1): 95-108.
[http://dx.doi.org/10.1016/j.pmr.2014.09.006] [PMID: 25479783]

[96]   Hostetler SG, Schwartz L, Shields BJ, Xiang H, Smith GA. Characteristics of pediatric traumatic amputations treated in hospital emergency departments: United States, 1990-2002. Pediatrics 2005; 116(5): e667-74.
[http://dx.doi.org/10.1542/peds.2004-2143] [PMID: 16263981]

[97]   Vollman D, Smith GA. Epidemiology of lawn-mower-related injuries to children in the United States, 1990-2004. Pediatrics 2006; 118(2): e273-8.
[http://dx.doi.org/10.1542/peds.2006-0056] [PMID: 16882772]

[98]   Edelstein J. Rehabilitation for children with limb deficiencies. Orthotics and Prosthetics in Rehabilitation, Butterword- Heinemann Oxford 2000.

[99]   Hall M, Wustrack R, Cummings D, Welling R Jr, Kaleta M, Koenig K Jr, *et al.* Morgan S Innovations in Pediatric Prosthetics JPOSNA 3(1)2021;

[100]   El-Sayed MM, Correll J, Pohlig K. Limb sparing reconstructive surgery and Ilizarov lengthening in fibular hemimelia of Achterman–Kalamchi type II patients. J Pediatr Orthop B 2010; 19(1): 55-60.
[http://dx.doi.org/10.1097/BPB.0b013e32832f5ace] [PMID: 19741550]

[101]   Sakkers R, van Wijk I. Amputation and rotationplasty in children with limb deficiencies: Current concepts. J Child Orthop 2016; 10(6): 619-26.
[http://dx.doi.org/10.1007/s11832-016-0788-7] [PMID: 27826906]

[102]   Geil M, Coulter C. Analysis of locomotor adaptations in young children with limb loss in an early prosthetic knee prescription protocol. Prosthet Orthot Int 2014; 38(1): 54-61.
[http://dx.doi.org/10.1177/0309364613487546] [PMID: 23685917]

[103]   Hall M, Cummings D, Welling R Jr, *et al.* Essentials of Pediatric Prosthetics 2020; 2(3).
[http://dx.doi.org/10.55275/JPOSNA-2020-168]

[104]   Verheul FJMG, Verschuren O, Zwinkels M, *et al.* Effectiveness of a crossover prosthetic foot in active children with a congenital lower limb deficiency. Prosthet Orthot Int 2020; 44(5): 305-13.
[http://dx.doi.org/10.1177/0309364620912063] [PMID: 32370612]

[105]   Jeans KA, Karol LA, Cummings D, Singhal K. Comparison of gait after Syme and transtibial amputation in children: factors that may play a role in function. J Bone Joint Surg Am 2014; 96(19): 1641-7.
[http://dx.doi.org/10.2106/JBJS.N.00192] [PMID: 25274789]

[106]   Michielsen A, Van Wijk I, Ketelaar M. Participation and quality of life in children and adolescents with congenital limb deficiencies: A narrative review. Prosthet Orthot Int 2010; 34(4): 351-61.
[http://dx.doi.org/10.3109/03093646.2010.495371] [PMID: 20704518]

[107]   Morris CD, Wustrack RL, Levin AS. Limb-Salvage Options in Growing Children with Malignant

Bone Tumors of the Lower Extremity. JBJS Rev 2017; 5(7): e7.
[http://dx.doi.org/10.2106/JBJS.RVW.16.00026] [PMID: 28742715]

[108] Weitao Y, Qiqing C, Songtao G, Jiaqiang W. Epiphysis preserving operations for the treatment of lower limb malignant bone tumors. Eur J Surg Oncol 2012; 38(12): 1165-70.
[http://dx.doi.org/10.1016/j.ejso.2012.05.005] [PMID: 22698890]

[109] Heare T, Hensley MA, Dell'Orfano S. Bone tumors: osteosarcoma and Ewing's sarcoma. Curr Opin Pediatr 2009; 21(3): 365-72.
[http://dx.doi.org/10.1097/MOP.0b013e32832b1111] [PMID: 19421061]

[110] Chimutengwende-Gordon M, Mbogo A, Khan W, Wilkes R. Limb reconstruction after traumatic bone loss. Injury 2017; 48(2): 206-13.
[http://dx.doi.org/10.1016/j.injury.2013.11.022] [PMID: 24332161]

[111] Schroeder JE, Mosheiff R. Tissue engineering approaches for bone repair: Concepts and evidence. Injury 2011; 42(6): 609-13.
[http://dx.doi.org/10.1016/j.injury.2011.03.029] [PMID: 21489529]

[112] Turcotte RE. Endoprosthetic replacements for bone tumors: review of the most recent literatüre 2007; 18(6): 572-8.
[http://dx.doi.org/10.1097/BCO.0b013e3282ef6eaf]

[113] Dotan A, Dadia S, Bickels J, *et al.* Expandable endoprosthesis for limb-sparing surgery in children: Long-term results. J Child Orthop 2010; 4(5): 391-400.
[http://dx.doi.org/10.1007/s11832-010-0270-x] [PMID: 21966302]

[114] Kocaoğlu S, Akdoğan E. Design and development of an intelligent biomechatronic tumor prosthesis. Biocybern Biomed Eng 2019; 39(2): 561-70.
[http://dx.doi.org/10.1016/j.bbe.2019.05.004]

[115] Cipriano CA, Gruzinova IS, Frank RM, Gitelis S, Virkus WW. Frequent complications and severe bone loss associated with the repiphysis expandable distal femoral prosthesis. Clin Orthop Relat Res 2015; 473(3): 831-8.
[http://dx.doi.org/10.1007/s11999-014-3564-3] [PMID: 24664193]

[116] Available from: https://www.maxonmotor.com/maxon/view/product/control/1-Q-EC-Verstaerker/367661 [Accessed: 12-May-2019].

[117] Kocaoğlu S, Akdoğan E. Pre-clinical validation and risk management of autonomous tumor prosthesis using FMEA approach. Periodicals of Engineering and Natural Sciences 2020; 8(2): 1152-64.

[118] Kocaoğlu S, Akdoğan E. Comparison of classification algorithms for detecting patient posture in expandable tumor prostheses. Adv Electr Comput Eng 2020; 20(2): 131-8.
[http://dx.doi.org/10.4316/AECE.2020.02015]

CHAPTER 4

Role of Nanomedicine in Ocular Parasitic Infections

Nagham Gamal Masoud[1], Nagwa Mostafa El-Sayed[2,*] and Manar Ezzelarab Ramadan[3]

[1] *Faculty of Medicine, Ain Shams University, Cairo, Egypt*

[2] *Department of Medical Parasitology, Research Institute of Ophthalmology, Giza, Egypt*

[3] *Department of Parasitology, Faculty of Medicine, Suez University, Suez, Egypt*

Abstract: Ocular parasites cause serious vision-threatening diseases. An early diagnosis and effective treatment are crucial to avoid side effects, such as blindness or eye removal. The first important step in diagnosing ocular parasite infections is to suspect them. Diagnosis is aided by ophthalmic examination, direct parasite identification in clinical samples and/or pathological lesions, immunoassays, and molecular methods. Despite this, ocular parasite infection diagnosis is fraught with difficulties in terms of sensitivity, specificity, and accuracy. The usage of nanoparticles may improve diagnosis by providing precise procedures for parasitic DNA, antigens, and antibodies detection in a variety of body specimens with fast, sensitive, and specific results. Low tolerability, long therapeutic duration, multiple adverse effects, and the emergence of medication resistance are all problems with existing anti-parasitic medications. Nanoparticles represent a promising way for the successful treatment of parasitic diseases by developing innovative drug carriers to target medications to infected sites while limiting high doses and adverse effects. They can also overcome the limitations of antiparasitic medications' low bioavailability, poor cellular permeability, non-specific distribution, and fast elimination from the body. The aim of the present chapter is to throw light on possible nanotechnology applications in ocular parasitic diseases caused by *Toxoplasma gondii*, *Acanthamoeba* spp. and *Toxocara* spp. with a focus on diagnosis, treatment, and vaccination.

Keywords: *Acanthamoeba* spp., Diagnosis, Eye infection, Nanotechnology, *Toxocara* Spp., *Toxoplasma gondii*, Treatment, Vaccination.

INTRODUCTION

Direct exposure to the environment makes the human eye susceptible to infections from a variety of microorganisms that are known to be major causes of

* **Corresponding author Nagwa Mostafa El-Sayed:** Department of Medical Parasitology, Research Institute of Ophthalmology, Giza, Egypt; E-mail: nag.elsaka@yahoo.com

Felipe López-Saucedo (Ed.)

ophthalmic diseases all over the world. Several parasitic infections have been associated with ocular lesions in either anterior or posterior segments of the eye. *Toxoplasma gondii, Toxocara* species, and *Onchorcerca volvulus* are the most common parasites that infect the ocular tissue(s) in the posterior segment of the eye. There are also case reports on ocular leishmaniasis, malaria, giardiasis, cysticercosis, hydatid cysts, and coenurosis [1]. *Acanthamoeba* spp. and *Loa Loa*, on the other hand, are the most prevalent parasites that impact the ocular tissue(s) in the anterior part of the eye. Moreover, *Microsporidia, Mansonella ozzardi, Thelazia, Gnanthostoma* spp., and *Angiostrongylus cantonensis* have all been documented in published case reports [2]. Infection with ocular parasites has been documented in a variety of geographic areas, largely based on the parasite's endemicity. The parasite's geographic spread, environmental pollution, and the patient's immune status all play a role in the prevalence of these infections.

These parasites may infect the eye directly through trauma or surgery, or indirectly through hematogenous transmission or dissemination from infected nearby tissues [2]. The resulting ophthalmic manifestations are linked to the causative agents and affected ocular tissues. Pathological lesions may result from direct damage caused by the presence of larvae or adult stages, parasite-released toxic products, or the host's immune response [1, 2]. Choroiditis, retinal vasculitis and hemorrhage, retinochoroiditis, detachment of retina, papilledema, orbital cysts, or optic nerve atrophy all affect the posterior segment, resulting in permanent retinal impairment and vision loss.

Clinical diagnosis frequently matches other viral and bacterial infections that induce visual morbidity. Suspecting ocular parasite infections is the most crucial step in diagnosing those [3]. A preliminary diagnosis is also aided by an ophthalmic examination, as well as the associated risk factors and the history of travel to endemic areas. The most common method for confirming the diagnosis is to use direct parasite detection in biological specimens. In ocular toxoplasmosis and toxocariasis, the detection of antigens/antibodies in aqueous humor, vitreous humor and sera samples frequently confirms the diagnosis. The use of molecular approaches, such as polymerase chain reaction (PCR), for the detection of parasite DNA, has improved diagnosis and species identification [1, 4]. Despite that, diagnosis of ocular parasitic infection faces many challenges regarding sensitivity, specificity, and accuracy.

Medical or surgical treatment is available for ocular parasitic infections. The severity of symptomatology, ocular inflammatory responses, affection of vision, macula affection, and presence of eye injuries are all factors that influence treatment. It is worth noting that the clinical response is the most important determinant of cure [5]. However, treating ocular parasitic infections exacerbates the challenge of limiting the effectiveness of current antiparasitic drugs, as well as their adverse effects and the possibility of the emergence of resistant strains. As a result, it is necessary to search for treatments that are both safer and more effective [6].

Medical nanotechnology can offer a new approach to diagnose and treat ocular parasitic diseases and to develop vaccinations against them. Use of nanoparticles (NPs) may provide precise procedures for parasitic DNA, antigen and antibody detection in a variety of body specimens with fast, sensitive, and specific results. Novel NP-related methods permitted the discovery of new target molecules avoiding cross-reactivity between parasites' antigens that are shared in antigenic epitopes. Moreover, NPs have promise for effective parasite disease therapy because they enable the development of innovative drug carriers or the delivery of new medications while overcoming high doses, low bioavailability, poor cellular permeability, non-specific distribution, and side effects that antiparasitic drugs have. In terms of the role of NPs in vaccine development, they can be used to carry whole or purified antigens, DNA, RNA, or act as adjuvants, enhancing uptake via antigen-presenting cells, producing specific antibodies (Abs), and eliciting the most effective T helper 1 (Th1) cells' immune response [7].

However, due to a lack of systematic studies in this field, nanotechnology has not been widely used in ocular parasitology. Therefore, this chapter clarified the expected role of nanomedicine in ocular parasitic infections caused by *Toxoplasma gondii*, *Acanthamoeba* spp. and *Toxocara* spp. with a focus on diagnosis, treatment, and vaccination.

ROLE OF NANOMEDICINE IN OCULAR PARASITES

Nanotechnology is the science dealing with tiny materials (10-100 nm) that can connect to certain molecules on the surface of cells or intracellularly modify a variety of physical, chemical, and biological characteristics [8]. Chemical and photochemical reactions being microwave-assisted, reverse micelles, thermal breakdown, as well as electrochemical, sonochemical [9], and biological techniques can all be used to create nanomaterials. Also, plant-mediated biological NP production is attracting attention owing to its simplicity and low cost [10].

Different nanomaterials have been used in applied parasitology. Antiparasitic effects of gold and silver NPs have been extensively researched. The conjugation of antiparasitic drugs with NPs has enhanced anti-amoebic [11], antitoxoplasmic [7], and anti-*Toxocara* [12] activities. The antiparasitic potential of metallic NPs is linked to several factors, including NPs interaction with microbial cell walls, toxic ion release, cell penetration, production of reactive oxygen species (ROS), and damage to DNA [13].

Liposomes NPs are sphere vesicles made up of phospholipid bilayers. They have become valuable drug delivery systems due to their ability to encapsulate hydrophilic or lipophilic medications with a higher rate of retention in the body and remain for a long time in circulation [14]. Also, they can prolong and modulate drug release as well as reduce drug dosage and frequency of administration [15].

Nanosuspension is a colloidal dispersion of submicron drug particles made from inert polymeric resins that are used for hydrophobic drug delivery. These formulations can improve ocular bioavailability by prolonging drug release, reducing irritation, and increasing precorneal residence time. The NPs' surface has a positive charge, which aids their adherence to the ocular surface. The use of nanosuspensions in ophthalmic formulations offers a promising way to solve the challenges of ocular drug delivery. Poorly soluble drugs can be given as nanosuspension because NPs have a long retention time in the ocular tissue [16].

Polymeric NPs with sizes ranging from 10 to 1000 nm are created from a variety of biocompatible and biodegradable colloids. They may contain synthetic polymers and act as drug carriers [17]. Polymeric NPs are classified into nanospheres and nanocapsules, which are polymeric or reservoir systems, respectively. In nanocapsules, the drug is contained in a cavity encircled by a polymer membrane, whereas in nanospheres, the drug is uniformly spread rather than contained in a cavity [18].

Chitosan is another form of NP. It has been shown to be an effective antiprotozoal agent against *Toxoplasma* and *Leishmania* and has mostly been used as a drug carrier [19]. Chitosan's antibacterial properties are mostly due to intrinsic characteristics, like physical state, positive charge density, and chelating capability [20]. It is known for being non-toxic and inexpensive, making it suited for use in underdeveloped countries.

Nanoemulsions are heterogeneous systems comprised of two forms of lipid NPs (oil in water with a water-based particle core, and water in oil with an oil-based particle core), enabling controlled drug and biomolecule release [21]. In addition, the solid lipid NPs with a solid lipid core have been widely investigated as a nanocarrier to target drugs to their sites of action due to their nontoxicity, high biocompatibility, and capability for large-scale production [22].

NPs are used in the diagnosis of different parasitic diseases to improve test's sensitivity and specificity. They have the potential to improve or replace currently existing and widely used diagnostic procedures [23, 24]. Nanomaterials may be used alone or have been used in drug delivery to target specific cells or intracellular organs. They can bind to their target within the cell or on its surface, resulting in high efficacy. Furthermore, they lead the drug to avoid non-specific binding, as well as the drug's side effects are reduced [7]. Monocytes, macrophages, dendritic cells, endothelial cells, and cancer cells are usually the targets of pharmaceutical drugs. Varied drug-loaded NPs have been created to combat infections caused by multidrug-resistant organisms, with the goal to increase the therapeutic effect by protecting the drug from metabolization [25], controlling drug release, enhancing drug permeability through membrane absorption, and decreasing the required drug doses [26]. Nanocarriers' uptake into cells is influenced by their size and charge, as well as hydrophobicity, which regulates nanocarrier absorption and distribution by promoting immune cell interactions, protein interactions, particle disposal, and protein charge [27].

The administration of the drugs' solutions to the posterior segment of the eye is quite challenging. Nanotechnology is currently being employed as an alternate approach for delivering long-acting medications to the eye in appropriate dosages, either in the anterior or posterior segment. NPs, liposomes, nanosuspensions, nanomicelles, nanocrystals, and dendrimers can be used as drug nanocarriers to the eyes for improving ocular bioavailability and minimizing irritation [28].

Furthermore, nanotechnology improves vaccine development by acting as adjuvants or carriers for a variety of antigens, promoting immunogenicity by increasing antigen availability to antigen-presenting cells and boosting Th1 response [29]. Nanoparticulate systems have been investigated for application in whole antigen vaccinations, protein, DNA, and RNA replicons [30, 31].

Role of Nanomedicine in Ocular Infection by *Acanthamoeba*

Acanthamoeba spp. are free-living organisms that can survive in a range of habitats, including water, soil, and air. *Acanthamoeba* alternates between two stages during its life cycle, trophozoite (active stage) and cyst (dormant stage) [32]. The trophozoite is irregularly shaped with locomotory pseudopods and acanthopodia that mediate adhesion to the corneal surface [33]. Under harsh environmental conditions, the trophozoite transforms into a spherical cyst with the formation of double-walled cysts. Cysts are extremely resistant to freezing, sterilization, and a variety of antimicrobial drugs [34]. Molecular techniques, especially the rRNA sequence, have been used to identify *Acanthamoeba* spp., which have been differentiated into twenty genotypes (T1-T20) [35]. T4 genotype isolates have been linked to most human infections because of their higher virulence, transmission rate, and their insusceptibility to therapeutic drugs [36].

Ocular infection by *Acanthamoeba* spp. affects the cornea causing painful *Acanthamoeba* keratitis (AK). Due to an increase in the number of AK reports each year, it has attracted worldwide attention [3, 37, 38]. Improper contact lens usage, lack of sterilization and continuous wearing, as well as corneal trauma and ocular exposure to contaminated water, are the main predisposing factors for this infection. Contact lens wearers are estimated to be associated with 81% of AK cases [3]. *Acanthamoeba* is commonly located in the epithelium of the cornea and can also infiltrate the beneath stroma and corneal nerves, causing neuritis and necrosis [33]. AK manifests as corneal ulcers, photophobia, inflammation and redness, stromal infiltration, edema, blurred stroma, severe eye pain, and ultimately vision loss [39]. Due to the existence of non-specific symptoms early in the infection, as well as the existence of microbial co-infections, misdiagnosis is common, leading to a delay in receiving proper treatment. As a result, reliable *Acanthamoeba* detection is critical for precise and rapid AK diagnosis to prevent visual impairments [40] and eye enucleation in severe uncontrolled infections.

The most common approach to identifying ocular *Acanthamoeba* infection is the microscopical examination of corneal scrapings followed by culturing. The detection has been stated to be improved by microscopic inspection of corneal smears stained with various staining techniques. However, such methods necessitate practical experience, and faulty results might arise if the corneal scraping sample is insufficient [32]. Culturing corneal scrapings take a long time, up to several weeks, to affirm *Acanthamoeba* outgrowth. As a result, very sensitive molecular techniques that amplify *Acanthamoeba* DNA are being used [3]. On the other hand, such diagnostic tools necessitate costly high-tech equipment and professional expertise, and they are only limited to specialized academic institutions.

To improve the diagnosis of AK, Toriyama *et al.* [41] studied the efficacy of fluorescent immunochromatographic assay (FICGA) employing fluorescent silica NPs for the rapid identification of *Acanthamoeba castellanii* in clinical samples. FICGA results were extremely similar to real-time PCR results, indicating high sensitivity. It looks to be beneficial for diagnosing AK in outpatient clinics because it is faster and easier than using traditional and molecular methods.

Effective PCR-based methods for molecular detection of *Acanthamoeba* spp. have been successfully established [3]. The sensitivity and specificity of PCR reactions can be affected by a variety of clinical and environmental contaminations. Furthermore, the heat transfer limitations in PCR thermal cyclers influence the efficiency of the reactions. In this regard, some studies have focused on using nanomaterials as PCR additives to enhance PCR efficiency, productivity, sensitivity, and specificity. Graphene oxide, copper oxide, and alumina NPs were used by Gabriel *et al.* [42] to improve PCR performance in the detection of pathogenic free-living amoebae using genus-specific probes. For the detection of *Acanthamoeba* spp., the optimum concentrations of graphene oxide, copper oxide, and alumina NPs were determined to be 0.4, 0.04, and 0.4 µg per mL, respectively. Nano-PCR provides an improved diagnostic method, owing to the efficacy of NPs at low concentrations and low prices, as well as their physical and chemical stability and ease of mixing with polymers. This is likely due to the increased thermal conductivity in PCR reactions involving NPs. These findings suggest that nano-PCR assay has potential in the clinical diagnosis of *Acanthamoeba* infection as well as for studying epidemiology and environmental monitoring of this infection.

Regarding treatment, *Acanthamoeba* infections are challenging to treat mainly due to the sturdy nature of cysts and failure to deliver the requisite lethal doses to the cornea. The resistance of *Acanthamoeba* cysts is considered the major obstacle in antimicrobial chemotherapy. No drug has been described as a single fully efficient therapy for AK because of the vast variety of drugs' susceptibility, pathogenicity, and virulence factors seen in different *Acanthamoeba* strains [33].

Current AK treatments are restricted to the use of diamidines and biguanides. In particular, the use of a pharmacological cocktail included chlorhexidine digluconate or polyhexamethylene biguanide combined with hexamidine or isethionate propamidine. However, these medications are not specific, and their prolonged application is highly toxic to the human eye [43]. Simultaneously, there is the risk of recurrence and the emergence of adverse effects [44]. To overcome these challenges, there is an urgent need for novel therapies that are less toxic to the eye and more effective against both trophozoite and cyst stages of *Acanthamoeba* spp [33, 45, 46]. The antiamoebic efficacy of the therapeutic medications was improved when they were conjugated with NPs that carry them to the cornea more efficiently with fewer adverse effects.

Conjugation of chlorhexidine with gold NPS displayed considerable amoebicidal activity against *A. castellanii* while also protecting host cells from cytotoxicity [47]. Also, the conjugation of guanabenz (antiparasitic agent) with silver and gold NPs enhanced the antiamoebic activity against *A. castellanii,* with more potent activity observed at lower effective concentrations than when the drug was used alone. During encystation and excystation experiments, guanabenz prevents interconversion between *A. castellanii* trophozoites and cysts. Additionally, pretreated *A. castellanii* by gold and silver conjugated guanabenz reduced the cytopathogenicity of host cells from 65% to 38% and 2%, accordingly [11].

Using amoebicidal and host cells' cytotoxicity assays, Anwar *et al.* [48] stated that amphotericin B and nystatin-coated silver NPs were more efficient against T4 genotype of *A. castellanii* than using drugs or silver NPs alone. On the other hand, fluconazole conjugation with silver NPs had a minimal antiamoebic effect. Moreover, pretreatment of amoeba by drug-coated silver NPs greatly reduced the cytotoxicity of the host cells. Because of their tiny size and higher loading abilities for drugs, silver NPs improved the antiamoebic properties of pharmaceuticals.

Medicinal plants are being studied extensively to produce alternative therapeutics against *Acanthamoeba* infections as phytochemicals derived from these plants have been shown to have amoebicidal effects against either trophozoite or cyst stages of *Acanthamoeba* [45, 46, 49]. Various plant extracts coated NPs have shown potent therapeutic values against *A. castellanii*. Panatieri *et al.* [50] demonstrated that the association of coumarin-rich n-hexane extract of *Pterocaulon balansae* to oil-in-water nanoemulsions induced an anti-amoebic effect equivalent to chlorhexidine against *Acanthamoeba* trophozoites. After 24 hours of incubation with nanoemulsion containing 1.25mg mL^{-1} of coumarin, the viability of trophozoites was reduced by 95%, a similar result to that shown with chlorhexidine. Similarly, at concentrations of 25 μg mL^{-1}, *Jatropha gossypifolia*-conjugated silver NPs demonstrated a stronger anti-amoebic effect, killing 74% of *Acanthamoeba* trophozoites [51].

In addition, purified natural compounds, like tannic acid [52], periglaucine A and betulinic acid [53], oleic acid [54], cinnamic acid [55], quercetin, kolavenic acid [10], and gallic acid [56] coated NPs have shown a step forward in the development of nanomedicine against challenging infections caused by *Acanthamoeba*.

Tannins are polyphenolic plant metabolites that have been shown to have anti-parasitic properties. Tannins can bind to proteins, nucleic acids, carbohydrates, as well as metal ions, forming insoluble complexes. Tannic acid is hydrolysable tannin having antimicrobial properties [57]. The results of Padzik *et al.* [52] revealed that the tannic acid-coated silver NPs had strong anti-amoebic activity against *Acanthamoeba* spp. clinical strains as well as they exhibited minimal cytotoxicity. Furthermore, silver NPs modified with tannic acid were well absorbed by trophozoites and did not cause encystation.

The viability of *A. castellanii* was affected by trans-cinnamic acid (CA), which is a natural organic compound in a wide range of plant species. Antiamoebic effect of CA was enhanced when it was conjugated to gold NPS (CA-AuNPs). *A. castellanii*-mediated host cells cytotoxicity was inhibited when amoebae were pretreated with CA-AuNPs [55]. Other natural anti-parasitic medications include periglaucine A (PGA) and betulinic acid (BA). Their conjugation with poly(D,L-lactide-*co*-glycolide) (PLGA) NPs demonstrated a considerable inhibitory effect on *Acanthamoeba* trophozoites survivability as well as a modest cysticidal effect after 72 hours [53].

Anwar *et al.* [10] studied the anti-acanthamoebic effects of natural compounds, like quercetin and kolavenic acid, isolated from *Polyalthia longifolia var pendula* and *Caesalpinia pulcherrima* extracts, as well as their nanoconjugates with silver NPs, against *A. castellanii* isolates of T4 genotype from keratitis patients. Such compounds inhibited the growth of *A. castellanii*, but Quercetin-coated silver NPS boosted anti-amoebic actions and impaired the amoebae's encystation and excystation activity. Also, both quercetins, kolavenic acid and nanoconjugates demonstrated no cytotoxic effect on human cells *in vitro*. Natural compounds could be nanoencapsulated in polymeric NPs, which could help with their delivery. Mahboob *et al.* [56] examined *in vitro* anti-acanthamoebic effect of *Leea indica* and its ingredient gallic acid. Gallic acid at a dose of 100μg mL^{-1} inhibited trophozoites and cysts by 83 and 69%, respectively. Encapsulated gallic acid in PLGA NPs inhibited *Acanthamoeba* trophozoites by 90% and reduced cytotoxic effect.

Acanthamoeba trophozoites are known for their ability to adhere to contact lens surfaces and their storing cases. Kusrini *et al.* [58] proposed that dysprosium-based NPs have been employed as effective anti-acanthamoebic medications, and they can be loaded upon contact lenses to treat *Acanthamoeba* keratitis. In this way, the drug can disperse and reach post-lens tear film, increasing the drug's residential duration upon eye surface up to thirty minutes compared to just two minutes with eye drops, thereby enhancing corneal drug bioavailability [59].

According to prior studies, most contact lens solutions are ineffective against *Acanthamoeba*. To improve their anti-amoebic activity, Padzik *et al.* [60] and Hendiger *et al.* [61] evaluated the anti-amoebic activity and cytotoxicity of varied contact lens solutions coated with Ag, AuNPs, as well as tannic acid-coated AgNPs against *Acanthamoeba* trophozoites. They found that combining contact lens solutions with NPs increased anti-amoebic activity while lowering toxicity to human cells, suggesting that this could be an effective method to prevent AK in contact lens wearers.

Role of Nanomedicine in Ocular Toxoplasmosis

Ocular toxoplasmosis is a contagious eye infection caused by *Toxoplasma gondii* (*T. gondii*), which is a parasite having different stages of development (tachyzoite, tissue cyst, and oocyst) and different strains. It is the most prevalent cause of eye inflammation worldwide, and it can be inherited or acquired [1]. Congenital infection occurs due to the passage of *Toxoplasma* organism through the placenta from newly infected mothers to fetuses [62]. While, acquired infection is commonly contracted by consuming raw or undercooked meat, raw fruits/vegetables, contaminated water, or dairy products, as well as contact with infected cats [63]. After *T. gondii* infection has been established in the small intestine, *Toxoplasma* tachyzoites invade the circulating blood as separate agents or within leukocytes and disseminate to different organs as well as the eye [64].

According to various histopathological and clinical findings, *T. gondii* tachyzoites penetrate the eye through the hematogenous dissemination into the retina via the choroidal vessels, which supply the photoreceptors in the outer retina, or via the retinal vessels entering at the optic nerve head, which supply the inner retina [65]. *T. gondii* can provoke a unilateral and unifocal retinochoroidal lesion usually associated with vitritis [66]. The severity of ocular toxoplasmosis is influenced by *T. gondii* strain and the host immune response. In immunocompetent individuals, the infection usually causes no symptoms and resolves without treatment within a few months. While in immunocompromised individuals, the parasite can reactivate and may result in permanent retinal damage and blindness [67].

The most prevalent cause of posterior uveitis is *Toxoplasma* retinochoroiditis. Recurrent incidences of toxoplasmic retinochoroiditis are caused by the disruption of dormant *Toxoplasma* cysts in the retina, which release active tachyzoites that trigger inflammation, necrosis, DNA damage in retinal cells, as well as hypersensitivity to *Toxoplasma* antigens [1, 68]. Consequently, intercellular adhesion molecule-1 (ICAM-1) has been found to be implicated in ocular toxoplasmosis pathogenic pathways [69]. Increased ICAM-1 production via *T. gondii* parasitized cells can actively enhance immunological responses throughout the early infections and reactivate retinochoroiditis attacks [70].

Ocular toxoplasmosis is diagnosed primarily by clinical manifestations. Even so, ocular toxoplasmosis-related symptoms are often atypical and misleading. Several techniques have advanced tremendously in the last years, allowing parasitologists to confirm the diagnosis in the most reported cases of ocular toxoplasmosis. Serological diagnosis using ELISA (enzyme-linked immunosorbent assay) to detect *Toxoplasma*-specific immunoglobulins (IgG & M) is the commonly used procedure. However, this diagnostic method may have concerns related to specificity and sensitivity, resulting in faulty outcomes [71, 72], particularly in immunocompromised individuals, HIV-positive, organ transplant, and diabetic patients. Moreover, the presence of specific antibodies can reveal an acquired infection but not necessarily a disease. Antibody levels decline after recovery, but they can remain for long periods of time, rendering them useless for evaluating treatment outcomes [73]. Antigen detection methods can help diagnose low titers in early acute or chronic infections and immunocompromised patients with low or absent antibody titers [74].

Other tests used for diagnosis of ocular toxoplasmosis include an indirect fluorescent antibody, latex agglutination, and indirect hemagglutination tests [75], i.e., immunoblotting, Goldman-Witmer coefficient, and molecular techniques to amplify *T. gondii* nucleic acids [76]. These techniques, on the other hand, take time, need costly devices, and demand well-trained professionals. To achieve the highest accuracy and sensitivity in ocular toxoplasmosis diagnosis, ophthalmologists examine aqueous humor using various techniques via anterior chamber puncture [77].

Nanotechnology can be used to develop new *T. gondii* diagnostic assays [24, 78 - 80]. NPs have the potential to increase test sensitivity and specificity, as well as enable rapid and early detection [23, 24]. Due to their wide surface area, the nanomaterial facilitates the binding of a large number of target molecules for ultra-sensitive detection [81]. Consequently, the use of nanomaterials enhances the identification of *T. gondii* DNA, antigens, or antibodies in various samples using fast and efficient handling techniques, displaying high accuracy and potentially serving as a screening and detection tool for various *T. gondii* strains and stages.

A dynamic flow immuno-chromatographic technique (DFICT) employing gold NPs attached with SPA (staphylococcal protein A) was devised to detect *T. gondii* infection. Nitrocellulose membrane strips were coated with recombinant antigens specific for *T. gondii* and SPA that react positively with serum anti-*Toxoplasma* IgG antibodies, resulting in gold-SPA antibody complex. Sensitivity and specificity of DFICT were nearly equivalent to those of the ELISA test. DFICT has the following advantages: fast, simple, does not require highly trained laboratory technicians or specialized equipment, and it only requires a small amount of serum sample without any sample processing [80].

Jiang *et al.* [79] used an electrochemical immunosensor with magnetic AuNPs and graphene sheets to detect *T. gondii* IgM antibodies. The goldmag increased the immunosensor's sensitivity by improving the reduction in H_2O_2. Throughout this, the immunosensor demonstrated a precise and accurate diagnosis of toxoplasmosis. Similarly, Li *et al.* [24] detected *T. gondii* IgM antibodies using lateral flow immuno-chromatographic (LFIA) assay. Using magnetic AuNPs coated with polymethacrylic acid and coupled with an anti-human IgM antibody, this LFIA achieved 100% sensitivity and specificity.

Hegazy *et al.* [23] developed an immunomagnetic bead-ELISA (IMB-ELISA) for the rapid detection of *T. gondii* circulating surface antigen 1 (SAG1) in human serum samples using IgG polyclonal antibody-coated magnetic microbead NPs. IMB-ELISA achieved better results for the diagnosis of human toxoplasmosis than sandwich ELISA by increasing binding capacity and antigen detection sensitivity. Compared to sandwich ELISA, the surface of the beads allows for much more antibodies to be active in the antigen reaction.

To detect specific IgG antibodies of *T. gondii* in sera samples, Wang *et al.* [78] modified the latex piezoelectric immunoagglutination assay (LPEIA) using AuNPs conjugated with antigens of *T. gondii* instead of latex. In addition, using parasite antigen-coated silica NPs on polymerized plasma film, Wang *et al.* [82] created a piezoelectric direct immunoassay. Yang *et al.* [83] also used photo-luminescent nanomaterials, like the quantum dot, to produce microarrays. These approaches exhibited high sensitivity and provided results comparable to those obtained by ELISA, and they are recommended as screening tests because they are rapid (just around twenty minutes) and need 10 times less antigen than ELISA.

A newly developed Nano-ELISA kit with excreted/secreted (E/S) antigens and conjugate of AuNPs was evaluated by Khodadadi *et al.* [84] for the diagnosis of toxoplasmosis. In the same serum samples, Nano-ELISA demonstrated a 93.33% higher sensitivity and specificity than conventional ELISA, demonstrating a significant increase by using AuNPs in the ELISA test, which allows more antibodies to join the antigen-antibody complex.

In addition, Miao *et al.* [85] used fluorescence resonance energy transfer to create quantum dot probe of nickel NPs having magnetic and fluorescent characteristics. Furthermore, employing nickel NPs, Xu *et al.* [86] produced molecular beacon probes. DNA probes were found to have high sensitivity, specificity, and speed in detecting *T. gondii* target DNA in both studies.

Ocular toxoplasmosis treatment is based on the location and size of *T. gondii*-induced lesions. Spontaneous recovery occurs within two months with little or no damage when the lesions are minimal and in the peripheral site [87]. Treatment is needed if the lesions affect the optic nerve and macula, and retinal vessel lesions cause impairment in the visual acuity, vitreoretinal traction, and hemorrhages in the vitreous and/or retina. Inappropriate treatment can result in sight loss. Antitoxoplasmic medications, topical steroids, systemic corticosteroids, and topical cycloplegic therapy are used to treat moderate to severe lesions in immunocompetent patients.

Corticosteroids are used to reduce choroid and retinal damage since the host immune response increases intraocular inflammation in response to *Toxoplasma* tachyzoites [88].

The treatment's goal is to reduce the threats of vision loss, repeated attacks, and severity of symptoms as well as to prevent more pathological lesions in the eye. Individuals with compromised immune systems, as well as pregnant women, need urgent treatment. Currently, the most effective antitoxoplasmic drugs are pyrimethamine and sulfadiazine, which act together to inhibit folic acid biosynthesis. Other medications include clindamycin, spiramycin, atovaquone, tetracyclines, and itraconazole. However, the major drawbacks of existing drug therapies are the emergence of resistance, limited effectiveness for *Toxoplasma* tissue cysts and the existence of side effects, including bone marrow suppression, hypersensitivity reaction, and teratogenic effects [89].

Anti-*Toxoplasma* medications must be safe, efficient against different strains and stages of *T. gondii,* and penetrate deeper into ocular tissues. There is currently no drug that offers all these properties. The use of nanotechnology in the treatment of toxoplasmosis is gaining a lot of interest by improving drug pharmacokinetic profiles, identifying specific targets (cells and receptors) related to clinical status, and selecting proper nanocarriers to achieve the desired response and minimize adverse effects [89].

Toxoplasma infection has been treated with a variety of nanomaterials. These nanomaterials may have antitoxoplasmic activity directly or indirectly as a drug carrier, allowing for more precise drug delivery and decreasing host toxic effects [7]. Gold, silver, and platinum NPs were tested for their anti-*T. gondii* activity by Adeyemi *et al.* [90]. *T. gondii* growth was inhibited by the NPs to a degree of more than 90%. The NPs did not affect the host cells. The parasite mitochondrial membrane, as well as parasite invasion, proliferation, and infectivity potential, were all affected by the NPs' anti-*T. gondii* effect. Vergara-Duque *et al.* [62] used fluorescence microscopy and scanning electron microscopy to examine the effects of AgNPs on *T. gondii* oocysts. AgNPs caused morphological changes in the structure of *T. gondii* oocysts at various exposure times during the treatments.

Costa *et al.* [91] studied the anti-*Toxoplasma* effects of biogenic AgNPs at various concentrations in trophoblast cells and villous explants. With the induction of inflammatory mediators in BeWo and HTR8/SVneo cells and independent of mediators in the chorionic villus, the biogenic AgNPs were able to minimize *T. gondii* replication. Owing to the wide surface area and small size structure of this compound, it has the potential to deliver pharmacological agents into the target tissue, resulting in fewer side effects with greater specificity as compared to traditional therapies. Moreover, biogenic NP synthesis provides the best efficiency, stability, and safety.

Chitosan and silver NPs, either alone or in combination, were found to be effective against experimental toxoplasmosis. Treatment by the combined NPs resulted in a considerable reduction in parasite load in the liver and spleen. In addition, these compounds caused *T. gondii* tachyzoites to become paralyzed and undergo morphological alterations. Furthermore, treated mice receiving AgNPs alone or combined with chitosan had higher levels of gamma interferon in their sera [19].

Oral administration of lactoferrin-loaded polymeric nanocapsules of alginate-chitosan to *T. gondii*-infected mice (BLF-NC) enhanced lactoferrin bioavailability in the liver and spleen, causing a decrease in parasite burden and inflammation and an increase in intracellular ROS production and nitric oxide and Th1 cytokines in mice sera, all of which led to *Toxoplasma* death and maintained mice's survival more than three weeks days post-infection [92].

Anti-*Toxoplasma* effects of chitosan NPs have been demonstrated by tachyzoites mortality and improved mice survival [93]. When chitosan NPs were used with other medications, they had almost the same effects. Spiramycin-loaded chitosan NPs reduced the death rate of mice infected by *Toxoplasma*. Furthermore, they reduced perivascular inflammatory cellular infiltration on a histopathological examination with significant improvement in the pathological changes caused by *Toxoplasma* infection in the eye [94].

Gold NPs play a role in photothermic treatments because of their biocompatibility. Their capacity for conjugation with anti- *T. gondii* antibodies and laser light absorption makes them effective in infection by *Toxoplasma* parasite. When gold nanorods ~39.5 nm and gold nanospheres ~20 nm conjugated with specific antibodies to *T. gondii* tachyzoites surface antigen were irradiated using laser, they enhanced parasite killing in a dose-dependent manner related to the levels of laser irradiation [95, 96]. Moreover, tachyzoites' infectivity was lowered after being pretreated with AuNPs coated with anti-*T. gondii* antibodies [96].

By using core-shell latex NPs (~213.4 nm), the growth of *T. gondii* tachyzoites was inhibited, and infected macrophage numbers in J744-A1 macrophage cell culture were reduced [97]. In addition, core lipid nanocapsules containing pyrimethamine were found to improve the survivability of mice infected intraperitoneally by tachyzoites of *T. gondii* RH strain [98].

Tachibana *et al.* [99] found that liposomes containing stearylamine (SA) and phosphatidylcholine (PC) liposomes, known as SA/PC-liposomes, had therapeutic and preventive effects against *T. gondii* infection. *In vitro* treatment of *Toxoplasma* tachyzoites by SA/PC-liposomes caused morphological alterations and destroyed 95% of tachyzoites. Giving SA/PC-liposomes to mice prior or post-infection increased their survival by 80 and 70%, respectively, over thirty days compared to controls.

Usnic acid, a naturally occurring compound with anti-*Toxoplasma* activity, is restricted in usage due to its toxicity and poor water solubility. Liposomes can help its delivery to intended targets. In intraperitoneally infected mice by *T. gondii* tachyzoites and treated with liposomal usnic acid (~130 nm), survival rates were greater than in mice treated by usnic acid alone [100]. Triclosan, another compound used for toxoplasmosis treatment, decreased *T. gondii* tachyzoites load *In vitro* and inhibited their replication through inhibition of the enzyme enoyl-acyl reductase carrier protein. But their therapeutic application is limited due to their low solubility following oral administration [101]. This can render drug efficacy more difficult. Triclosan's activity can be increased by liposomes. Triclosan-liposomal treatment of *Toxoplasma*-tachyzoites infected mice showed no alternations in mice behavior compared to infected untreated mice and significantly decreased the tachyzoites burden in peritoneum fluids and liver specimens in comparison to Triclosan-treated mice [102]. Furthermore, triclosan-liposomal treatment lowered the mice infectivity by *Toxoplasma* cysts (Me49 strain) to 70%, while triclosan alone reduced the infectivity of *T. gondii* cysts by 20%. By transmission electron microscopy, bradyzoites found within triclosan-liposomal treated cysts had a partly destroyed wall, membrane disruption, and widespread vacuolization [103].

Despite the possible effectiveness of atovaquone against *Toxoplasma* infection, it has poor solubility and tissue bioavailability. As a result, several scientists tested atovaquone's bioavailability and anti-*Toxoplasma* activity after coating it with NPs. Conjugation of atovaquone with nanosuspensions increased its therapeutic level in the blood and protected infected mice against reactivated toxoplasmosis by 100%, resulting in the absence of *Toxoplasma* organisms and inflammatory foci in their brains and livers [104]. It is well-recognized that sodium dodecyl sulfate (SDS) improves molecule transport through epithelial barriers and stabilizes drug-loaded NPs. In a model of *Toxoplasma* reactivation, administration of atovaquone nanosuspension incorporated into SDS increased bioavailability and decreased inflammation, parasitic load, and *T. gondii* DNA concentration in the mice's brains [105].

In addition, Sordet *et al.* [106] stated that encapsulating atovaquone in polymeric nanocapsules of polylactic acid (206 nm) enhanced atovaquone activity, resulting in increased survival and reduced the parasite number in murine models of acute and chronic toxoplasmosis.

Sulfadiazine (SDZ) is a toxoplasmosis medication drug used in conjunction with pyrimethamine. However, its high doses cause serious side effects. Dendrimers may be used as drug carriers that can increase sulfadiazine efficacy while also lowering toxoplasmosis prescription doses. An *in vitro* study by Prieto *et al.* [107] determined the effect of sulfadiazine-dendrimers complexes, SDZ-DG4 (cationic dendrimer) and SDZ-DG4.5 (anionic dendrimer) complexes, on Vero cells infected by *T. gondii* RH strain over 4 hours. Although both complexes decreased Vero cells infection, SDZ-DG4 at 0.03 μM recorded the best effect by 60%, indicating that SDZ–DG4 complex is a promising antitoxoplasmic therapy candidate.

Immunological effects of varied *T. gondii* antigens and a variety of adjuvants have been used in several animal studies [108]. Significant applications of nanotechnology currently aim at the advancement of anti-*Toxoplasma* vaccines. Nanomaterials may be used as carriers of *T. gondii* antigens or genetic materials (DNA/RNA). The use of NPs has significant consequences for generating an effective immune response to *T. gondii* infections [109 - 111].

Ducournau *et al.* [111] tested the protective effect of DGNP/TE vaccine against *T. gondii* infection, comprising maltodextrin NPs (DGNP) coated to total extract (TE) of *T. gondii* proteins. DGNP/TE vaccine elicited a Th1-cellular response driven by the release of IL-12 and IFN-γ via antigen-presenting cells after two intranasal or intradermal injections. Intradermally vaccinated animals' Th1 responses tended to be controlled by IL-10 secretion, but intranasally vaccinated animals' Th1 responses were not. When compared to non-vaccinated animals, intranasal vaccination resulted in a significant reduction in brain cysts. Therefore, intranasal administration of the DGNP/TE vaccine provided excellent protection against latent and congenital toxoplasmosis. Similarly, nasal vaccination of mouse model given TE of *T. gondii* antigen from virulent strain encapsulated in porous maltodextrin-based lipid core NPs (DGNP/TE) showed a protective effect. DGNP/TE (~88.4 nm) enhanced delivery of antigens into dendritic cells and macrophages, stimulating various cytokines production, like TNF-α, IL-12p40, IL-6 and IL-1. Furthermore, this NP protected the murine model from acute and chronic *T. gondii* infection by activating the humoral and cellular immune responses [110].

Vaccination of C57BL/6 mice by *T. gondii* profiling, a potent stimulant, encapsulated in liposomes coating in oligomannose (TgPF-OML) reduced *Toxoplasma* cysts in mouse brains and increased survival rates. Moreover, it enhanced IgG and IFN-γ production while lowering IL-10 in comparison to mice that only received TgPF or were not vaccinated [112]. Moreover, Chen *et al.* [113] used liposome-encapsulated dense granule protein-4 recombinant antigen (pGRA4-liposomes) and pGRA4 alone to immunize mice before infection with *T. gondii* ME49 and RH strain. Immunization with pGRA4-liposomes increased serum IgG2a levels and production of IFN-γ and IL-2 via splenocytes stimulation. When compared to immunization with only pGRA4, this vaccine protected mice by reducing *Toxoplasma* cysts and increasing rates of survival.

Regarding the development of DNA vaccines against *T. gondii* infection, Chen *et al.* [114] demonstrated that immunizing mice with liposomes encapsulated surface antigen (SAG1) and rhoptry antigen (ROP1) DNA elicited cellular response driven by the release of IL-2 and IFN-γ, as well as a humoral immune response resulting in the production of anti-*T. gondii* antibodies. It is known that intracellular proliferation of *T. gondii* is regulated by ROP18, a rhoptry kinase that is a major *Toxoplasma* virulence factor [115]. Intranasal administration of recombinant protein of ROP18 (rROP18) DNA in nanospheres to mouse model increased humoral immune response, resulting in the production of IgG2a and IgA [116].

Following infection by *T. gondii* RH-strain, immunizing mouse model with rhomboid 4 and dense granule antigen (GRA1) DNA plus nano-adjuvant calcium phosphate NPs raised IFN-γ and IgG2a/IgG1 ratio, decreased parasite numbers, and improved rates of survival [117]. Also, encapsulating DNA-GRA1plasmid in 400 nm chitosan NPs improved cellular absorption and protected DNA from degradation. Regardless, this NP elicited a predominant Th2 response in C3H/HeN mice by increasing specific IgG production that is linked to toxoplasmosis susceptibility [30].

The DNA-GRA1 plasmid was encapsulated in 400nm chitosan NPs, which increased cellular absorption and protected the DNA from degradation. Regardless, by enhancing specific IgG production, this NP triggered a significant Th2 response in C3H/HeN mice, which is connected to illness risk. Vaccines developed from liposomes with plasmids carrying DNA from various *T. gondii* antigens, like matrix antigen 1 (MAG1), dense granule antigens (GRA1,4,6,7), and microneme antigen 3 to immunized infected mice enhanced IgG and IFN-γ production. At the same time, immunized mice with MAG1 and GRA7 DNA liposomes only showed a higher IgG2/IgG1 ratio, indicating a Th1 immune response [118 - 120].

According to Chahal *et al.* [31], RNA replicon encapsulated in modified dendrimer NPs that encodes GRA6, SAG1, SAG2A, ROP18, ROP2A, and apical membrane antigen-1, can deliver self-replicating RNA to cells' cytoplasm, avoiding nuclease action. Moreover, the single dose of modified dendrimer NPs-RNA immunization protected mice from *T. gondii* infection and stimulated CD8+ T lymphocyte proliferation, as well as mild IFN-γ and IL-2 expression. As a result, modified dendrimer NPs-RNA is a promising vaccine as it avoids the risk of mutagenic integration with the host DNA.

Role of Nanomedicine in Ocular Toxocariasis

Ocular toxocariasis is an infectious disease provoked by larvae of *Toxocara canis* (*T. canis*) and *Toxocara cati*, which are roundworms that colonize dogs and cats' small intestines, penetrating the posterior segment of the eye. Infected dogs/cats excrete large numbers of eggs in their feces, contaminating the soil, water supplies, and raw vegetables. Humans are infected by ingesting embryonated *Toxocara* eggs. The parasite does not mature into an adult stage and persists as L2 larvae, which can migrate to the eyes and cause ocular toxocariasis, a common cause of posterior and diffuse uveitis [121, 122]. Ocular toxocariasis is more common in children. It is frequently unilateral in 90% of cases. According to estimates, ocular toxocariasis causes five to twenty percent of uveitis-related blindness [122, 123].

The degree of visual deterioration relies on the larval location and the level of the fibrotic granulomatous response. Recorded pathological lesions in ocular toxocariasis are posterior pole granuloma, peripheral chorioretinal granuloma, chronic endophthalmitis, and panuveitis. Scleritis, vitreous opacities, yellowish-white intraretinal lesions, papilledema, and tractional retinal detachment, as well as diffuse unilateral subacute neuropathy, are among the other pathologies [122].

Patients with ocular toxocariasis need to be diagnosed as soon as possible so that they can be properly managed and treated. Currently, clinical diagnosis of ocular toxocariasis is based on fundoscopic inspection to observe the moving larva underneath the retina and detect the typical pathologies [124, 125]. For laboratory diagnosis, immunoassays can identify specific immunoglobulins versus excretory/secretory antigens of *Toxocara* in blood and ocular fluids samples [122].

Cross-reactivity with so many other nematode antigens, like *Ascaris lumbricoides*, difficulty distinguishing between active and past infections and low concentration of TES antigens in sera samples in cases of active ocular toxocariasis, are the main limitations of ELISA and monoclonal/polyclonal antibodies-based assay for

TES antigens identification. The use of western blot (WB) for identifying anti-*Toxocara* IgG has considerably improved immunodiagnosis [126].

Nanobodies (Nbs) can be used as an alternative to monoclonal antibodies as they are cost-effective and more stable immunoreagents. Nbs are easy to identify and may be generated in *Escherichia coli* using standardized conditions of protein production, reducing variance seen with traditional poly and monoclonal antibodies [127]. Nbs can identify cryptic epitopes on their cognate antigen because of their small size [128]. Nbs can also be used as modular structures by incorporating a tag or enzyme label, which can be used to improve test sensitivity [129].

Morales-Yanez *et al.* [130] created a diagnostic approach with a highly sensitive electrochemical readout based on single-domain antigen-binding fragments (Nbs) from camel heavy-chain antibodies. This method has been shown to have high specificity and sensitivity with no cross-reactivity with other nematodes larval antigens. It employs Nbs as a binder for TES antigens found in the serum of mice infected with *T. canis* eggs 3 days after infection. Nbs can identify common epitopes on multiple components of TES antigen due to the presence of disulphide linkages in the target antigen, which are required for these Nbs to recognize epitopes.

Morales-Yánez *et al.* [131] assessed the electrochemical magnetosensor assay performance for recognition of TES antigens in sera of Ecuadorian children. Nb-based electrochemical assay revealed that 38% of the analyzed samples were positive for TES antigens recognition at a concentration of 2.1 ng mL^{-1}. Furthermore, there was no cross-reaction between the electrochemical assay's positive results and the existence of other helminth infections. Nb-based electrochemical assay allows for highly sensitive quantification of TES antigens in sera, which could aid in the diagnosis of active toxocariasis in humans. Similarly, Jofre *et al.* [132] created a microfluidic immunosensor for the detection of anti-*T. canis* IgG antibodies electrochemically. Core-shell gold-ferric oxide NPs and ordered mesoporous carbon (CMK-8) in chitosan were utilized to improve the sensor's specificity and sensitivity. The electrochemical immunosensor was demonstrated to be a promising technique for identifying anti-*T. canis* IgG antibodies, with a total assay time of 20 minutes, 0.10 ng mL^{-1} detection limit, and coefficients variance less than 6% for intra- and inter-assay.

Medawar *et al.* [133] designed a microfluidic immunosensor for quantifying anti-*T. canis* IgG antibodies. TES antigens were immobilized covalently on to 3-aminopropyl-functionalized silica-NPs in a non-competitive immunoassay. Serum antibodies interacted with TES antigens and then were quantified by a second antibody labeled by cadmium selenide zinc sulphide quantum dots. The assay detection limit was calculated to be 0.12 ng mL^{-1} and less than 6% coefficients variance of intra- and inter-assay. The proposed immunosensor might be used to detect toxocariasis promptly.

Ocular toxocariasis treatment is influenced by the severity of ocular manifestations, presence of ocular inflammations, involvement of macula and impaired vision [5]. Ocular toxocariasis can be treated with medications or surgery. Ophthalmologists tend to use steroids and antihelminthic medications for medical care, although there are no established limits for dosages, durations, or methods of administration [122]. Corticosteroids are the most common medication for ocular toxocariasis because they suppress inflammation, prohibit vitreous opacity and tractional retinal detachment. Despite this, corticosteroids are ineffective in treating structural retinal complications [125].

Many ophthalmologists recommend administering antiparasitic drugs in conjunction with steroids, whilst others recommend them when the response to steroids is ineffective. Anthelmintic drugs can cause intraocular inflammations as a result of a hypersensitivity reaction to the death of *Toxocara* larva within the eye, which can result in complete eye damage [134]. Albendazole and diethylcarbamazine, the two most often used drugs, have an anti-*Toxocara* larval effect and can penetrate blood-brain barriers. Anthelmintic medications, including thiabendazole, mebendazole, and tinidazole, are very successful at keeping *Toxocara* larvae from proceeding to the neurotropic phase of infection [121, 134].

Albendazole is a water-insoluble, low-permeability medication that is partially absorbed from the gastrointestinal tract following oral administration. This property is a major disadvantage during the treatment of ocular toxocariasis. Moreover, it has several side effects, like acute and granulomatous hepatitis [135, 136]. Incorporating ABZ into lipid-based nanocarrier devices would not only improve treatment effectiveness by lowering the therapeutic dosage, but it would also reduce side effects on the host. Following oral administration of ABZ solid lipid NPs, the number of *Toxocara* larvae in the liver, brain, lungs, and kidneys was reduced in a mouse model [137].

To treat larval toxocariasis, Velebný *et al.* [138] undertook therapeutic trials on *Toxocara*-infected mice with fenbendazole and albendazole entrapped in liposomes and on combined administration of immunostimulator glucan. The results revealed that FBZ and ABZ administered in liposomal formulations had higher larvicidal efficacy, and the potentiation of their effect by co-administration of liposomal glucan was greatly enhanced.

In vitro anti-*Toxocara* effects of ZnO and iron oxide NPs were studied by Dorostkar *et al.* [12]. *Toxocara vitulorum* worms were incubated for 24 hours with various NP concentrations. Both NPs significantly reduced worm mobility, increased mortality rate, and elevated malondialdehyde and nitric oxide (NO) levels in a time and concentration-dependent manner as compared to the control group. The activity of superoxide dismutase was increased at low concentrations of NPs but decreased at higher concentrations. Anti-*Toxocara* effect of metal oxide NPs is mediated by the activation of oxidative/nitrosative stress, thereby exerting their cytotoxic effects.

In a trial to develop a vaccine, Malheiro *et al.* [139] studied the effect of DNA vector plasmid (pcDNA3-CpG) and plasmid expressing murine IL-12 (pcDNA-IL-12) plus adjuvant of gold microparticles against *Toxocara* infection. The vaccines were given to a mouse model of toxocariasis via a percutaneous route. Vaccination by pcDNA-IL12 resulted in a persistent reduction in blood/bronchoalveolar eosinophilia, while airway hyperresponsiveness was prevented by pcDNA3-CpG vaccination. In pcDNA-IL-12-vaccinated mice, type-1 immune response was the most dominant, as seen by the higher IFN-γ/IL-4 ratio, lower levels of IgG1, and high IgG2a/IgG1 ratio. These findings demonstrate that the produced vaccines have different therapeutic advantages towards toxocariasis, implying that they may be combined in future therapeutic interventions.

As evident from the preceding discussion, several investigations are on the way related to the development of nanodiagnostic assays, nanomedicines, and nanovaccines against ocular parasite infections, as accurate diagnosis and timely treatment can reduce eye morbidity (Fig. **1**).

CONCLUSION

Ocular parasitic infections are a leading cause of ocular morbidity, not because they are incurable, but because of latency or misdiagnosis, mostly due to a lack of awareness. Untreated severe infections can result in visual deterioration and blindness. Ocular parasitic lesions should be diagnosed and treated early to minimize ocular morbidity.

Considering the importance of ocular parasites and their widespread prevalence around the world, developing new diagnostic and successful therapeutic approaches with few side effects is considered one of the most critical vision-protection priorities. Advanced studies and research are, nevertheless, required to produce effective, low-cost medications to combat these pathogenic disorders.

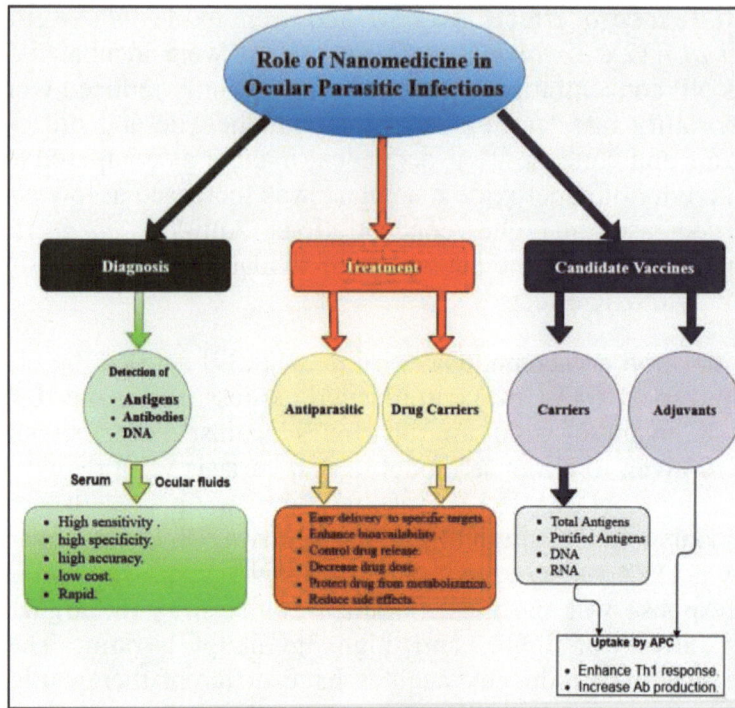

Fig. (1). The role of nanomedicine in ocular parasite infections. Nanomaterials can be used for diagnosis, treatment, and vaccinations. Nanomaterials can be utilized in diagnosis to detect parasites' antigens, antibodies, or DNA in a variety of specimens, with high sensitivity, and specific approaches. Nanomaterials can act directly as antiparasitic or indirectly as drug carriers, enhancing drug delivery to target sites with minimal toxic effects. Nanomaterials aid in vaccine production by functioning as adjuvants or carriers of antigens and DNA/RNA, increasing immunogenicity and boosting Th1 response.

The diversity of nanomaterials gives the advantage for application in diagnosis, treatment, and vaccinations against numerous infectious diseases, including ocular parasitic diseases. Nanomaterials may operate directly on microorganisms or act as carriers. Drug delivery to specific targets (particularly intracellular parasites) improves the bioavailability and stability of drugs while also regulating the release, increasing activity, preventing degradation, and lowering toxicity. Immunization using different types of nanomaterials loaded with parasitic antigens or DNA in experimental models can enhance the internalization and antigen delivery in macrophages and dendritic cells, activate cellular immune response, induce the inflammatory cytokines and specific IgG production, reduce parasitic burden, and increase survival rates.

CONSENT FOR PUBLICATION

Not applicable.

CONFLICT OF INTEREST

The author declares no conflict of interest, financial or otherwise.

ACKNOWLEDGEMENT

Declared none.

REFERENCES

[1] El-Sayed NM, Safar EH. Characterization of the parasite-induced lesions in the posterior segment of the eye. Indian J Ophthalmol 2015; 63(12): 881-7.
[http://dx.doi.org/10.4103/0301-4738.176028] [PMID: 26862090]

[2] El-Sayed NM, Safar EH, Issa RM. Parasites as a cause of keratitis: Need for increased awareness. Aperito J Ophthalmol 2015; 1: 103.

[3] El-Sayed NM, Younis MS, Elhamshary AM, Abd-Elmaboud AI, Kishik SM. *Acanthamoeba* DNA can be directly amplified from corneal scrapings. Parasitol Res 2014; 113(9): 3267-72.
[http://dx.doi.org/10.1007/s00436-014-3989-3] [PMID: 24951167]

[4] Saki J, Eskandari E, Feghhi M. Study of toxoplasmosis and toxocariasis in patients suffering from ophthalmic disorders using serological and molecular methods. Int Ophthalmol 2020; 40(9): 2151-7.
[http://dx.doi.org/10.1007/s10792-020-01393-6] [PMID: 32424529]

[5] Woodhall D, Starr MC, Montgomery SP, *et al.* Ocular toxocariasis: epidemiologic, anatomic, and therapeutic variations based on a survey of ophthalmic subspecialists. Ophthalmology 2012; 119(6): 1211-7.
[http://dx.doi.org/10.1016/j.ophtha.2011.12.013] [PMID: 22336630]

[6] Rathinam SR, Annamalai R, Biswas J. Intraocular parasitic infections. Ocul Immunol Inflamm 2011; 19(5): 327-36.
[http://dx.doi.org/10.3109/09273948.2011.610024] [PMID: 21970664]

[7] Assolini JP, Concato VM, Gonçalves MD, *et al.* Nanomedicine advances in toxoplasmosis: diagnostic, treatment, and vaccine applications. Parasitol Res 2017; 116(6): 1603-15.
[http://dx.doi.org/10.1007/s00436-017-5458-2] [PMID: 28477099]

[8] Salata OV. Applications of nanoparticles in biology and medicine. J Nanobiotechnology 2004; 2(1): 3.
[http://dx.doi.org/10.1186/1477-3155-2-3] [PMID: 15119954]

[9] Maribel G, Guzman Dille J, Godet S. Synthesis of silver nanoparticles by chemical reduction method and their antimicrobial activity. Int J Chem Biomolec Engin 2009; 2(3): 104-11.

[10] Anwar A, Ting ELS, Anwar A, *et al.* Antiamoebic activity of plant-based natural products and their conjugated silver nanoparticles against Acanthamoeba castellanii (ATCC 50492). AMB Express 2020; 10(1): 24.
[http://dx.doi.org/10.1186/s13568-020-0960-9] [PMID: 32016777]

[11] Anwar A, Mungroo MR, Anwar A, Sullivan WJ Jr, Khan NA, Siddiqui R. Repositioning of Guanabenz in conjugation with gold and silver nanoparticles against pathogenic amoebae *Acanthamoeba castellanii* and *Naegleria fowleri*. ACS Infect Dis 2019; 5(12): 2039-46.
[http://dx.doi.org/10.1021/acsinfecdis.9b00263] [PMID: 31612700]

[12] Dorostkar R, Ghalavand M, Nazarizadeh A, Tat M, Hashemzadeh MS. Anthelmintic effects of zinc oxide and iron oxide nanoparticles against *Toxocara vitulorum*. Int Nano Lett 2017; 7(2): 157-64.
[http://dx.doi.org/10.1007/s40089-016-0198-3]

[13] Dhakal Y, Meena RS, Kumar S. Effect of INM on nodulation, yield, quality and available nutrient status in soil after harvest of green gram. Legume Res 2016; 39(4): 590-4.

[14]   Ahsan F, Rivas IP, Khan MA, Torres Suarez AI. Targeting to macrophages: role of physicochemical properties of particulate carriers—liposomes and microspheres—on the phagocytosis by macrophages. J Control Release 2002; 79(1-3): 29-40.
[http://dx.doi.org/10.1016/S0168-3659(01)00549-1] [PMID: 11853916]

[15]   Treiger Borborema SE, Schwendener RA, Osso Junior JA, de Andrade Junior HF, do Nascimento N. Uptake and antileishmanial activity of meglumine antimoniate-containing liposomes in *Leishmania (Leishmania)* major-infected macrophages. Int J Antimicrob Agents 2011; 38(4): 341-7.
[http://dx.doi.org/10.1016/j.ijantimicag.2011.05.012] [PMID: 21783345]

[16]   Mandal AK. Silver nanoparticles as drug delivery vehicle against infections. Glob J Nano 2017; 3(2): 555607.

[17]   Banik BL, Fattahi P, Brown JL. Polymeric nanoparticles: the future of nanomedicine. Wiley Interdiscip Rev Nanomed Nanobiotechnol 2016; 8(2): 271-99.
[http://dx.doi.org/10.1002/wnan.1364] [PMID: 26314803]

[18]   Kumar R, Sahoo GC, Pandey K, *et al.* Development of PLGA–PEG encapsulated miltefosine based drug delivery system against visceral leishmaniasis. Mater Sci Eng C 2016; 59: 748-53.
[http://dx.doi.org/10.1016/j.msec.2015.10.083] [PMID: 26652429]

[19]   Gaafar MR, Mady RF, Diab RG, Shalaby TI. Chitosan and silver nanoparticles: Promising anti-*toxoplasma* agents. Exp Parasitol 2014; 143: 30-8.
[http://dx.doi.org/10.1016/j.exppara.2014.05.005] [PMID: 24852215]

[20]   Kong M, Chen XG, Xing K, Park HJ. Antimicrobial properties of chitosan and mode of action: A state of the art review. Int J Food Microbiol 2010; 144(1): 51-63.
[http://dx.doi.org/10.1016/j.ijfoodmicro.2010.09.012] [PMID: 20951455]

[21]   Pavoni L, Pavela R, Cespi M, *et al.* Green micro- and nanoemulsions for managing parasites, vectors and pests. Nanomaterials (Basel) 2019; 9(9): 1285.
[http://dx.doi.org/10.3390/nano9091285] [PMID: 31505756]

[22]   Kamboj S, Bala S, Nair AB. Solid lipid nanoparticles: an effective lipid based technology for poorly water soluble drugs. Int J Pharm Sci Rev Res 2010; 5: 78-90.

[23]   Hegazy S, Farid A, Rabae I, El-Amir A. Novel IMB-ELISA assay for rapid diagnosis of human toxoplasmosis using SAG1 antigen. Jpn J Infect Dis 2015; 68(6): 474-80.
[http://dx.doi.org/10.7883/yoken.JJID.2014.444] [PMID: 25866114]

[24]   Li X, Zhang Q, Hou P, *et al.* Gold magnetic nanoparticle conjugate-based lateral flow assay for the detection of IgM class antibodies related to TORCH infections. Int J Mol Med 2015; 36(5): 1319-26.
[http://dx.doi.org/10.3892/ijmm.2015.2333] [PMID: 26329478]

[25]   Moen MD, Lyseng-Williamson KA, Scott LJ. Liposomal amphotericin B: a review of its use as empirical therapy in febrile neutropenia and in the treatment of invasive fungal infections. Drugs 2009; 69(3): 361-92.
[http://dx.doi.org/10.2165/00003495-200969030-00010] [PMID: 19275278]

[26]   Verma D, Verma S, Blume G, Fahr A. Particle size of liposomes influences dermal delivery of substances into skin. Int J Pharm 2003; 258(1-2): 141-51.
[http://dx.doi.org/10.1016/S0378-5173(03)00183-2] [PMID: 12753761]

[27]   Saptarshi SR, Duschl A, Lopata AL. Interaction of nanoparticles with proteins: relation to bio-reactivity of the nanoparticle. J Nanobiotechnology 2013; 11(1): 26.
[http://dx.doi.org/10.1186/1477-3155-11-26] [PMID: 23870291]

[28]   Omerović N, Vranić E. Application of nanoparticles in ocular drug delivery systems. Health Technol (Berl) 2020; 10(1): 61-78.
[http://dx.doi.org/10.1007/s12553-019-00381-w]

[29]   Torres-Sangiao E, Holban A, Gestal M. Advanced nanobiomaterials: Vaccines, diagnosis and

treatment of infectious diseases. Molecules 2016; 21(7): 867.
[http://dx.doi.org/10.3390/molecules21070867] [PMID: 27376260]

[30]   Bivas-Benita M, Laloup M, Versteyhe S, *et al.* Generation of *Toxoplasma gondii* GRA1 protein and DNA vaccine loaded chitosan particles: preparation, characterization, and preliminary *in vivo* studies. Int J Pharm 2003; 266(1-2): 17-27.
[http://dx.doi.org/10.1016/S0378-5173(03)00377-6] [PMID: 14559390]

[31]   Chahal JS, Khan OF, Cooper CL, *et al.* Dendrimer-RNA nanoparticles generate protective immunity against lethal Ebola, H1N1 influenza, and *Toxoplasma gondii* challenges with a single dose. Proc Natl Acad Sci USA 2016; 113(29): E4133-42.
[http://dx.doi.org/10.1073/pnas.1600299113] [PMID: 27382155]

[32]   El-Sayed NM, Hikal WM. Several staining techniques to enhance the visibility of *Acanthamoeba* cysts. Parasitol Res 2015; 114(3): 823-30.
[http://dx.doi.org/10.1007/s00436-014-4190-4] [PMID: 25346196]

[33]   Lorenzo-Morales J, Khan NA, Walochnik J. An update on *Acanthamoeba* keratitis: diagnosis, pathogenesis and treatment. Parasite 2015; 22: 10.
[http://dx.doi.org/10.1051/parasite/2015010] [PMID: 25687209]

[34]   Visvesvara GS, Moura H, Schuster FL. Pathogenic and opportunistic free-living amoebae: *Acanthamoeba* spp., *Balamuthia mandrillaris*, *Naegleria fowleri*, and *Sappinia diploidea*. FEMS Immunol Med Microbiol 2007; 50(1): 1-26.
[http://dx.doi.org/10.1111/j.1574-695X.2007.00232.x] [PMID: 17428307]

[35]   Martín-Pérez T, Criado-Fornelio A, Martínez J, Blanco MA, Fuentes I, Pérez-Serrano J. Isolation and molecular characterization of *Acanthamoeba* from patients with keratitis in Spain. Eur J Protistol 2017; 61(Pt A): 244-52.
[http://dx.doi.org/10.1016/j.ejop.2017.06.009] [PMID: 28756938]

[36]   Fabres LF, Maschio VJ, Santos DL, *et al.* Virulent T4 *Acanthamoeba* causing keratitis in a patient after swimming while wearing contact lenses in Southern Brazil. Acta Parasitol 2018; 63(2): 428-32.
[http://dx.doi.org/10.1515/ap-2018-0050] [PMID: 29654672]

[37]   Randag AC, van Rooij J, van Goor AT, *et al.* The rising incidence of *Acanthamoeba* keratitis: A 7-year nationwide survey and clinical assessment of risk factors and functional outcomes. PLoS One 2019; 14(9): e0222092.
[http://dx.doi.org/10.1371/journal.pone.0222092] [PMID: 31491000]

[38]   Liu HY, Chu HS, Wang IJ, Chen WL, Hou YC, Hu FR. Clinical features and outcomes of *Acanthamoeba* keratitis in a tertiary hospital over 20- year period. J Formos Med Assoc 2020; 119(1): 211-7.
[http://dx.doi.org/10.1016/j.jfma.2019.04.011] [PMID: 31076316]

[39]   Lorenzo-Morales J, Martín-Navarro CM, López-Arencibia A, Arnalich-Montiel F, Piñero JE, Valladares B. *Acanthamoeba* keratitis: an emerging disease gathering importance worldwide? Trends Parasitol 2013; 29(4): 181-7.
[http://dx.doi.org/10.1016/j.pt.2013.01.006] [PMID: 23433689]

[40]   Claerhout I, Goegebuer A, Van Den Broecke C, Kestelyn P. Delay in diagnosis and outcome of *Acanthamoeba* keratitis. Graefes Arch Clin Exp Ophthalmol 2004; 242(8): 648-53.
[http://dx.doi.org/10.1007/s00417-003-0805-7] [PMID: 15221303]

[41]   Toriyama K, Suzuki T, Inoue T, *et al.* Development of an immunochromatographic assay kit using fluorescent silica nanoparticles for rapid diagnosis of *Acanthamoeba* keratitis. J Clin Microbiol 2015; 53(1): 273-7.
[http://dx.doi.org/10.1128/JCM.02595-14] [PMID: 25392356]

[42]   Gabriel S, Rasheed AK, Siddiqui R, Appaturi JN, Fen LB, Khan NA. Development of nanoparticle-assisted PCR assay in the rapid detection of brain-eating amoebae. Parasitol Res 2018; 117(6): 1801-11.

[http://dx.doi.org/10.1007/s00436-018-5864-0] [PMID: 29675682]

[43]   Chomicz L, Conn DB, Padzik M, *et al.* Emerging threats for human health in Poland: Pathogenic isolates from drug resistant *Acanthamoeba* Keratitis monitored in terms of their *in vitro* dynamics and temperature adaptability. BioMed Res Int 2015; 2015: 1-8.
[http://dx.doi.org/10.1155/2015/231285] [PMID: 26682216]

[44]   Rezaeian M, Farnia S, Niyyati M, Rahimi F. Amoebic keratitis in Iran (1997–2007). Iran J Parasitol 2007; 2(3): 1-6.

[45]   El-Sayed NM, Ismail KA, Ahmed SAEG, Hetta MH. *In vitro* amoebicidal activity of ethanol extracts of *Arachis hypogaea* L., *Curcuma longa* L. and *Pancratium maritimum* L. on *Acanthamoeba castellanii* cysts. Parasitol Res 2012; 110(5): 1985-92.
[http://dx.doi.org/10.1007/s00436-011-2727-3] [PMID: 22146994]

[46]   Kadry GM, Ismail MAM, El-Sayed NM, El-Kholy HS, El-Akkad DMH. *In vitro* amoebicidal effect of *Aloe vera* ethanol extract and honey against *Acanthamoeba spp.* cysts. J Parasit Dis 2021; 45(1): 159-68.
[http://dx.doi.org/10.1007/s12639-020-01292-8] [PMID: 33746401]

[47]   Aqeel Y, Siddiqui R, Anwar A, Shah MR, Khan NA. Gold nanoparticle conjugation enhances the antiacanthamoebic effects of chlorhexidine. Antimicrob Agents Chemother 2016; 60(3): 1283-8.
[http://dx.doi.org/10.1128/AAC.01123-15] [PMID: 26666949]

[48]   Anwar A, Siddiqui R, Hussain MA, Ahmed D, Shah MR, Khan NA. Silver nanoparticle conjugation affects antiacanthamoebic activities of amphotericin B, nystatin, and fluconazole. Parasitol Res 2018; 117(1): 265-71.
[http://dx.doi.org/10.1007/s00436-017-5701-x] [PMID: 29218442]

[49]   Dodangeh S, Niyyati M, Kamalinejad M, Lorenzo-Morales J, Haghighi A, Azargashb E. The amoebicidal activity of *Ziziphus vulgaris* extract and its fractions on pathogenic *Acanthamoeba* trophozoites and cysts. Trop Biomed 2017; 34(1): 127-36.
[PMID: 33592990]

[50]   Panatieri LF, Brazil NT, Faber K, *et al.* Nanoemulsions containing a coumarin-rich extract from *Pterocaulon balansae* (*Asteraceae*) for the treatment of ocular *Acanthamoeba* Keratitis. AAPS PharmSciTech 2017; 18(3): 721-8.
[http://dx.doi.org/10.1208/s12249-016-0550-y] [PMID: 27225384]

[51]   Borase HP, Patil CD, Sauter IP, Rott MB, Patil SV. Amoebicidal activity of phytosynthesized silver nanoparticles and their *in vitro* cytotoxicity to human cells. FEMS Microbiol Lett 2013; 345(2): 127-31.
[http://dx.doi.org/10.1111/1574-6968.12195] [PMID: 23746354]

[52]   Padzik M, Hendiger EB, Chomicz L, *et al.* Tannic acid-modified silver nanoparticles as a novel therapeutic agent against *Acanthamoeba.* Parasitol Res 2018; 117(11): 3519-25.
[http://dx.doi.org/10.1007/s00436-018-6049-6] [PMID: 30112674]

[53]   Mahboob T, Nawaz M, Tian-Chye T, Samudi C, Wiart C, Nissapatorn V. Preparation of poly (dl-lactide-co-glycolide) nanoparticles encapsulated with periglaucine A and betulinic acid for *in vitro* anti-*Acanthamoeba* and cytotoxicity activities. Pathogens 2018; 7(3): 62.
[http://dx.doi.org/10.3390/pathogens7030062] [PMID: 30012991]

[54]   Anwar A, Abdalla SAO, Aslam Z, Shah MR, Siddiqui R, Khan NA. Oleic acid–conjugated silver nanoparticles as efficient antiamoebic agent against *Acanthamoeba castellanii.* Parasitol Res 2019; 118(7): 2295-304.
[http://dx.doi.org/10.1007/s00436-019-06329-3] [PMID: 31093751]

[55]   Anwar A, Siddiqui R, Shah MR, Khan NA. Gold nanoparticle conjugated cinnamic acid exhibit antiacanthamoebic and antibacterial properties. Antimicrob Agents Chemother 2018; 62(9): e00630-18.
[http://dx.doi.org/10.1128/AAC.00630-18] [PMID: 29967024]

[56]   Mahboob T, Nawaz M, de Lourdes Pereira M, *et al.* PLGA nanoparticles loaded with Gallic acid- a constituent of Leea indica against *Acanthamoeba triangularis.* Sci Rep 2020; 10(1): 8954.
[http://dx.doi.org/10.1038/s41598-020-65728-0] [PMID: 32488154]

[57]   Khan NS, Ahmad A, Hadi SM. Anti-oxidant, pro-oxidant properties of tannic acid and its binding to DNA. Chem Biol Interact 2000; 125(3): 177-89.
[http://dx.doi.org/10.1016/S0009-2797(00)00143-5] [PMID: 10731518]

[58]   Kusrini E, Sabira K, Hashim F, *et al.* Design, synthesis and antiamoebic activity of dysprosium-based nanoparticles using contact lenses as carriers against *Acanthamoeba* sp. Acta Ophthalmol 2021; 99(2): e178-88.
[http://dx.doi.org/10.1111/aos.14541] [PMID: 32701190]

[59]   Xu J, Xue Y, Hu G, *et al.* A comprehensive review on contact lens for ophthalmic drug delivery. J Control Release 2018; 281: 97-118.
[http://dx.doi.org/10.1016/j.jconrel.2018.05.020] [PMID: 29782944]

[60]   Padzik M, Hendiger EB, Żochowska A, *et al.* Evaluation of *in vitro* effect of selected contact lens solutions conjugated with nanoparticles in terms of preventive approach to public health risk generated by *Acanthamoeba* strains. Ann Agric Environ Med 2019; 26(1): 198-202.
[http://dx.doi.org/10.26444/aaem/105394] [PMID: 30922053]

[61]   Hendiger EB, Padzik M, Żochowska A, *et al.* Tannic acid-modified silver nanoparticles enhance the anti-*Acanthamoeba* activity of three multipurpose contact lens solutions without increasing their cytotoxicity. Parasit Vectors 2020; 13(1): 624.
[http://dx.doi.org/10.1186/s13071-020-04453-z] [PMID: 33353560]

[62]   Vergara-Duque D, Cifuentes-Yepes L, Hincapie-Riaño T, Clavijo-Acosta F, Juez-Castillo G, Valencia-Vidal B. Effect of silver nanoparticles on the morphology of Toxoplasma gondii and Salmonella braenderup. J Nanotech 2020; pp. 1-11.

[63]   El-Sayed NM, Ramadan ME, Masoud NG. A Step Forward towards Food Safety from Parasite Infective agents 2021.https://www.springer.com/gp/book/9783030506711.
[http://dx.doi.org/10.1007/978-3-030-50672-8_40]

[64]   Dubey JP. Toxoplasmosis of animals and humans. 2nd ed. Boca Raton: CRC Press Inc 2016; p. 313.
[http://dx.doi.org/10.1201/9781420092370]

[65]   Feustel SM, Meissner M, Liesenfeld O. *Toxoplasma gondii* and the blood-brain barrier. Virulence 2012; 3(2): 182-92.
[http://dx.doi.org/10.4161/viru.19004] [PMID: 22460645]

[66]   Harrell M, Carvounis PE. Current treatment of *toxoplasma* retinochoroiditis: an evidence-based review. J Ophthalmol 2014; 2014: 1-7.
[http://dx.doi.org/10.1155/2014/273506] [PMID: 25197557]

[67]   Machala L, Kodym P, Malý M, Geleneky M, Beran O, Jilich D. Toxoplasmosis in immunocompromised patients. Epidemiol Mikrobiol Imunol 2015; 64(2): 59-65.
[PMID: 26099608]

[68]   El-Sayed NM, Aly EM. *Toxoplasma gondii* infection can induce retinal DNA damage: an experimental study. Int J Ophthalmol 2014; 7(3): 431-6.
[PMID: 24967186]

[69]   El-Sayed NM, Ismail KA. Role of intracellular adhesion molecules-1 (ICAM-1) in the pathogenesis of toxoplasmic retinochoroiditis. Journal of Molecular Pathophysiology 2012; 1(1): 37-42.
[http://dx.doi.org/10.5455/jmp.20120307042400]

[70]   Nagineni CN, Detrick B, Hooks JJ. *Toxoplasma gondii* infection induces gene expression and secretion of interleukin 1(IL-1), IL-6, granulocyte-macrophage colony, and intracellular adhesion molecule-1 by human retinal pigment epithelial cells. Infect Immun 2000; 68(1): 407-10.
[http://dx.doi.org/10.1128/IAI.68.1.407-410.2000] [PMID: 10603418]

[71] El-Sayed NM, Abdel-Wahab MM, Kishik SM, Alhusseini NF. Do we need to screen Egyptian voluntary blood donors for toxoplasmosis? Asian Pac J Trop Dis 2016; 6(4): 260-4. [http://dx.doi.org/10.1016/S2222-1808(15)61027-1]

[72] Zhang N, Wang S, Wang D, *et al.* Seroprevalence of *Toxoplasma gondii* infection and risk factors in domestic sheep in Henan province, central China. Parasite 2016; 23: 53. [http://dx.doi.org/10.1051/parasite/2016064] [PMID: 27882868]

[73] Bruschi F, Castagna B. The serodiagnosis of parasitic infections. Parassitologia 2004; 46(1-2): 141-4. [PMID: 15305704]

[74] Garcia LS. Tissue protozoa: Toxoplasma gondii.Diagnostic Medical Parasitology. 4th ed. Am Soc Microbiol 2007; pp. 130-41.

[75] Liu Q, Wang ZD, Huang SY, Zhu XQ. Diagnosis of toxoplasmosis and typing of *Toxoplasma gondii*. Parasit Vectors 2015; 8(1): 292. [http://dx.doi.org/10.1186/s13071-015-0902-6] [PMID: 26017718]

[76] Fekkar A, Bodaghi B, Touafek F, Le Hoang P, Mazier D, Paris L. Comparison of immunoblotting, calculation of the Goldmann-Witmer coefficient, and real-time PCR using aqueous humor samples for diagnosis of ocular toxoplasmosis. J Clin Microbiol 2008; 46(6): 1965-7. [http://dx.doi.org/10.1128/JCM.01900-07] [PMID: 18400917]

[77] Greigert V, Di Foggia E, Filisetti D, *et al.* When biology supports clinical diagnosis: review of techniques to diagnose ocular toxoplasmosis. Br J Ophthalmol 2019; 103(7): 1008-12. [http://dx.doi.org/10.1136/bjophthalmol-2019-313884] [PMID: 31088793]

[78] Wang H, Lei C, Li J, Wu Z, Shen G, Yu R. A piezoelectric immunoagglutination assay for *Toxoplasma gondii* antibodies using gold nanoparticles. Biosens Bioelectron 2004; 19(7): 701-9. [http://dx.doi.org/10.1016/S0956-5663(03)00265-3] [PMID: 14709388]

[79] Jiang S, Hua E, Liang M, Liu B, Xie G. A novel immunosensor for detecting *toxoplasma gondii*-specific IgM based on goldmag nanoparticles and graphene sheets. Colloids Surf B Biointerfaces 2013; 101: 481-6. [http://dx.doi.org/10.1016/j.colsurfb.2012.07.021] [PMID: 23010058]

[80] Jiang W, Liu Y, Chen Y, *et al.* A novel dynamic flow immunochromatographic test (DFICT) using gold nanoparticles for the serological detection of *Toxoplasma gondii* infection in dogs and cats. Biosens Bioelectron 2015; 72: 133-9. [http://dx.doi.org/10.1016/j.bios.2015.04.035] [PMID: 25978441]

[81] Cao-Milán R, Liz-Marzán LM. Gold nanoparticle conjugates: recent advances toward clinical applications. Expert Opin Drug Deliv 2014; 11(5): 741-52. [http://dx.doi.org/10.1517/17425247.2014.891582] [PMID: 24559075]

[82] Wang H, Li J, Ding Y, Lei C, Shen G, Yu R. Novel immunoassay for *Toxoplasma gondii*-specific immunoglobulin G using a silica nanoparticle-based biomolecular immobilization method. Anal Chim Acta 2004; 501(1): 37-43. [http://dx.doi.org/10.1016/j.aca.2003.09.018] [PMID: 18761119]

[83] Yang H, Guo Q, He R, *et al.* A quick and parallel analytical method based on quantum dots labeling for ToRCH-related antibodies. Nanoscale Res Lett 2009; 4(12): 1469-74. [http://dx.doi.org/10.1007/s11671-009-9422-7] [PMID: 20652102]

[84] Khodadadi A, Madani R, Hoghooghi Rad N, Atyabi N. Development of Nano-ELISA method for serological diagnosis of toxoplasmosis in mice. Arch Razi Inst 2021; 75(4): 419-26. [PMID: 33403837]

[85] Miao H, Xu SC, Yang YQ, *et al. Toxoplasma gondii* DNA sensor based on a novel Ni-magnetic sensing probe. Adv Mat Res 2010; 152-153: 1510-3. [http://dx.doi.org/10.4028/www.scientific.net/AMR.152-153.1510]

[86] Xu S, Zhang C, He L, *et al.* DNA detection of *Toxoplasma gondii* with a magnetic molecular beacon probe via CdTe@Ni quantum dots as energy donor. J Nanomater 2013; 2013: 1-6.
[http://dx.doi.org/10.1155/2013/473703]

[87] Butler NJ, Furtado JM, Winthrop KL, Smith JR. Ocular toxoplasmosis II: clinical features, pathology and management. Clin Exp Ophthalmol 2013; 41(1): 95-108.
[http://dx.doi.org/10.1111/j.1442-9071.2012.02838.x] [PMID: 22712598]

[88] Holland GN, Lewis KG. An update on current practices in the management of ocular toxoplasmosis. Am J Ophthalmol 2002; 134(1): 102-14.
[http://dx.doi.org/10.1016/S0002-9394(02)01526-X] [PMID: 12095816]

[89] El-Sayed NM. Recent advances in the treatment of toxoplasmosis. Frontiers in Clinical Drug Research - Anti-Infectives, Atta-ur-Rahman (Ed). Bentham Science, Vol 6, 2020; pp. 127–164.
[http://dx.doi.org/10.2174/9789811425745120060006]

[90] Adeyemi OS, Murata Y, Sugi T, Kato K. Inorganic nanoparticles kill *Toxoplasma gondii* via changes in redox status and mitochondrial membrane potential. Int J Nanomedicine 2017; 12: 1647-61.
[http://dx.doi.org/10.2147/IJN.S122178] [PMID: 28280332]

[91] Costa IN, Ribeiro M, Silva Franco P, *et al.* Biogenic silver nanoparticles can control *Toxoplasma gondii* infection in both human trophoblast cells and villous explants. Front Microbiol 2021; 11: 623947.
[http://dx.doi.org/10.3389/fmicb.2020.623947] [PMID: 33552033]

[92] Anand N, Sehgal R, Kanwar RK, Dubey ML, Vasishta RK, Kanwar JR. Oral administration of encapsulated bovine lactoferrin protein nanocapsules against intracellular parasite *Toxoplasma gondii*. Int J Nanomedicine 2015; 10: 6355-69.
[PMID: 26504384]

[93] Teimouri A, Jafarpour Azami S, Keshavarz H, *et al.* Anti-*Toxoplasma* activity of various molecular weights and concentrations of chitosan nanoparticles on tachyzoites of RH strain. Int J Nanomedicine 2018; 13: 1341-51.
[http://dx.doi.org/10.2147/IJN.S158736] [PMID: 29563791]

[94] Etewa SE, El-Maaty DAA, Hamza RS, *et al.* Assessment of spiramycin-loaded chitosan nanoparticles treatment on acute and chronic toxoplasmosis in mice. J Parasit Dis 2018; 42(1): 102-13.
[http://dx.doi.org/10.1007/s12639-017-0973-8] [PMID: 29491568]

[95] Pissuwan D, Valenzuela SM, Miller CM, Cortie MB. A golden bullet? Selective targeting of *Toxoplasma gondii* tachyzoites using antibody-functionalized gold nanorods. Nano Lett 2007; 7(12): 3808-12.
[http://dx.doi.org/10.1021/nl072377+] [PMID: 18034505]

[96] Pissuwan D, Valenzuela SM, Miller CM, Killingsworth MC, Cortie MB. Destruction and control of *Toxoplasma gondii* tachyzoites using gold nanosphere/antibody conjugates. Small 2009; 5(9): 1030-4.
[http://dx.doi.org/10.1002/smll.200801018] [PMID: 19291731]

[97] Leyke S, Köhler-Sokolowska W, Paulke BR, Presber W. Effects of nanoparticles in cells infected by *Toxoplasma gondii*. E-Polymers 2012; 12(1): 647-63.
[http://dx.doi.org/10.1515/epoly.2012.12.1.647]

[98] Pissinate K, dos Santos Martins-Duarte É, Schaffazick SR, *et al.* Pyrimethamine-loaded lipid-core nanocapsules to improve drug efficacy for the treatment of toxoplasmosis. Parasitol Res 2014; 113(2): 555-64.
[http://dx.doi.org/10.1007/s00436-013-3715-6] [PMID: 24292545]

[99] Tachibana H, Yoshihara E, Kaneda Y, Nakae T. Protection of *Toxoplasma gondii*-infected mice by stearylamine-bearing liposomes. J Parasitol 1990; 76(3): 352-5.
[http://dx.doi.org/10.2307/3282665] [PMID: 2352064]

[100] Si K, Wei L, Yu X, *et al.* Effects of (+)-usnic acid and (+)-usnic acid-liposome on *Toxoplasma gondii*.

Exp Parasitol 2016; 166: 68-74.
[http://dx.doi.org/10.1016/j.exppara.2016.03.021] [PMID: 27004468]

[101] Vandhana S, Deepa PR, Aparna G, Jayanthi U, Krishnakumar S. Evaluation of suitable solvents for testing the anti-proliferative activity of triclosan - a hydrophobic drug in cell culture. Indian J Biochem Biophys 2010; 47(3): 166-71.
[PMID: 20653288]

[102] El-Zawawy LA, El-Said D, Mossallam SF, Ramadan HS, Younis SS. 2015.Preventive prospective of triclosan and triclosan-liposomal nanoparticles against experimental infection with a cystogenic ME49 strain of Toxoplasma gondii. Acta Trop 2015; 141: 103–111.

[103] El-Zawawy LA, El-Said D, Mossallam SF, Ramadan HS, Younis SS. Triclosan and triclosan-loaded liposomal nanoparticles in the treatment of acute experimental toxoplasmosis. Exp Parasitol 2015; 149: 54-64.
[http://dx.doi.org/10.1016/j.exppara.2014.12.007] [PMID: 25499511]

[104] Dunay IR, Heimesaat MM, Bushrab FN, *et al.* Atovaquone maintenance therapy prevents reactivation of toxoplasmic encephalitis in a murine model of reactivated toxoplasmosis. Antimicrob Agents Chemother 2004; 48(12): 4848-54.
[http://dx.doi.org/10.1128/AAC.48.12.4848-4854.2004] [PMID: 15561866]

[105] Shubar HM, Lachenmaier S, Heimesaat MM, *et al.* SDS-coated atovaquone nanosuspensions show improved therapeutic efficacy against experimental acquired and reactivated toxoplasmosis by improving passage of gastrointestinal and blood–brain barriers. J Drug Target 2011; 19(2): 114-24.
[http://dx.doi.org/10.3109/10611861003733995] [PMID: 20367080]

[106] Sordet F, Aumjaud Y, Fessi H, Derouin F. Assessment of the activity of atovaquone-loaded nanocapsules in the treatment of acute and chronic murine toxoplasmosis. Parasite 1998; 5(3): 223-9.
[http://dx.doi.org/10.1051/parasite/1998053223] [PMID: 9772721]

[107] Prieto MJ, Bacigalupe D, Pardini O, *et al.* Nanomolar cationic dendrimeric sulfadiazine as potential antitoxoplasmic agent. Int J Pharm 2006; 326(1-2): 160-8.
[http://dx.doi.org/10.1016/j.ijpharm.2006.05.068] [PMID: 16920292]

[108] Liu Q, Singla LD, Zhou H. Vaccines against *Toxoplasma gondii* : Status, challenges and future directions. Hum Vaccin Immunother 2012; 8(9): 1305-8.
[http://dx.doi.org/10.4161/hv.21006] [PMID: 22906945]

[109] El Bissati K, Zhou Y, Dasgupta D, *et al.* Effectiveness of a novel immunogenic nanoparticle platform for *Toxoplasma* peptide vaccine in HLA transgenic mice. Vaccine 2014; 32(26): 3243-8.
[http://dx.doi.org/10.1016/j.vaccine.2014.03.092] [PMID: 24736000]

[110] Dimier-Poisson I, Carpentier R, N'Guyen TTL, Dahmani F, Ducournau C, Betbeder D. Porous nanoparticles as delivery system of complex antigens for an effective vaccine against acute and chronic *Toxoplasma gondii* infection. Biomaterials 2015; 50: 164-75.
[http://dx.doi.org/10.1016/j.biomaterials.2015.01.056] [PMID: 25736506]

[111] Ducournau C, Moiré N, Carpentier R, *et al.* Effective nanoparticle-based nasal vaccine against latent and congenital toxoplasmosis in sheep. Front Immunol 2020; 11: 2183.
[http://dx.doi.org/10.3389/fimmu.2020.02183] [PMID: 33013917]

[112] Tanaka S, Kuroda Y, Ihara F, *et al.* Vaccination with profilin encapsulated in oligomannose-coated liposomes induces significant protective immunity against *Toxoplasma gondii.* Vaccine 2014; 32(16): 1781-5.
[http://dx.doi.org/10.1016/j.vaccine.2014.01.095] [PMID: 24530937]

[113] Chen R, Lu S, Tong Q, *et al.* Protective effect of DNA-mediated immunization with liposome-encapsulated GRA4 against infection of *Toxoplasma gondii.* J Zhejiang Univ Sci B 2009; 10(7): 512-21.
[http://dx.doi.org/10.1631/jzus.B0820300] [PMID: 19585669]

[114] Chen H, Chen G, Zheng H, Guo H. Induction of immune responses in mice by vaccination with Liposome-entrapped DNA complexes encoding *Toxoplasma gondii* SAG1 and ROP1 genes. Chin Med J (Engl) 2003; 116(10): 1561-6.
[PMID: 14570624]

[115] El Hajj H, Lebrun M, Arold ST, Vial H, Labesse G, Dubremetz JF. ROP18 is a rhoptry kinase controlling the intracellular proliferation of *Toxoplasma gondii*. PLoS Pathog 2007; 3(2): e14.
[http://dx.doi.org/10.1371/journal.ppat.0030014] [PMID: 17305424]

[116] Nabi H, Rashid I, Ahmad N, *et al*. Induction of specific humoral immune response in mice immunized with ROP18 nanospheres from *Toxoplasma gondii*. Parasitol Res 2017; 116(1): 359-70.
[http://dx.doi.org/10.1007/s00436-016-5298-5] [PMID: 27785602]

[117] Rahimi MT, Sarvi S, Sharif M, *et al*. Immunological evaluation of a DNA cocktail vaccine with co-delivery of calcium phosphate nanoparticles (CaPNs) against the *Toxoplasma gondii* RH strain in BALB/c mice. Parasitol Res 2017; 116(2): 609-16.
[http://dx.doi.org/10.1007/s00436-016-5325-6] [PMID: 27909791]

[118] Hiszczyńska-Sawicka E, Akhtar M, Kay GW, *et al*. The immune responses of sheep after DNA immunization with, *Toxoplasma gondii* MAG1 antigen—with and without co-expression of ovine interleukin 6. Vet Immunol Immunopathol 2010; 136(3-4): 324-9.
[http://dx.doi.org/10.1016/j.vetimm.2010.03.018] [PMID: 20409592]

[119] Hiszczyńska-Sawicka E, Olędzka G, Holec-Gąsior L, *et al*. Evaluation of immune responses in sheep induced by DNA immunization with genes encoding GRA1, GRA4, GRA6 and GRA7 antigens of *Toxoplasma gondii*. Vet Parasitol 2011; 177(3-4): 281-9.
[http://dx.doi.org/10.1016/j.vetpar.2010.11.047] [PMID: 21251760]

[120] Hiszczyńska-Sawicka E, Li H, Xu J, *et al*. Induction of immune responses in sheep by vaccination with liposome-entrapped DNA complexes encoding *Toxoplasma gondii* MIC3 gene. Pol J Vet Sci 2012; 15(1): 3-9.
[http://dx.doi.org/10.2478/v10181-011-0107-7] [PMID: 22708351]

[121] El-Sayed NM, Ramadan ME. Toxocariasis in children: an update on clinical manifestations, diagnosis and treatment. J Pediatr Infect Dis 2017; 12(4): 222-7.
[http://dx.doi.org/10.1055/s-0037-1603496]

[122] El-Sayed NM, Masoud NG. Ocular toxocariasis: a neglected parasitic disease in Egypt. Bull Natl Res Cent 2019; 43(1): 146.
[http://dx.doi.org/10.1186/s42269-019-0185-8]

[123] Arevalo JF, Espinoza JV, Arevalo FA. Ocular Toxocariasis. J Pediatr Ophthalmol Strabismus 2013; 50(2): 76-86.
[http://dx.doi.org/10.3928/01913913-20120821-01] [PMID: 22938514]

[124] Cortez RT, Ramirez G, Collet L, Giuliari GP. Ocular parasitic diseases: a review on toxocariasis and diffuse unilateral subacute neuroretinitis. J Pediatr Ophthalmol Strabismus 2011; 48(4): 204-12.
[http://dx.doi.org/10.3928/01913913-20100719-02] [PMID: 20669882]

[125] Ahn SJ, Ryoo NK, Woo SJ. Ocular toxocariasis: clinical features, diagnosis, treatment, and prevention. Asia Pac Allergy 2014; 4(3): 134-41.
[http://dx.doi.org/10.5415/apallergy.2014.4.3.134] [PMID: 25097848]

[126] Magnaval JF, Malard L, Morassin B, Fabre R. Immunodiagnosis of ocular toxocariasis using Western-blot for the detection of specific anti- *Toxocara* IgG and CAP™ for the measurement of specific anti- *Toxocara* IgE. J Helminthol 2002; 76(4): 335-9.
[http://dx.doi.org/10.1079/JOH2002143] [PMID: 12498639]

[127] Voskuil JLA. Commercial antibodies and their validation. F1000 Res 2014; 3: 232.
[http://dx.doi.org/10.12688/f1000research.4966.1] [PMID: 25324967]

[128] De Genst E, Silence K, Decanniere K, *et al*. Molecular basis for the preferential cleft recognition by

dromedary heavy-chain antibodies. Proc Natl Acad Sci USA 2006; 103(12): 4586-91.
[http://dx.doi.org/10.1073/pnas.0505379103] [PMID: 16537393]

[129] Schumacher D, Helma J, Schneider AFL, Leonhardt H, Hackenberger CPR. Nanobodies: chemical functionalization strategies and intracellular applications. Angew Chem Int Ed 2018; 57(9): 2314-33.
[http://dx.doi.org/10.1002/anie.201708459] [PMID: 28913971]

[130] Morales-Yanez FJ, Sariego I, Vincke C, Hassanzadeh-Ghassabeh G, Polman K, Muyldermans S. An innovative approach in the detection of *Toxocara canis* excretory/secretory antigens using specific nanobodies. Int J Parasitol 2019; 49(8): 635-45.
[http://dx.doi.org/10.1016/j.ijpara.2019.03.004] [PMID: 31150611]

[131] Morales-Yánez F, Trashin S, Sariego I, *et al.* Electrochemical detection of *Toxocara canis* excretory-secretory antigens in children from rural communities in Esmeraldas Province, Ecuador: association between active infection and high eosinophilia. Parasit Vectors 2020; 13(1): 245.
[http://dx.doi.org/10.1186/s13071-020-04113-2] [PMID: 32398157]

[132] Jofre CF, Regiart M, Fernández-Baldo MA, Bertotti M, Raba J, Messina GA. Electrochemical microfluidic immunosensor based on TES-AuNPs@Fe$_3$O$_4$ and CMK-8 for IgG anti-Toxocara canis determination. Anal Chim Acta 2020; 1096: 120-9.
[http://dx.doi.org/10.1016/j.aca.2019.10.040] [PMID: 31883578]

[133] Medawar V, Messina GA, Fernández-Baldo M, Raba J, Pereira SV. Fluorescent immunosensor using AP-SNs and QDs for quantitation of IgG anti-*Toxocara canis*. Microchem J 2017; 130: 436-41.
[http://dx.doi.org/10.1016/j.microc.2016.10.027]

[134] Schneier AJ, Durand ML. Ocular Toxocariasis. Int Ophthalmol Clin 2011; 51(4): 135-44.
[http://dx.doi.org/10.1097/IIO.0b013e31822d6a5a] [PMID: 21897146]

[135] Choi GY, Yang HW, Cho SH, *et al.* Acute drug-induced hepatitis caused by albendazole. J Korean Med Sci 2008; 23(5): 903-5.
[http://dx.doi.org/10.3346/jkms.2008.23.5.903] [PMID: 18955802]

[136] Marin Zuluaga JI, Marin Castro AE, Perez Cadavid JC, Restrepo Gutierrez JC. Albendazole-induced granulomatous hepatitis: a case report. J Med Case Reports 2013; 7(1): 201.
[http://dx.doi.org/10.1186/1752-1947-7-201] [PMID: 23889970]

[137] Kudtarkar A, Shinde U, Bharkad GP, Singh K. Solid lipid nanoparticles of albendazole for treatment of *Toxocara Canis* infection: In-vivo efficacy studies. Nanosci Nanotechnol Asia 2017; 7(1): 80-91.
[http://dx.doi.org/10.2174/2210681206666160726164457]

[138] Velebný S, Hrčková G, Tomašovičová O. *Toxocara canis* in mice: effect of stabilised liposomes on the larvicidal efficacy of fenbendazole and albendazole. Helminthol 2000; 37(4): 195-8.

[139] Malheiro A, Aníbal FF, Martins-Filho OA, *et al.* pcDNA-IL-12 vaccination blocks eosinophilic inflammation but not airway hyperresponsiveness following murine *Toxocara canis* infection. Vaccine 2008; 26(3): 305-15.
[http://dx.doi.org/10.1016/j.vaccine.2007.11.023] [PMID: 18083279]

CHAPTER 5

Anticancer Delivery: Nanocarriers and Nanodrugs

Hatice Feyzan Ay[1], Zeynep Karavelioglu[1,*], Rabia Yilmaz-Ozturk[1], Hilal Calik[1] and Rabia Cakir-Koc[1,2]

[1] *Department of Bioengineering, Faculty of Chemical and Metallurgical Engineering, Yıldız Technical University, Istanbul, Turkey*

[2] *Turkish Biotechnology Institute, Health Institutes of Turkey (TUSEB), Istanbul, Turkey*

Abstract: Cancer is a disease in which cells grow uncontrollably and spread to different tissues. Existing treatment methods developed for cancer do not allow this disease to be completely cured, and these methods have various side effects. The search for effective cancer treatment has encouraged scientists to produce new ideas with nanotechnological methods. With the help of nanotechnological methods, which are becoming more popular day by day, the material is reduced to nano size, where it shows quantum effect, and gains unique physicochemical, mechanical, and biological properties. Thanks to the large surface area of the nanocarriers, more drug loading can be achieved on the unit surface, and their easy modification procedures enable these materials to be conjugated with biological molecules to become more specific structures. Due to the several advantages of nanocarriers, such as different synthesis methods, being open to modification, and relatively easy production, these materials can provide effective delivery of cancer drugs and even increase their efficacy. Moreover, there are also many nanodrugs approved for different routes of administration. Thanks to all these features, nanocarriers are promising ways to develop new drug formulations for cancer treatment. In this chapter, the anticancer activity of nanocarriers synthesized by different methods is clarified. Besides, the effects of the nanocarriers on different types of cancer, the targeting strategies of nanocarriers, and the effects of their size, surface charge, and shape, on their anticancer activity are summarized.

Keywords: Active targeting, Anticancer effect, Antitumor effect, Cancer, Cancer cells, Cancer treatment, Carbon nanotubes, Chemotherapy, Drug delivery, Graphene, Lipid nanocarriers, Magnetic nanoparticles, Metallic nanoparticles, Nanocarriers, Nanodrugs, Nanoparticles, Nanotechnology, Passive targeting, Polymeric nanocarriers, Targeted delivery.

* **Corresponding author Zeynep Karavelioglu:** Department of Bioengineering, Faculty of Chemical and Metallurgical Engineering, Yıldız Technical University, Istanbul, Turkey; E-mail: zeynep.karaveli@gmail.com

Felipe López-Saucedo (Ed.)

INTRODUCTION

Cancer, a disease characterized by uncontrolled cell proliferation, replicative immortality, and cell death resistance, is the second cause of death worldwide [1]. According to the estimates of the GLOBOCAN 2020 represented by the International Agency for Research on Cancer (IARC), in 2020, there were 19,3 million new diagnosed cases and 10 million cancer-related deaths worldwide [2].

Surgical operations, radiation therapy, and chemotherapy are conventionally used to treat cancer, but these methods have side effects on healthy cells [3, 4]. The action mechanism of radiotherapy is based upon the destruction of cancer cells by DNA damage. It is desirable for healthy tissues to be unharmed or have minimal damage; however, healthy cells can be exposed to radiation doses, albeit at low levels, and this causes damage [4].

Cancer immunotherapy, which is an alternative to these conventional therapy methods, is based on the principle of stimulating the immune system to recognize, target, and destroy cancer cells [5]. Nevertheless, the usage of this therapy is limited due to problems, such as the possibility of immune-related side effects and poor specificity in tumor cell targeting [6].

Another strategy used for cancer treatment is gene therapy. Gene therapy is an approach that enables a gene or gene product to be selectively delivered to a specific cell or tissue with minimum toxicity. Using this therapy, the expression of the defective gene can be blocked, or a normal gene expression can be achieved [7]. Gene therapy depends on gene delivery vectors, and they are classified into two categories according to vectors: viral vectors and non-viral vectors. The drawbacks of viral vectors, such as limited cloning capacity, complicated production, immune response triggering potential, and the risk of insertional mutagenesis, have led researchers to develop non-viral systems [7]. The non-viral delivery systems are less efficient than viral vectors but have advantages, such as flexibility and safety. The recent advancements in non-viral gene therapy are based on nanoparticle technologies [8].

In addition to these treatments, photodynamic therapy (PDT) is an important approach based on the destruction of tumors by the generation of singlet oxygen and reactive oxygen species (ROS) in cancer cells. Although PDT provides many advantages in tumor targeting and minimal toxicity, some challenges limit its utilization [9].

Chemotherapy is one of the most common methods used against cancer. Chemotherapeutic agents, such as etoposide (ETO), docetaxel (DTX), doxorubicin (DOX), cisplatin, and paclitaxel (PTX), have shown remarkable potential in clinical studies and have enhanced the survival rate of cancer patients. However, cancer cells may develop drug resistance against chemotherapeutic agents, reduce drugs' effectiveness and make treatment less effective or ineffective [10, 11]. Drug resistance causes a reduction in a drug's efficacy and potency to deliver therapeutic benefits, and thus it poses a significant barrier to disease treatment and patient survival [12]. Also, these chemotherapeutic agents can cause some serious side effects. Generally, chemotherapy is not specific to cancer cells, so it can damage healthy tissues. Therefore, it has dose-limiting effects, and as a result, the required critical dose cannot be reached, and the effectiveness of the drug may decrease [13]. Also, low water solubility, lack of stability, rapid metabolism, unfavorable pharmacokinetic properties, and non-selective drug distribution are other disadvantages of many chemotherapeutics [14].

Recently, nano-based drug delivery systems have become one of the most promising strategies to eliminate the side effects of chemotherapeutics and to increase the effectiveness of existing cancer treatment methods [15]. Nanoparticles (NPs) are defined as materials ranging from 1 to 1000 nm in length or 1 to 100 nm in diameter, but particles less than 200 nm in size are preferred for nanomedicine [16, 17]. Since most anticancer drugs are hydrophobic, they have some drawbacks, such as low solubility and weak bioavailability [15]. Nanosizing of drugs overcomes these drawbacks and provides benefits, such as increasing targeting ability, drug stability, and dissolution rate, besides reducing toxicity, drug resistance, and the required dose [18]. They have many advantages compared to free drugs due to their properties, such as continuous and slow release, increased half-life of drugs, and ensuring effective dose at low concentrations [13, 19]. These advantages make nanocarriers promising vehicles for anticancer drug delivery.

In this chapter, the features and targeting strategies of nanocarriers, the properties of the polymer-based, lipid-based, and inorganic NPs and their anticancer research in the literature, and finally, nano-drugs approved by various administrations for cancer treatment will be summarized.

NANOCARRIERS FOR ANTICANCER DRUG DELIVERY

Nanocarrier vehicles, which have an important place for anticancer studies, can effectively reduce the toxic effect of chemotherapeutics by modifying their pharmacokinetic properties, and thus alleviate the therapeutic dosage limitation [20]. In order to reduce the disadvantages of free drugs and increase their effectiveness, nano formulations of chemotherapeutic drugs have been developed.

Doxil®, Abraxane®, and Onivyde® are approved nanoformulations of chemotherapeutic drugs, such as DOX, PTX, and irinotecan [13].

Nanocarriers can be produced using many different synthesis methods and material types, and they can show anticancer efficacy *via* active and passive targeting strategies, as shown in Fig. (**1**). Nanocarriers have an important role in the production of nanoformulations of anticancer drugs due to their small size and the unique properties of the materials from which they are produced [21]. Therefore, the polymer-based, lipid-based, and inorganic nanocarriers provide important contributions due to the properties of their materials, the characteristics of additional substances, the modifications during their production, or their spontaneously possessed anticancer activities. In this section, these nanocarriers are explained in detail, and their applications in anticancer studies are discussed.

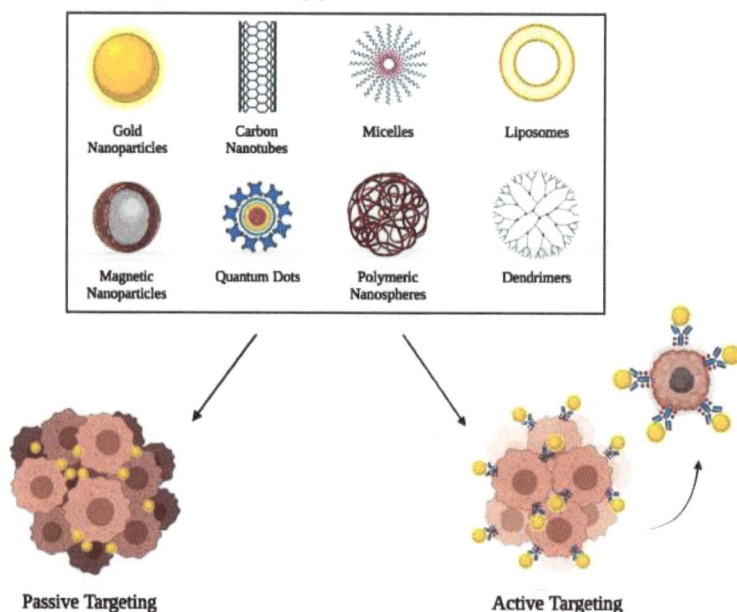

Fig. (1). Different types of nanocarriers and their cancer-targeting mechanisms. Created with Biorender.

Polymer-based NPs as Nanocarriers in Cancer Therapy

Polymer-based NPs can be easily synthesized in a wide variety of sizes by many different synthesis methods and are widely used as nanocarriers in cancer research. In general, it is stated that the size range of polymer-based nanocarriers varies between 1 and 1000 nm [22]. There are a small number of Food and Drug Administration (FDA)-approved formulations of polymer-based nanocarriers for clinical use. In addition to this, many polymer-based nanocarriers are currently being investigated in clinical trials [23].

The general advantages of polymer-based NP systems are the capability to control NP properties, increase the bioavailability of therapeutic agents, and improve their therapeutic index [23]. In addition, polymer-based NPs are drawing attention in cancer therapy because they are generally biodegradable, biocompatible, and their structures can be modified.

Polymer-based NPs can be classified as natural, synthetic, or semi-synthetic according to the type of polymer used, and nanocarriers can gain many different properties depending on the polymer. The most used natural polymers for the synthesis of polymer-based NPs are chitosan, alginate, hyaluronic acid (HA), albumin, gelatin, and dextran. On the other hand, poly(ethylene glycol) (PEG), poly(lactic acid) (PLA), poly(lactic-*co*-glycolic acid) (PLGA), poly(e-caprolactone) (PCL), polyglutamic acid (PGA), polycaprolactone (PCL), polyvinyl alcohol (PVA), poly(alkyl cyanoacrylate), thiolated poly (methacrylic acid), and poly(DL-lactide-*co*-glycolide) (PLGH) are common types of synthetic polymers [24, 25].

According to their preparation forms, polymer-based nanocarriers are classified into 7 groups as polymeric micelles, polymeric NPs, hydrogels, dendrimers, polymersomes, polymer-drug conjugates, and albumin-bound NPs. Depending on the preparation method of polymer-based NPs, drugs can be physically attached to the polymer matrix or covalently bound [26].

This section summarizes the features and applications of various polymer-based nanocarriers, primarily polymeric NPs, polymeric micelles, hydrogels, and dendrimers as potential anticancer nanocarriers.

Polymeric Nanoparticles

Among polymeric NPs, the colloidal NPs may carry therapeutic agents through loading, adsorbing, or in a conjugated system in the polymer matrix [27]. They can be synthesized with several methods listed, such as emulsification (solvent displacement or diffusion), nanoprecipitation, microfluidic techniques, and ionic gelation method. The sizes of NPs can be controlled according to the synthesis method. While polymeric NP sizes are defined as less than 100 nm in theory, polymeric NPs up to 300 nm size can be accepted as a drug carriers in studies [28]. In addition, it has been reported that polymeric NPs with sizes ranging from 100 to 200 nm have an enhanced permeability and retention (EPR) effect in tumor tissues [29].

Polymeric NPs have drawn attention as drug carriers in cancer research due to their many advantageous properties. One of the advantages of polymeric NPs is that polymeric NPs are readily modulated and have simple and controllable synthesis parameters. Modulation of features, such as chemical composition and surface charge of polymeric NPs, provides controlled-sustained drug release and increased loading capacity, circulation time, and half-life [22]. In addition, due to their modifiable structures, NPs can deliver drugs to target sites while increasing their uptake into the cell, thereby reducing the *via*bility of cancer cells. Martínez-Relimpio *et al.* developed folate-targeted NPs based on bovine serum albumin (BSA) and alginate (ALG) for PTX delivery *via* active targeting [30]. It was reported that folate receptor-targeted PTX-loaded BSA/ALG NPs showed greater intracellular uptake than PTX-loaded BSA/ALG NPs and cell *via*bility was significantly decreased on MCF-7 and MDA-MB-231 breast cancer cell lines and HeLa cervical cancer cells. As a result, the anticancer agent was effectively delivered to cancer cells, and the effectiveness of the anticancer agent was increased due to active targeting.

In addition to all these advantages, polymeric NPs increase the effectiveness of drugs while reducing their cytotoxic effects on normal cells and tissues [27]. In a study by Kucuksayan *et al.*, DOX and epoxomicin-loaded PLGA NPs showed less cytotoxic effects compared to DOX and epoxomicin on human umbilical vein endothelial cells while showing an increased anticancer effect on MCF-7 cancer cells [30].

In addition, polymeric NPs protect the encapsulated therapeutic agents from drug efflux pumps that cause drug resistance in cancer and accelerate their entry through cell membranes thanks to their nano size. For instance, Wu *et al.* developed multifunctional polymeric NPs encapsulating P-glycoprotein (P-gp) small interfering RNA (siRNA) and the positively charged aggregation-induced emission fluorogen (AIEgen) Py-TPE and conjugated PEG-*b*-poly(5-mthy--5-propargyl1,3-dioxan-2-one)-*g*-PTX (PMP), to form (Py-TPE/siRNA@PMP), as a delivery system against drug resistance in ovarian cancer. Results showed that Py-TPE/siRNA@PMP was a promising agent for the effective treatment of PTX-resistant cells and overcoming the multi drug-resistance in ovarian cancer [31].

These advantages of polymeric NPs have increased the importance of *in vitro*, *in vivo*, and clinical studies on their use as drug carriers in cancer therapy [26]. On the other hand, there are some disadvantages associated with the use of polymeric NPs. For example, toxicity may develop due to increased particle aggregation.

Besides, it is difficult to obtain monodisperse formulations during synthesis, there are limitations in industrial production, and the solvents that are used in production can exhibit toxic effects [22].

Polymeric Micelles

Polymeric micelles are smaller than 100 nm in size and consist of phospholipids and polymers. They form spontaneously in an aqueous solution and are frequently used for the delivery of hydrophobic drugs due to their amphiphilic properties as a consequence of their hydrophobic core and hydrophilic shell [24]. Hydrophobic core-forming polymers generally used in micelles are PLA, PLGA, PCL, poly(ethylene-*co*-propylene-*co*-ethylene oxide) (PEO-*b*-PPO-*b*-PPO), poly(ethylene-*co*-propylene oxide) (PEO-*b*-PPO), poly(D,L-lactide) (PDLLA), and poly(hydroxybutyrate) (PHB) [32]. Polymeric micelles can be combined through physical, chemical, or electrostatic interactions with therapeutic agents [33].

The critical micelle concentration (CMC) is defined as the concentration of amphiphilic substance or surfactant at which micelles begin to form [34]. The amphiphilic copolymer used for micelle formation provides better stability. For this reason, polymeric micelles are more preferred than conventional micelles. In addition, polymeric micelles provide a slow and controlled release, high drug loading capacity, and increased EPR effect compared to conventional micelles due to their controllable and versatile chemical compositions [22]. In a study published by González-Pastor *et al.*, two types of micelles based on Pluronic® F127 hybrid dendritic-linear-dendritic block copolymers modified with polyester or poly(esteramide) dendrons derived from 2,2'-bis(hydroxymethyl)propionic acid and 2,2'-bis(glycyloxymethyl)propionic acid were separately loaded with cisplatin (CDDP) and chloroquine. The encapsulation efficiency of chloroquine was found to be higher compared to cisplatin for both micelles. It has been reported that both micelles loaded with chloroquine showed a continuous release profile for 4 days, and the dual treatment of chloroquine-loaded micelles and CDDP exhibited a higher antitumor effect on A549 lung cancer cells while having good biocompatibility with normal cells [35]. Rapid renal excretion can be prevented by the size of the polymeric micelles; thus, high accumulation can occur in tumor tissues with the effect of EPR [36]. Gao *et al.* reported a study based on HA-e--zinc protoporphyrin micelles synthesized at 40 nm sizes and showed an antitumor effect on *in vitro* C26 mouse colon cancer cells and *in vivo* mouse sarcoma S180 solid tumor model by strongly accumulating in the tumor with the EPR effect [37].

Polymeric micelles can be separated in several different ways. Conventional polymeric micelles are formed by using natural or synthetic diblock or triblock copolymers to load various therapeutic agents/molecules [32]. Pluronic micelles are copolymers consisting of a central hydrophobic polyoxypropylene (poly(propylene oxide)) (PEO-*b*-PPO-*b*-PEO) chain surrounded by two hydrophilic polyoxyethylene (poly(ethylene oxide)) chains, and these are a group of micelle drug carriers generally 40 nm in size [34].

Other groups of polymeric micelles are pH-sensitive and acid-degradable polymeric micelles that release the loaded therapeutic substances into the environment by interacting with intracellular acidic compartments, such as lysosomes and endosomes [25]. In a study, Cavalcante *et al.* loaded DOX into two different pH-sensitive micelles based on distearyl phosphatidylethanolamine (DSPE), DSPE-PE and DSPE-PEG/OA, and then investigated their antitumor activity against 4T1 mouse breast cancer cells *in vitro* and 4T1 tumor-bearing BALB/c mice *in vivo* [38]. It was stated that PEG/OA6/DOX micelles showed tumor growth inhibition of 81.3% in tumor models, while free DOX and pH-insensitive micelles showed 50% and 38% tumor growth inhibition, respectively. The results revealed that the drug-releasing micelles exhibited high antitumor activity in response to acidic tumor pH.

The polymeric shells of the micelles prevent non-specific interactions with biological components. However, the uncertain tissue distribution of polymeric micelles can be mentioned as a disadvantage. Also, various polymer micelles can control the induction of the release of therapeutic agents into the environment, but sometimes polymeric micelles can release the active components easily and quickly. To eliminate this disadvantage, cross-linked biodegradable micelles are formed. Huang *et al.* developed an intelligent drug delivery system. These amphiphilic star copolymer cross-linked micelles are responsive to pH/redox stimuli for the controlled release of DOX [39]. It has been reported that these developed star copolymer cross-linked micelles had good stability and pH/reduction sensitivity. Also, while it released less DOX to normal tissues, rapid DOX release in tumor tissues and/or lower pH value conditions were observed.

Hydrogels

Hydrogels are three-dimensional reticulate structures formed by crosslinking polymer chains [22]. Hydrogels are generally considered nano-sized hydrogels when they are smaller than 200 nm [40]. These structures have a hydrophilic nature, adjustable physical and chemical properties, and are formed in a water-rich environment. Hydrogels can be prepared through reversible interactions, by self-assembly, by adding non-reversible covalent bonds, by irreversible chemical reactions, and by Ultraviolet (UV)/photopolymerization [33]. Polymeric materials are preferred for hydrogel formation since their molecular weights and also their physical and chemical properties can be adjusted, and they can develop a response mechanism against some stimuli [33]. Mechanical properties, water content, flexibility, and dynamic structures of natural polymers attract attention for the formation of hydrogels that structurally and biologically mimic the cellular environment [41]. However, the properties of natural polymers can be manipulated in a limited way, and natural polymers can perform limited degradation. Synthetic polymers, on the other hand, can increase biodegradation or allow the addition of chemical moieties that change the hydrogel property [41]. To take advantage of the good properties of natural polymers and synthetic polymers, their combinations are often used.

Since hydrogels are thermodynamically compatible with water, they show properties, such as swelling or gelling in aqueous environments [33]. The gelation process depends on the time and concentration of each component, and it can be triggered by several factors, such as pH, temperature, or light. The degree of swelling of hydrogels is defined by their amount of water uptake and polymer-water interactions. Hydrogels with hydrophilic groups show swelling caused by polymer-water interaction forces; ionic hydrogels show expansion due to the repulsion of charges [41].

The advantages of hydrogels can be listed as biocompatibility, biodegradability, drug loading ability, and controlled drug release [40]. Omtvedt *et al.* developed a hydrogel-based drug delivery system by loading PTX into alginate functionalized with beta-cyclodextrins (β-CyD). As a result, the final hydrogel structure showed a very slow release profile and had a cytotoxic effect on PC-3 prostate cancer cells compared to free PTX and unfunctionalized hydrogel [42].

Hydrogels attract attention in cancer treatment because they form stable complexes with nucleic acids and can easily cross the cell membrane. Hydrogel-based polymeric systems can extend the physical stability of chemotherapeutic agents or nucleic acids for months [22]. It has been shown that hydrogels have good stability due to being resistant to enzymatic degradation and temperature, and also cause tumor inhibition in many different types of cancer in many studies [43].

One of the most striking features of hydrogels is the ability to respond to a stimulus, such as degradation or conformational changes that provide a very high drug delivery efficiency for cancer treatments [40]. Hydrogels can be stimulated by several factors, such as temperature, light, pH, redox potential, magnetic field, and ultrasound [44]. In a study by Lu *et al.*, irinotecan (CPT-11) was loaded on epidermal growth factor receptor-targeted cetuximab (CET)-conjugated graphene oxide (GO) (GO-CET/CPT11) for pH-sensitive drug release for the treatment of glioblastoma [45]. The gene system used in the study was prepared together with GO-CET/CPT1, thermosensitive polymer chitosan-*g*-poly(*N*-isopropy lacrylamide) (CPN) obtained with stomatin-like protein 2 (SLP2) short hairpin RNA (shRNA) and CPN@GO-CET/CPT11. The @shRNA hydrogel was synthesized, and the drug and shRNA-loaded hydrogel exhibited a high antitumor activity while exhibiting minimal systemic toxicity.

The combined application of hydrogels with therapeutic agents for anticancer drug delivery is an effective treatment strategy to increase tumor inhibition in cancer treatments [24]. Xu *et al.* reported a study based on Arg-Trp-D-Arg-An-Arg-(D-Lys-D-Leu-D-Ala-D-Lys-D-Leu-D-Ala-D-Lys)$_2$ (RWrNR-kla) structure that was obtained by conjugation of apoptotic peptide kla and RWrNR ligand. Then, this structure was loaded to acid-sensitive nanogels (RWrNR-kla /poly--arginine/carboxymethyl chitosan) consisting of poly-L-arginine and carboxymethyl chitosan. They reported that RWrNR-kla/poly-L-arginine/carboxymethyl chitosan nanogels effectively stimulated apoptosis *in vivo* with regular cytosolic distribution and showed synergistic anticancer effect and effectively inhibited tumor growth with minimal adverse effects [46].

Dendrimers

Dendrimers are highly branched polymers with a three-dimensional structure that can be regulated in terms of mass, size, form, and surface chemistry, and show high structural homogeneity [47]. The most commonly used dendrimers for the delivery of cancer therapeutics are poly(ethyleneimine) (PEI), poly(propylene-imine) (PPI), poly-L-lysine (PLL), poly(amidoamine) (PAMAM), carbosilane dendrimers, phosphorus dendrimers, and glycodendrimers [48]. Active functional groups on the surface of dendrimers can bind to a surface by conjugation, while drugs can be loaded into the interior part. Dendrimers can release loaded substances as a response to certain stimuli in a controlled manner [48]. These stimuli can originate from the tumor environment, such as acid, enzyme, redox potentials, or external stimuli, such as light or temperature [49]. Hu *et al.* formed His-PAMAM-ss-PEG-Tf dendrimers by adding histidine (His) for increased pH and redox sensitivity and transferrin (Tf) for active tumor targeting and loading DOX for antitumor effect. According to the results, His-PAMAM-ss-PEG-Tf dendrimers showed pH and redox double sensitivity and higher cellular uptake and anticancer activity on HepG2 liver cancer cells than other control groups [50].

The advantageous properties of dendrimers make them remarkable nanocarriers in drug delivery in cancer treatments. These notable features are their nano sizes ranging from 1-20 nm, the controllability of the surface functional groups on their structures, high branching abilities, and high penetrating capacity into cells caused by their well-defined molecular weights [23]. Dendrimers can also be preferred due to their monodispersity, long therapeutic agent retention, high water solubility, and low side effects [22]. Alfei *et al.* loaded gallic acid (GA) into 2,2-bis(hydroxymethyl) propionic acid in dendrimer structures with a polyester-based inner matrix based on tree-like repetitions of units of 64 peripheral hydroxyl groups [51]. In the results of the study, they found the loading and encapsulation efficiencies of dendrimers to be 74.1% and 148.4%, respectively. Results showed that 42% of GA was sustainably released from the dendrimers for 24 hours, and the dendrimer structure increased the water solubility of GA by 6.1-8.5 times and reported that GA-loaded dendrimers showed high anticancer activity in neuroblastoma cells.

The encapsulation/conjugation of both hydrophilic and lipophilic therapeutic agents to dendrimers and their easy surface decoration draw attention for their use as carriers in cancer-specific targeted therapies [25, 26]. With these advantages, dendrimers have been preferred as carriers of genetic materials, such as nucleic acids and small molecules used in cancer treatments [47]. Hu *et al.* synthesized HA-conjugated PAMAM dendrimers to target CD44 receptors in cancer cells and loaded 4G/DNA and 5G/DNA into HA-PAMAM dendrimers for gene delivery to cancer cells. As a result, increased anticancer activity was shown on HeLa, Bel-7402, and HepG2 cancer cells, depending on increasing concentrations of PAMAM 4G/DNA and PAMAM 5G/DNA dendrimers [37].

The complex synthesis processes of dendrimers, high production costs caused by multi-step synthesis methods, and poor escape ability from immune cells are among the disadvantages of these nanocarrier systems [25]. In addition, the presence of amine groups in dendrimers limits the clinical applications of dendrimers. Cationic amine groups make dendrimers toxic, and therefore, the toxicity problem can be overcome by modification procedures, such as PEGylation or acetylation [36].

Polymer-Drug Conjugates

High molecular weight polymer-drug conjugates are obtained by direct coupling of the therapeutic agent to the ligands by covalent conjugation [22]. In general, polymers conjugated with therapeutic agents are PEG, PEI, PCL, PLA, PLGA, dextran, HA, and poly (acrylic acid) (PAA) [52]. Besides, drug conjugates that are based on PEG and *N*-(2-hydroxypropyl)methacrylamide (HPMA) polymers have been approved by the FDA [53]. Polymer-drug conjugates aim to improve drug delivery to the desired site without affecting its solubility, stability, or biodegradability [24]. In addition, direct conjugation of drugs on polymeric carriers enables more effective treatment *via* high drug loading capacity and increased blood circulation time due to filtration in the kidneys [52]. In the study of Arkaban *et al.*, $CoFe_2O_4$@PAA-FA-DOX NPs were obtained by conjugating

$CoFe_2O_4$@PAA NPs with folic acid (FA) and DOX for cancer treatment [54]. The encapsulation and loading efficiencies of NPs were found to be $80.00\pm5.3\%$ and $53.33\pm3.5\%$, respectively, and it was observed that DOX was released faster at pH 5.4 than at pH 7.4. Consequently, $CoFe_2O_4$@PAA-FA-DOX NPs have been reported to exert a more effective therapeutic effect compared to DOX and $CoFe_2O_4$@PAA-FA-DOX NPs on 4T1 cancer cells.

In many studies on polymer-drug conjugates, advantages, such as cytotoxic effects on cancer cells, a synergistic effect caused by the two agents, high selectivity towards cancer microenvironment, increased drug half-life, and long-term and controlled release of the drug, have been demonstrated [55]. Sun *et al.* conjugated PTX and gemcitabine (GEM) chemotherapeutics with sulfobetaine zwitterionic polymer to obtain a 46 nm zwitterionic polymer-drug conjugate (ZPDC) nanoformulation and reported that ZPDCs showed a high therapeutic effect with high cellular uptake and effective tumor accumulation on MIA PaCa-2 human pancreatic cancer cells *in vitro* [56]. They also stated that the ZPDCs were highly biocompatible and had a long blood circulation capacity. By triggering the covalent bonds formed between drug molecules and polymer in the presence of certain stimuli, such as pH, enzymes, temperature, and light, the polymer can release the drug only in the tumor tissue [57]. This causes the tumor tissue to be exposed to the most therapeutically effective concentration of the drug. Du *et al.* designed a degradable polymer-drug formulation based on polydopamine-chlorambucil conjugate NPs. This formulation was produced to be sensitive to intracellular pH and redox stimulation with high drug loading capacity for breast cancer treatment [58]. Results showed that polydopamine-chlorambucil conjugate NPs had 3.6 times longer blood circulation time than free chlorambucil and showed 4.1 times stronger selective tumor deposition with photothermal therapy than control.

Covalent linkers can be used to form a conjugate with a polymer and genetic materials, such as siRNA molecules, for efficient release. In a study by Wen *et al.*, they synthesized tumor tissue=sensitive HA-P-PEI/siRNA delivery system by conjugating matrix metalloproteinase-2 (MMP-2) with HA-PEI for PD-L1-siRNA delivery [59]. They stated that HA-P-PEI/siRNA NPs caused less cytotoxicity than PEI, showed a high cellular uptake and down-regulated PD-L1 expression on NCI-H1975 non-small cell lung cancer cells. Coating of polymer-drug conjugates with PEG is important because of its advantages, such as increasing surface area, recognition of some functional groups, and targeting cancer cells. Grigelotto *et al.* investigated PEG conjugates loaded with PTX and FA in two different (FA)/PEG ratios as polymer-drug conjugates for selective targeting that can cause therapeutic effects against cancer cells [60]. In the study, the cytotoxicity of polymer-drug conjugates was investigated on positive (HT-29) and negative

(HCT-15) FA receptor (FR)-cells, which are colon cancer cell lines, and it was reported that PTX-PEG-FA and PTX-PEG-(FA)3 conjugates showed 4 times and 28 times greater therapeutic activity in HT-29 cells, respectively. It was also reported in the study that both conjugates had a better pharmacokinetic profile in mice and effectively inhibited the migration and invasion of HT-29 cells.

In addition to all these advantages, the disadvantage is that polymer-drug conjugates are mostly passively targeted to tumor tissues. In addition, when compared to nanocarriers, polymer-drug conjugates require a synthesis method that includes more chemical processes due to their nature. The *in vivo* characterization of the stability, release, metabolism, excretion, and toxicity of polymer-drug conjugates may also be more difficult than other nanocarriers [57]. Besides, since polymer-drug conjugates are accepted by the FDA as a new type of drug, the toxicity and pharmacokinetics of their metabolites are being studied in detail.

Nanoparticle Albumin-bound

Albumin is a protein in the blood and circulation [22]. Albumin protein is also a natural carrier of endogenous hydrophobic molecules, such as vitamins and hormones [24]. Albumin is a stable, soluble, acidic, hydrophobic, non-toxic, and biodegradable protein [61].

Albumin-bound NPs (Nab) facilitate the transport and dilution of drugs. It shows high conjugation capacity due to its high binding availability and long half-life (19 days). Albumin specifically binds to two different proteins overexpressed in tumor tissues [61, 62]. The binding of albumin to the glycoprotein receptor gp60 in endothelial cells stimulates the activation of the intracellular protein caveolin-1 and the transcytosis of NPs around the cell membrane [62]. In addition, albumin can selectively bind to the extracellular matrix glycoprotein SPARC, which is not found in normal tissue but is over-expressed in tumor tissue [61].

Albumin has attracted attention as a carrier of hydrophobic chemotherapeutics, especially PTX and DTX, and can bind water-insoluble molecules in a reversible non-covalent manner. Cancer therapeutics can be covalently attached to albumin or can be encapsulated with albumin. The methods that enable it to be used as a nanocarrier by encapsulation can be summarized as desolvation, thermal gelation, nano-spray drying, emulsification, and self-assembly techniques [61]. The disadvantages of the albumin-bound NP group are that they can create immunogenicity and exhibit poor metabolic stability *in vivo*. In the study by Chen *et al.*, Nab-PTX-PAs were obtained by loading PTX palmitate into albumin NPs, and this formulation reduced toxicity compared to commercial Abraxane in normal organs, providing a sustained release of PTX in tumor tissue [63]. In addition, it has been reported that it exhibits better antitumor activity by inhibiting the proliferation of tumor cells more effectively in *in vitro* and *in vivo* 4T1 mouse breast cancer models.

Commercially available PTX is formulated with organic solvents, such as ethanol. New formulations with PTX have been sought to overcome the toxicity caused by the formulation with organic solvents and to improve clinical use. For these reasons, one of the formulations developed was Nab-PTX (nano-albumin-linked PTX) Abraxane® (Abraxis®), produced in 130 nm size. Abraxane®, which is a Nab-PTX formulation, was approved by The European Medicines Agency (EMA) and FDA for the treatment of metastatic breast cancer in 2005, metastatic non-small cell lung cancer in 2012, and metastatic pancreas patients in 2013, and it has been reported to be effective in patients with taxane resistance [62, 64].

Polymersomes

Polymersomes, one of the less-studied polymeric NPs, are self-assembled artificial vesicles with amphiphilic block copolymer membranes [23]. Few polymers, such as poly(ethylene glycol)-polycaprolactone (PEG-PCL) and poly(ethylene glycol)-polylactic acid (PEG-PLA) can be self-assemble as biodegradable polymers in nanocarrier systems have been reported [65]. They generally provide very high stability and substance retention efficiency while causing a local response. In a study, polypeptide-blocked poly(L-glutamic acid)-*block*-poly(L-phenylalanine) (PGlu-*b*-PPhe) polymersomes containing 117 L-glutamic acid and 165 L-phenylalanine were synthesized and their size was approximately 200-400 nm with varying parameters [66]. Cytotoxicity of these polymersomes was investigated on Caco-2 colon cancer cells and HEK-293 embryonic kidney cells. While the cytotoxic effect was observed on Caco-2 cells, no cytotoxic effect was observed in HEK-293 cells at any concentration value for 48 hours.

Polymersomes have significant advantages compared to liposomes, such as tunable physicochemical features, improved pharmacokinetic behaviors, and improved membrane stability to prevent drug release [67]. However, while hydrophobic substances can be loaded into the aqueous interior of polymersomes, only weak bases of the hydrophilic agents can be loaded into them [67]. In a study, PTX, a hydrophobic anticancer drug, and DOX, a hydrophilic anticancer drug, were co-encapsulated using the PEG-PLA (OL2) double-block copolymer and inert copolymer PEG-PBD (PEG-butadien; OB18) as the drug delivery system [68]. It has been reported that the polymersomes exhibited improved pharmacodynamics, accumulated in tumors, and caused double cell death in tumors than free drugs.

In the study of Colley *et al.*, pH-sensitive poly-2-(methacryloyloxy) ethyl phosphorylcholine (PMPC)-poly-2-(diisopropylamine) ethyl methacrylate (PDPA) polymersomes were loaded with PTX and DOX, and their effect was studied in normal oral cells and head and neck squamous cell carcinoma (HNSCC) in *in vitro* 2D and 3D cell cultures [69]. Obtained results showed that polymersomes loaded with PTX and DOX showed a significant synergistic effect and anticancer effect on oral head and neck squamous cancer cells compared to normal oral cells over 24 hours.

In a study, the anticancer drug DOX was loaded into poly(trimethylene carbonate)-b-poly(L-glutamic acid) (PTMC-*b*-PGA) block copolymer polymersomes by nanoprecipitation method [70]. Aqueous solutions of these polymersomes were stable for more than 6 months without nanostructure degradation or drug precipitation, and colloidally stable for at least 10 hours in serum. In the study, the loading content of DOX in PTMC-*b*-PGA vesicles was determined as 23% at pH 7.4 and 47% at pH 10.5%. Release kinetics of DOX-loaded PTMC-*b*-PGA vesicles was examined; when DOX was loaded into the vesicles at pH 7.4, 80% was released within 10 hours, and when DOX was loaded at pH 10.5, only 20% was released after 24 hours. Thus, it was shown that polymersomes improved the pharmacodynamic properties of the anticancer agents they encapsulated.

In another study, the efficacy of DOX-loaded polymersomes was examined against breast cancer, both hydrophilic DOX-HCl and hydrophobic DOX, were encapsulated into amphiphilic polyphosphazenes with high loading and encapsulation efficiency [65]. The therapeutic efficacy of these polymersomes was investigated on MCF-7 and MCF-7/Adr breast cancer cells *in vitro*, and *in vivo* experiments were also carried out. It was reported that they showed high efficacy by overcoming drug resistance in resistant breast cancer cells. In addition, it was stated that the toxicity of both DOX·HCl and hydrophobic DOX were significantly reduced with polymersomes while their therapeutic efficacy did not change.

Lipid-based Nanocarriers for Anticancer Drug Delivery

Lipids, which have self-convening properties, are used to create a wide range of colloidal drug carriers with various structures [71]. Lipids generally have a significant role in drug delivery, especially for drugs with poor permeability [72]. Lipid-based drug delivery systems have the potential to improve the bioavailability and solubility of drugs that are poorly water-soluble [73, 74]. Lipid-based nanocarriers are made from physiological lipids; therefore, they are well-tolerated and generally non-toxic in comparison to inorganic or polymeric

NPs [75, 76]. Lipid-based NPs offer properties including biodegradability, biocompatibility, high physical stability, high loading capacity, easy preparation, low cost of production, and easily large scaled up. They can deliver hydrophobic and hydrophilic drugs by encapsulating them and enhancing the drug action time due to their longer half-life and controlled drug release capacity [77, 78]. Various kinds of Lipid-based NPs have been developed for drug delivery, and the most common of these are liposomes, niosomes, solid lipid NPs, nanostructured lipid nanocarriers, lipid micelles, and nano-emulsions.

In this section, the anticancer activities of various lipid-based nanocarriers, including liposomes, niosomes, solid lipid NPs, nanostructured lipid nanocarriers, lipid micelles, and nanoemulsions, will be summarized.

Liposomes

One of the most effective nano-carriers among lipid-based NP systems is liposomes for anticancer drug delivery. Liposomes, first identified as phospholipid spherules by Dr. Bangham in 1961, were then accepted for use in chemotherapeutic drug delivery by Gregoriadis in 1974 [79]. Liposomes are closed spherical lipid bilayer vesicles occurring by self-assembling in the aqueous phase [80]. Liposomes have phospholipid layers, which consist of biocompatible, biodegradable, non-immunogenic, and non-toxic materials, and due to the amphiphilic nature of phospholipids, they are utilized for better targeting of both hydrophilic and hydrophobic drugs in drug delivery. They can also quickly move through vascular pores and accumulate in tumors due to their small particle size [81]. Liposomes have important advantages in drug delivery that include raising the drug solubility, better-targeting drug delivery, decreasing the toxic effect of drugs, and increasing the circulation half-life of active pharmaceutical ingredients. Moreover, liposomal drug formulations have the potential to enhance effectiveness and reduce the side effects of chemotherapeutic drugs [82].

Liposomes are generally formed from phospholipids and cholesterol [14], and can be zwitterionic, positively or negatively charged, or uncharged, depending on the charge of the polar head and type of phospholipids [83]. For instance, liposomes made from phosphatidylserine (PS) or phosphatidylglycerol (PG) have a negative net charge, whereas phosphatidylcholine (PC) or phosphatidylethanolamine (PE) have a neutral net charge [84].

Liposomes are produced using two different types of lipids. These lipids can be natural or synthetic double-chain polar lipids and sterols, such as cholesterol [83]. Cholesterol affects liposome size, permeability, and fluidity, resulting in modulating the release of hydrophilic molecules from liposomes [85]. The addition of cholesterol to liposomes enhances the stability of the liposomal membrane in biological fluids while reducing the fluidity of the liposomal membrane bilayer and the permeability of water-soluble molecules from the liposomal membrane. Also, apart from cholesterol, hydrophilic groups, such as PEG, can be conjugated to the surface of liposomes. PEG conjugation or PEGylation enables liposomes to circulate for extended periods and enhances liposome accumulation into the tumor and their circulation half-life [86].

Liposomal anticancer drugs are the first nano-based formulations to be approved by the FDA for cancer treatment. There are many liposomal drugs approved, including Doxil® (DOX), DaunoXome® (daunorubicin), and DepoCyt® (cytarabine) as shown in Table **1** [87]. The PEGylated liposomal-based formulation of Doxil® was approved by FDA for AIDS-related Kaposi's sarcoma, ovarian cancer, breast cancer, and multiple myeloma [88].

Liposomes are effective in reducing the toxic effects of the drugs in healthy tissues by delivering the drugs to the target site. Liposomes can be more targeted by various agents or stimuli. For instance, ligand-targeted liposomes, formed using a variety of ligands, including transferrin, mannose, folate, peptide, and antibody [89], have a higher uptake by the target tissues. Except for ligands, liposomes can also be targeted by developing them to be sensitive to certain stimuli. If the stimuli-responsive liposomes are not exposed to any stimulation at the target site, they do not release their cargo [90]. In a stimulus-sensitive system, stimuli, such as pH, temperature, enzymes, redox potential, and electrolyte concentration can induce the release of drugs [89]. pH-sensitive liposomes trigger drug release by destabilization of the lipid bilayer after reaching the cancer cells by increasing targeting [91].

Yang *et al.* (2017) reported that pH-sensitive PEGylated liposomes displayed more intracellular accumulation and higher anti-proliferative effect, in comparison to non-pH-sensitive liposomes and the free drug [92]. Pourhassan *et al.* (2017) reported therapeutic effects of the platinum drug encapsulated in the secretory phospholipase A2 (sPLA2)-sensitive liposomes. In the study, the sPLA2-sensitive liposomes were loaded with oxaliplatin (L-OHP) and evaluated for their antiproliferative activities against HT-29 and Colo 205 human colon carcinoma cell lines. Results showed that sPLA2-sensitive liposomal L-OHP formulation showed an extremely cytotoxic effect on cancer cells by inhibiting the cell growth of these cells [93]. In another study, PTX-loaded liposomes (Ptx-Glu6-FA-Lip) were modified with FA and glutamic hexapeptide (Glu6) ligands and evaluated for bone metastatic breast cancer treatment. The results showed that thanks to their high targeting capacity, PTX-Glu6-FA-Lip accumulated more in bone metastatic lesions compared to normal bone tissue. It was also noted that Ptx-Glu6-FA-Lip has been shown to be in a higher concentration in tumor tissue compared to PTX-Glu6-Lip. In conclusion, it has been reported that the PTX-Glu6-FA-Lip is effective in preventing the metastasis of bone cancer while reducing the toxic effect on healthy tissues [94]. In another study, curcumin was loaded into cationic liposomes and then complexed with the signal transducer and activator of transcription 3 (STAT3) siRNA and evaluated against the A431 epidermoid cancer cells. It was found that delivering a curcumin-loaded liposome-siRNA combination together inhibited cell proliferation more effectively compared to individual treatments of curcumin and free STAT3 siRNA [95].

Niosomes

Niosomes, also called nonionic surfactant vesicles, are produced by the self-assembly of cholesterol and nonionic surfactants, like alkyl-ether, esters, and amides [96]. Niosomes are spheroidal structures that can be in uni- or multilamellar form. Niosomes consist of nonionic surfactant bilayers in aqueous media. They have several advantageous properties, such as low cost, high chemical stability, and ease of production [97]. Niosomes increase drug availability at the target site by preventing the early release of the drug into circulation [98]. In addition to all these advantages, enhanced penetration into the skin, good adhesion ability, and long-term drug release are among the other advantages of niosomes. Niosomes are quite similar to liposomes. However, the bilayers of niosomes are composed of nonionic surfactants, which distinguishes them from liposomes. Compared to other vesicular systems, they exhibit better chemical stability, biodegradability, biocompatibility, and non-immunogenicity, and have lower toxicity, depending on the structure of the surfactant [96, 97].

In a study, Curcumin (C) and DOX (D) were encapsulated into the polyethylene glycolated niosomes (PEGNIO) prepared with DSPE-PEG (2000). PEGNIO/D-C forms were modified with the tumor-homing peptide tLyp-1. It was found that PEGNIO/D-C/t-Lyp-1 was more effective on U87 human glioblastoma cells in comparison to free D-C and PEGNIO/D-C [99]. In another study, Tymoquinone (TQ), TQ-loaded niosome (Nio/TQ), Carum, and Carum-loaded niosome (Nio/Carum) were compared for their efficiency on MCF-7 cancer cells. The antitumor effects of TQ, Nio/TQ, Carum, Nio/Carum, and blank niosome were investigated on cancer cells. It was shown that Nio/TQ exhibited a more antitumor activity than free TQ; similarly, Nio/Carum exhibited a more prominent antitumor activity compared to free Carum. Moreover, in TQ and Carum-loaded niosomes, the cancer cell migration was inhibited. Besides, Nio/TQ showed better antitumor activity and cell migration inhibition compared to Nio/Carum [100]. Rajput *et al.* developed multilamellar gold (Au) niosomes containing TQ and Akt-siRNA (siRNA-Nio-Au-TQ) and tested them on tamoxifen-resistant MCF-7 and T-47D breast cancer cells and BALB/c mice. When compared with free TQ and Nio-A--TQ, the siRNA-Nio-Au-TQ formulation has been reported to reduce the *via*bility of cancer cells. On the other hand, siRNA-Nio-Au-TQ decreased tumor volumes in the mice more than the other drugs [101]. In another study, niosomes were obtained with hydrophobic molecules, curcumin and letrozole, using Span family surfactants. Letrozole and curcumin were co-loaded into the niosomes in the presence (LC-FA-S80-10) and absence (LC-S80-10) of FA, and LC-FA-S80-10 exhibited higher toxicity in breast cancer cells. Also, it was found that the curcumin/letrozole co-loaded niosomes showed a high biocompatible profile on HEK-293 normal cells while inducing considerable apoptosis on cancer cells, compared to free drug molecules [102].

Solid Lipid Nanoparticles

In the early 1990s, solid lipid NPs (SLNs) were developed to overcome the constraints of nanocarriers, like liposomes and polymeric NPs [21]. The SLNs are safer and more effective in comparison to other NPs; also, since they are made from lipids, they are solid at room temperature and body temperature [80]. SLNs are submicron-sized (50 to 1000 nm) and less toxic due to their biocompatible compositions, and they consist of biodegradable materials that can be loaded with lipophilic and hydrophilic drugs [103, 104]. They are also ideal carrier systems for drugs with poor water solubility [105]. They have several advantageous properties, including small size, large surface area, and the capacity for high drug stacking [106]. Besides, high physical stability, controlled drug release, and low-cost scale-up are one of the advantages of SLNs [99]. Moreover, they can increase the antitumor activity in cancer cells while decreasing toxicity in normal cells [107].

SLNs are produced by entrapping a drug into a biocompatible lipid core and a surfactant at the external shell [105]. Common constituents used in the production of SLNs are lipids, emulsifiers, co-emulsifiers, and water [104]. SLNs can be prepared with a variety of lipids, including lipid acids, mono-, di-, or triglycerides, or glyceride mixtures, and can be stabilized with biocompatible surfactants chosen based on the desired ionic or nonionic properties. Lipids used in the preparation of SLNs are hard fats, glyceryl behenate, cetyl palmitate, triglycerides (such as tristearin, tripalmitin, and trilaurin), and lipid acids (such as stearic acid and palmitic acid). Also, emulsifiers, such as poloxamer 188, polysorbate 80, lecithin, and sodium glycocholate are used in formulations [108]. Surfactants with amphiphilic structures keep their hydrophobic groups away from water, while polar heads direct them towards the aqueous phase [109]. Amphiphilic molecules, such as monoacylglycerides of long-chain fatty acids, phospholipids, some esters, poloxamers, and polysorbates, can be used as surfactants. Besides, bile salts, such as taurodeoxycholate, or alcohols, such as butanol and ethanol, are used as co-surfactants to improve stability [110]. It has been suggested that some surfactants increase the degradation of SLNs, while others slow their enzymatic degradation by stabilizing SLNs [104].

There are many studies on the antitumor activities of drugs loaded by SLNs. Xu *et al.* reported that Taxotere®-loaded SLN (tSLN) prepared with galactosylated dioleoylphosphatidyl ethanolamine was found to be more cytotoxic than Taxotere® and non-targeted SLNs (nSLNs) against Bel-7402 hepatocellular carcinoma cells and had better antitumor activity on the murine model [111]. The study based on DOX-loaded SLNs demonstrated that *in vitro* anticancer efficiency of drug-loaded SLNs was higher than the free drug against MCF-7 breast cancer and HL-60 promyelocytic leukemia cells [112]. In the study on the HT-29 colorectal cancer cells, it was also found that DOX-loaded SLNs had a higher half-life than free drug, and the required dose to inhibit cell growth was lower [113]. Wang *et al.* stated that the combination of resveratrol-SLNs (Res-SLNs) showed a greater inhibition on MDA-MB-231 cells, and the IC50 value was lower for Res-SLNs compared to the free drug [114]. In another study on breast cancer, curcumin-SLNs (Cur-SLNs) were found to be cytotoxic against SKBR3 human breast cancer cells, and Cur-SLNs caused a higher apoptotic effect on cancer cells than free drug [115]. In another study, Tf-modified DOX and pEGFP loaded SLN (T-SLN/DE) were evaluated for anticancer activity and gene transfection efficacy. Results indicated that T-SLN/DE showed a better antitumor effect [116].

Nanostructured Lipid Nanocarriers

Nanostructured lipid nanocarriers (NLCs) are modified SLNs that consist of a lipidic phase containing both solid (fat) and liquid (oil) lipids, contrary to SLNs, which contain only solid fat. NLCs are produced by a solid lipid matrix composed of a combination of solid and liquid lipids and an aqueous phase that includes a surfactant mixture [117 - 119]. Since NLCs contain both solid and liquid lipids in their structure, this situation causes defections in the crystal structure that result in increased drug loading capacity [120]. NLCs have been created to overcome the disadvantages associated with SLN, since they have a better drug loading capacity and increased stability to prevent drug expulsion during storage. However, no significant difference was noticed among the biocompatibility of SLNs and NLCs [117, 121].

NLCs can be categorized as imperfect type, amorphous type, and multiple types based on the lipid mix composition. NLCs with the imperfect crystal type are produced by mixing a variety of liquid lipids with solid lipids, and as a result, imperfections occur in the crystal structure. These defects allow high drug loading into NLCs. The amorphous type NLCs are occurred by combining solid lipids with specific liquid lipids. The produced lipid matrix is solid and in an amorphous state. Also, because there is no crystallization, the expulsion of the drug is prevented [122]. Thus, drugs stay embedded in the amorphous matrix. This makes amorphous NLCs more advantageous than imperfect NLCs [123]. Multiple NLCs are oil-in-fat-in-water nanocarriers (o/f/w) that occur from a solid lipid matrix containing multiple liquid oil nano-compartments. The solid matrix prevents drug leakage, while the oil-based nano-compartments enhance the capacity of the drug loading [124].

In a study, HA-coated and PTX-loaded nanostructured lipid carriers (HA-NLCs) were evaluated for their cytotoxicity and antitumor activity against B16 melanoma, CT26 mouse colon, and HCT116 human colon cancer cells and B16-bearing Kunming mice model. Results showed that the IC50 value of HA-NLC was considerably lower than the IC50 value of PTX for three cancer cell lines. Moreover, tumor volume was considerably smaller in B16-bearing Kunming mice treated with HA-NLC; however, mice had lower body weights compared to treatment with PTX, indicating that HA-NLC had fewer side effects [125]. In another study, resveratrol-loaded NLCs (RSV-NCLs) were obtained using stearic acid as a solid lipid and oleic acid as a liquid lipid. For better targeting, RSV-NCLs were modified with FA (RSV-FA-NLCs). Results showed that RSV-F--NLCs had a smaller particle size and high anticancer effect compared to RSV-NLCs. Also, these two RSV-NCLs formulations exhibited a higher anti-proliferative effect than free RSV on MCF-7 and A549 cells [126]. In another study, gefitinib-NLC (NANOGEF) combination was prepared using stearic acid (solid lipid), sesame oil (liquid lipid), and sodium lauryl sulfate and tween 80 (surfactants). NANOGEF displayed a higher anticancer efficacy on HCT116 cells when compared to GEF alone [127]. Taratula *et al.* developed two NLCs loaded with DOX or PTX. They modified them with siRNA molecules that were targeted to MRP1 mRNA and BCL2 mRNA, respectively. Then, they conjugated them with luteinizing hormone-releasing hormone (LHRH). Compared to LHRH-dru--NLC and the free drug, the LHRH-drug-NLC-siRNA reduced the expression of both BCL2 and MRP1, and it induced cell death against lung cancer [128].

Lipid Micelles

Lipid micelles occur inside the aqueous media when the micellar concentration reaches above the critical level. These lipid micelles are self-assembled spherical structures inside a solvent [129]. Hence, lipid micelles are described as monolayer nanostructures of 7 - 35 nm in size that comprise PEG-conjugated phospholipids or lysolipids [130]. These micelles consist of a single micelle layer consisting of hydrophilic (polar) heads making an outer shell and a hydrophobic (non-polar) tail region. Thanks to the biocompatible and biodegradable properties of lipids, lipid micelles have low toxicity [129]. In addition, thanks to their hydrophobic/hydrophilic structure, their use as drug delivery systems to increase drug availability and increase the solubility of water-insoluble drugs attracts attention. However, their application poses some obstacles, such as limited chemical flexibility and structural instability [25, 131].

Phospholipid micelles are promising tools for improving the therapeutic effectiveness of anticancer drugs and for enhancing drug accumulation in tumor tissues due to their EPR effect and ligand-mediated active targeting properties [132]. PEG-PE complex creates lipid-core micelles instead of liposomes when their concentration exceeds a critical limit. PEG-PE micelles exhibit more stable characteristics due to hydrophobic interactions between double acyl chains of phospholipids, so the drug loading efficiency of PEG-PE micelles correlates with the hydrophobicity of a drug [133]. The micelles are typically prepared with PEGylated phospholipids for utilization in drug delivery. As another example of lipid micelles, the sterically stabilized phospholipid simple micelles (SSMs) that are effective in increasing the solubility of water-insoluble drugs have a central lipid core that allows hydrophobic compounds to be efficiently solubilized. Also, it is possible to increase the drug solubility by increasing the lipid concentration of SSMs. They are composed of the PEGylated phospholipid DSPE-PEG$_{2000}$ [130]. PEG-DSPE, also known as PEG-lipid micelles, is made from conjugates of PEG and DSPE. Also, more efficient uptake by tumor cells was reported for PEG-lipid micelles. Moreover, anticancer drugs loaded into PEG-DSPE micelles have been displayed to improve their biological activity against a variety of cancer cell lines [134].

Nanoemulsions

Nanoemulsions, which are also known as mini emulsions, ultrafine emulsions, and submicron emulsions, are oil-water emulsions that have oil-in-water or water-in-oil forms [135]. Nanoemulsion is a colloidal dispersion of two immiscible liquids that are thermodynamically unstable, consisting of a dispersed liquid phase and another liquid phase forming the dispersion medium [136]. They are solid spheres being amorphous, lipophilic, and negative-charged. Nanoemulsions improve the therapeutic effectiveness, physical stability, and bioavailability of drugs while reducing their adverse effects and toxicities [137]. In addition, they exhibit advantages, such as dissolving hydrophobic substances and providing protection in cases of hydrolysis and enzymatic degradation. Moreover, they have advantages over large particle-sized emulsions due to a large surface area and free energy [135].

Nanoemulsions include two immiscible liquids and emulsifying agents, such as surfactant or co-surfactant [137]. The oil phase may include triglycerides, free fatty acids, or vegetable oils and mineral oils. The oils are generally selected depending on drug solubility, so oil phases that provide high drug loading are preferred. Commonly used surfactants are spans, tweens, polysaccharides, phospholipids, and amphiphilic proteins. Co-surfactants or co-solvents are used in combination with a surfactant because nanoemulsion production requires extra-low negative interfacial tension. PEG, propylene glycol, ethanol, ethylene glycol, glycerin, and propanol are generally used as co-surfactants or co-solvents [15, 138].

There are three different types of nanoemulsions according to their types of production: oil in water nanoemulsion in which oil is spread in the continuous aqueous phase (o/w), water in oil nanoemulsion in which water is spread in the continuous oil phase (w/o), and bi-continuous (multiple) nanoemulsions [137]. The solubility of the surfactant determines the type of nanoemulsion; if the surfactant is water-soluble, it forms o/w nanoemulsion, while if the surfactant is oil soluble, it forms w/o nanoemulsion [138]. Nanoemulsion formulations are developed according to the physicochemical properties of the drug. Lipophilic drugs are loaded into the oil phase of o/w nanoemulsions, while hydrophilic drugs are loaded into the aqueous phase of w/o emulsions. The drug's solubility in the oil or aqueous phase is a significant criterion for the selection of oils and water, respectively [139].

In a study, oil-in-water (o/w) nanoemulsion was loaded with Lapachol (NE-LAP). It was observed that the anticancer activity of NE-LAP was higher than the free drug against 4T1 mouse and MDA-MB-231 human breast cancer cells. In addition, NE-LAP and NE-blank nanoemulsions did not show any toxic effects on normal cells [140]. In another study, in order to encapsulate DTX, a water-i--garlic-oil nanoemulsion (DTX-NEGO) was produced, and antitumor activity and cardiotoxicity were evaluated on Ehrlich ascites carcinoma (EAC)-bearing mice. DTX-NEGO healed the oxidative stress in the mice's cardiac tissue and enhanced the mean survival time of mice. It has been suggested that DTX-NEGO enhances the antitumor activity and can eliminate its cardiotoxicity [141]. In a study, tocotrienols and caffeic acid were loaded into water-in-oil-in-water (w/o/w) multiple nanoemulsions with cisplatin. This formulation caused cell death and enhanced ROS formation in A549 and HepG2 cancer cells [142].

Inorganic Nanocarriers for Anticancer Delivery

Recently, the interest in utilizing inorganic nanocarriers has increased in anticancer delivery applications due to their significant features, such as easy functionalization, biocompatibility, and availability of synergistic applications [143]. Particularly, the use of inorganic nanocarriers in different treatment strategies such as drug delivery, photothermal therapy (PTT), photodynamic therapy (PDT), magnetic hyperthermia therapy (MHT), and their combination improves the therapeutic effects of cancer treatment. PDT is a technique based on the generation of ROS after light irradiation in the presence of photosensitizer molecules and subsequently, it induces cancer cell death [144]. PTT is based on heat-induced apoptotic cell death that arises from absorbing light in the near-infrared (NIR) or visible region by nanocarriers. On the other hand, MHT is a strategy that involves the production of heat by magnetic materials in the presence of a magnetic field. Since tumor cells are far more sensitive to heat when compared to healthy cells, PTT and MHT enable tumors to be destroyed without harming surrounding tissue [145].

In this section, a summary of the properties of inorganic nanocarriers, including metallic nanocarriers, mesoporous silica NPs, and carbon-based nanocarriers, and their applications as potential anticancer delivery vesicles are included.

Metallic Nanocarriers

Metallic nanocarriers have emerged as attractive candidates for anticancer drug or gene delivery purposes because of their capability to be customized in size and shape, as well as their biocompatibility. Metal nanocarriers are usually modified with polymers or biological substances, such as peptides, antibodies, *etc.* to improve their biocompatibility, solubility, and biorecognition [146]. They have great potential to be used in various anticancer therapies, such as chemotherapy, PDT, and hyperthermia. Silver NPs, gold nanocarriers, platinum NPs, and iron oxide NPs are some of the most common metallic nanocarriers that have received significant interest as anticancer agents and drug carriers in cancer treatment [147].

Gold Nanocarriers

Gold NPs (AuNPs) have outstanding optical, magnetic, and plasmonic characteristics, and also high surface area, good biocompatibility, and a multipurpose structure, allowing them to be a suitable carrier for anticancer drug delivery [148]. In recent years, different gold-based carrier systems, including spherical NPs, nanoclusters, nanoshells, nanorods, *etc.*, have been developed. Two gold nanoshell-based anticancer agents, Auroshell® and Aurolase®, are now in clinical Phase I and human pilot trials for cancer therapy, respectively [149]. AuNPs have many attractive properties, including small size and ease of penetration throughout the body, as well as deposition in tumors thanks to the EPR effect, and high capacity to actively target cancer cells [150]. AuNPs have also been examined as photosensitizer carriers and photothermal agents owing to their surface plasmon resonance (SPR) effect, which occurs when NP surfaces are exposed to a certain light spectrum and lead to the oscillation of electrons on the surface of gold [149, 151].

There are numerous studies that employ AuNPs as chemotherapeutic drug carriers. For instance, Safwat *et al.* loaded 5-Fluorouracil, an antimetabolite drug used to treat cancer, to AuNPs with a size between 9-17 nm and found that AuNPs-5-Fluorouracil increased the anticancer efficacy of the free drug 2-folds against colorectal cancer tissues, induced apoptosis, and interrupted the colorectal cell cycle while minimizing the complications of the drug [152]. Yahyaei *et al.* conjugated AuNPs with multiple chemotherapeutic drugs, capecitabine, tamoxifen, and PTX, and found that PTX conjugated AuNPs showed greater anticancer effect on gastric and breast cancer cells than PTX alone. Surprisingly, the capecitabine and tamoxifen conjugated AuNPs exhibited no toxicity on gastric and breast cancer cells owing to the drug inactivation after conjugation [153].

The surface of AuNPs can be modified to offer controlled release of anticancer agents by external stimulation, such as heat, pH, and light. For instance, Taghdisi *et al.* designed pH-sensitive AuNPs loaded with polyvalent aptamer and daunorubicin for the targeted treatment of leukemia cells. Aptamer-drug-AuNPs complex released the daunorubicin faster in acidic conditions, showed higher anticancer activity on leukemia cells than the drug alone, and also reduced the cytotoxic effect of the drug in healthy cells [154]. Zhang *et al.* developed a three-dimensional mesoporous gold nano-network with a size of 165 nm for both photothermal therapy and chemotherapy against breast cancer. The DOX-loaded gold network was capable of the controlled temperature-dependent release of drugs based on the melting of lauric acid and releasing from the system when subjected to laser or heating. The results showed that the network displayed strong photothermal activity when exposed to 808 nm NIR laser irradiation, increased the temperature to 43 °C (the melting point of lauric acid) in 2.5 minutes, and had high cytotoxic activity on 4T1 cancer cells under the laser irradiation [155].

In a study by Zhu *et al.*, biomimetic gold nanocages coated with the red blood cell membrane and anti-Epcam antibodies were designed for both targeted delivery of PTX and hyperthermia therapy. This combined therapy showed a significant anticancer effect against breast cancer in comparison to free drug, thanks to the immune escape ability and stability provided by red blood cells, antibodies targeting the EpCam proteins overexpressed in the cell, and the photothermal activity of gold [156]. Furthermore, the photothermal and anti-angiogenic characteristics of AuNPs enable synergistic therapy to improve the anticancer activity of chemotherapeutics. In particular, the natural anti-angiogenic effect of

AuNPs is based on the inhibition of vascular endothelial growth factor, and thus, prevention of blood flow into the tumor site [157]. For example, You *et al.* developed polydopamine-coated gold nanostars for targeted and synergistic treatment to eliminate drug resistance in breast cancer. Nanostars have demonstrated considerable toxic effects in folate receptor-positive breast cancer cells and drug-resistant cancer cells in combination with photothermal heating and, above all, they prevented the proliferation and tube development of human umbilical vein endothelial cells by blocking vascular endothelial growth factor-mediated angiogenesis [158].

Silver Nanoparticles

Silver NPs (AgNPs) have been increasingly studied in medicine as drug carriers since they have simple production techniques, versatile surface chemistry, excellent surface/volume ratio, and high functionalization performance [159]. AgNPs can be produced by physical, chemical, and biological techniques. Physical production methods are fast and non-hazardous; however, they have many concerns, such as poor efficiency, high energy consumption, and the need for expensive equipment. Chemical methods offer some advantages, such as easy manufacturing, low cost, and high efficiency, but they have agglomeration issues and require the use of hazardous chemicals [160]. Biological techniques, or "green synthesis," involve the reduction of silver ions. AgNPs are simple and cost-effective solutions without the need for dangerous chemicals that are environmentally harm-free. Bacteria, fungi, algae, plants, vitamins, or amino acids, may be employed to create green AgNPs with a defined shape and size [161].

Several anticancer researches involving green synthesized AgNPs from various sources have been published in the literature. For instance, Almalki *et al.* investigated the anticancer activity of AgNPs biosynthesized from Bacillus sp. on MCF-7 cells and found that AgNPs resulted in a 15% reduction in cell *via*bility, significantly decreased cell migration, and promoted apoptosis [162]. Majeed *et al.* obtained AgNPs from the Bacillus cereus strain and coated them with BSA due to their immune evading property and multifunctional binding ability. BSA-AgNPs improved anticancer activity by reducing the inhibitory effector concentration of NPs on breast, colon, and bone cancer cells [163].

Biological AgNPs synthesized by using a fungus, Botryosphaeria rhodina, exhibited significant anticancer activity at low concentrations against human lung cancer cells by inducing DNA fragmentation, suggesting that the toxicity may be caused by biomolecules in fungi as well as AgNPs [164]. Silver nanocolloids made from a marine sponge, Haliclona exigua, extract resulted in the formation of flower-shaped nanocolloids with a diameter ranging from 100-120 nm. Almost all concentrations of silver nanocolloids demonstrated greater anticancer effects on oral cancer cells after 48 hours when compared to chemically produced AgNPs [165].

AgNPs synthesized from plant-based materials may include a variety of active compounds with anticancer potential because plants contain numerous phytochemicals, such as flavonoids and tannins [161]. Anticancer effects of green synthesized AgNPs using various plant-based materials have been reported against cervical cancer [166], gastric cancer [167], colon cancer [168], lung cancer [169], glioblastoma [170], prostate cancer [171], and many other cancers. In the study of Pei *et al.*, AgNPs with a diameter of 6-45 nm that were produced from Coptis Chinensis have been found to exert antiproliferative activity on the A549 cell line in a dose-dependent manner based on the mechanism of activating intrinsic apoptotic cell death [172]. Furthermore, Erdogan *et al.* produced AgNPs from *Cynara scolymus* extract for synergistic therapy against MCF7 cells and found that AgNPs exhibited antiproliferative activity on breast cancer cells after 24h when combined with PDT. This combined treatment enhanced ROS generation, increased mitochondrial apoptosis in cancer cells, and reduced cell migration in comparison to AgNPs or PDT alone [173].

Platinum Nanoparticles

Platinum-based chemotherapeutics, such as cisplatin or carboplatin, play a significant role in cancer treatment today; however, their drawbacks, like toxicity and drug resistance, limit their applications. Therefore, drug delivery systems based on platinum NPs (PtNPs) have emerged as promising solutions to the limitations of platinum drugs [174]. PtNPs can be utilized in cancer treatment with a similar strategy to cisplatin, inducing DNA damage in cancer cells [187, 188]. PtNPs have been found to perform better than cisplatin in terms of anticancer activity and tumor selectivity [175].

There are few studies on the use of PtNPs in targeted therapy [175], combined therapy with PTT or RT [176, 177], and electrodynamic therapy [178]. In a study, mesoporous PtNPs modified with PEG and loaded with DOX (PEG@Pt/DOX) were developed for tumor chemo-photothermal therapy against DOX-resistant breast cancer cells. PEG@Pt/DOX internalized by cancer cells after 1 h and released drug into the cell nucleus after 24 h. However, when only DOX treatment was applied, they obtained approximately 50% cell *via*bility, indicating that only DOX in PEG@Pt/DOX is insufficient to kill cancer cells. Therefore, the authors also investigated the synergistic effect of PEG@Pt and found that cell *via*bility was drastically reduced after using an 808 nm laser when compared to DOX or PTT alone [179].

Ma *et al.* created ultrasmall PtNPs (4-6 nm) consisting of triphenyl phosphonium, which is a mitochondria-targeting peptide ligand, for targeted photothermal treatment. Triphenyl phosphonium-PtNPs effectively generated thermal effects in cells upon NIR irradiation, reduced cell *via*bility to 30% in liver cancer cells, and showed higher cellular internalization and mitochondria localization than the control group. Moreover, triphenyl phosphonium-PtNPs successfully suppressed tumor development at a greater rate than control and exhibited no systemic toxicity, tissue necrosis, or damage *in vivo* owing to its ultrasmall size [180]. In another study, PtNPs were examined for dual cancer therapy against malignant melanoma cancer cells using PTT and X-ray therapy. The melanoma cell *via*bility was around 40-50% after X-ray radiation was applied alone, and when the 808-nm laser was applied before the X-ray, melanoma cell viability decreased to 13-29% after 72 h, suggesting that the combined therapy is more effective to kill the cancer cells [177].

Several studies have shown that PtNPs produced using green synthesis methods are potential agents against various tumor types. For instance, Ullah *et al.* found that PtNPs prepared with Maytenus royleanus extract have a significant anticancer effect against the A549 cells [181]. Alshatwi *et al.* found that biosynthesized PtNPs using tea polyphenol enhanced the number of cells in the cell death phase and led to the death of SiHa cervical cancer cells [182]. Recently, Aygun *et al.* suggested that PtNPs produced from *Nigella sativa L.* extract demonstrated dose-dependent toxicity in cervical and breast cancer cells [183]. Finally, Baskaran *et al.* indicated that PtNPs biosynthesized using Streptomyces sp. showed an antiproliferative effect on MCF-7 cells [184].

Iron Oxide Nanoparticles

Iron oxide NPs (IONPs) are innovative tools for cancer diagnostic and therapeutic applications. IONPs-based anticancer carriers increase the effectiveness of anticancer agents due to their magnetic properties. However, bare IONPs have undesirable characteristics, such as poor stability, low magnetism due to oxidation, low bioavailability, and aggregation. Functionalization of IONPs with polymers, peptides, or other biomolecules overcomes these problems and provides new features to NPs, such as active targeting, solubility, and extended circulation time [185]. Recently, Norouzi *et al.* examined the anticancer activity of salinomycin-loaded PEI-PEG magnetic IONPs on glioblastoma cells. PEI-PE--IONPs showed no toxicity and high cellular uptake on endothelial and glioblastoma cells, and IONPs increased the anticancer activity of salinomycin after 48 h treatment. Moreover, NPs exhibited a high permeability in *in vitro* model of the blood-brain barrier using magnetic targeting and increased the penetration of salinomycin into the brain [186].

Particularly, superparamagnetic IONPs (SPIONS) have been frequently used as drug carriers in cancer therapy and MHT [201]. NanoTherm®, aminosilane-coated IONPs were approved in 2010 by EMA for the treatment of recurrent glioblastoma (GBM) and by FDA in 2018 for clinical testing for the treatment of prostate cancer [187].

Khan *et al.* developed a drug carrier SPION formulation of curcumin (SP-CUR) for a combined chemotherapy strategy [188]. The combination of SP-CUR and GEM inhibited the proliferation of pancreatic cancer cells, restricted *in vivo* tumor growth and metastasis, and improved survival rate by inhibiting the activation of CXCR4/CXCL-12/SHH signaling, which is a critical pathway that promotes cancer metastasis and growth. Moreover, they revealed that SPION increased the bioavailability of GEM and reduced the chemoresistance of pancreatic cancer cells to GEM.

Cancer-targeted SPIONs can agglomerate in the tumor site because of the combination of the EPR effect, magnetic field, and active targeting. Fang *et al.* have developed SPIONs modified from HA and loaded with DOX for the purpose of targeted cancer therapy [189]. They observed that SPIONs improved the therapeutic efficacy of DOX by enhancing internalization by CD44-expressing breast cancer cells and DOX release. Also, the employment of the magnetic field improved the retention of the NPs in the tumor site and allowed them to hinder tumor growth and decrease the adverse effects of the drug.

SPIONs can develop a potential hyperthermia effect in tumor tissues by producing heat in an AC magnetic field. Jordan *et al.* investigated the anticancer effect of aminosilane-coated and carboxydextran-coated SPIONS against glioblastoma multiforme in a rat tumor model and found that thermotherapy with aminosilane-coated NPs increased the survival rate 4.5-fold compared to control [190]. This approach, combined with mild radiotherapy treatment, was investigated in clinical trials. 41 patients with recurrent glioblastoma were treated with magnetic fluid for six weeks and radiotherapy before or after thermal therapy. Treatment caused a significant increase in the overall survival of patients and no serious side effects were observed [191].

MHT, in combination with chemotherapy, provides a more effective cancer therapy approach by addressing the problems, such as side effects of hyperthermia and drug resistance [192]. For instance, DOX-loaded and FA-coated multifunctional PLA-PEG-SPIONs were found to be promising nanocarriers for both targeted chemotherapy and MH against cervix and colon cancer cells [192]. Quinto *et al.* developed phospholipid-PEG-coated SPIONs for delivering DOX to the tumor site as well as generating hyperthermia by applying an external magnetic field [193]. The electrostatic interactions between DOX and phospholipids were exploited to produce water solubility and high drug loading capacity. SPIONs released drugs continuously for 72 hours and elevated local temperatures to apoptotic levels. DOX-loaded SPIONs caused the death of HeLa cells equivalent to free drug, and SPION-induced hyperthermia increased the anticancer activity of DOX.

SPIONs have previously been reported to be a promising and successful platform for siRNA-based anticancer treatment. For instance, Galactose (Gal) and PEI-modified SPIONs have been shown to deliver therapeutic siRNA to liver cancer with high efficiency [194]. Gal was employed as a targeting agent to minimize off-target effects by binding selectively with the asialoglycoprotein receptor on the hepatocellular carcinoma cell surface. In comparison to free siRNA, the Gal-PEI-SPIO platform increased siRNA accumulation in the tumor site, shielded the siRNA from degradation, and therefore extended the siRNA's half-life. Gal-PE--SPIO was deposited in the tumor, and as a consequence, reduced tumor growth *in vivo*.

Mesoporous Silica Nanocarriers

Mesoporous silica nanocarriers have become successful anticancer delivery systems in recent years thanks to their significant features, including high surface area, large pore volume, great loading efficiency, good biocompatibility, and tunable pore structure. They have been utilized over the years to enhance chemotherapeutic activity by increasing drug solubility, reducing adverse effects, allowing multidrug delivery, and improving cytotoxic efficiency [195]. Additionally, they can be easily modified by various polymers and proteins, enabling controlled, targeted, and stimuli-responsive release of chemotherapeutic drugs into tumor sites [196]. For instance, magnetic-field controlled drug delivery to C26 cancer cells was achieved using epirubicin-loaded mesoporous silica NPs combined with SPIONs. *In vitro* and *in vivo* studies revealed that the nanocarriers with a size of 18-20 nm had better cellular uptake and greater tumor suppression compared to the free drug when subjected to a magnetic field [197].

Modification of mesoporous silica NPs by responsive polymers forms responsive or smart mesoporous silica NPs, which are sensitive to specific biological stimuli, including pH, temperature, and ultrasound. Since most cancer tissues possess lower pH values and have higher temperature values than normal tissues, smart mesoporous silica NPs are promising carriers that have the potential to be utilized for passive tumor tissue targeting and enhancing drug efficiency [198]. In addition, multi-targeted mesoporous silica NPs enable the delivery of anticancer agents to target tissues and even organelles. For instance, the conjugation of HA and triphenylphosphine to DOX-loaded mesoporous silica NPs allowed endocytosis mediated by the CD44 receptor and mitochondrial targeting in gastric cancer cells [199]. The degradation of HA by hyaluronidase initiated the drug release into the environment, indicating that the produced MSNPs are an enzyme-responsive anticancer drug carrier.

In another study, ultrasound-responsive mesoporous silica NPs were decorated with PEG using thermoresponsive linkers, which trigger the detachment of PEG chains from the surface by the effect of increasing temperature after the application of ultrasound [200]. The dissociation of PEG chains exposed the positively charged surface of mesoporous silica NPs, allowing the nanocarrier to be internalized by osteosarcoma cancer cells and improving its anticancer effect. In a study by Kundu *et al.*, umbelliferone drug-loaded mesoporous silica NPs coated with FA for targeted therapy and pH-responsive PAA for controlled release have been found to exhibit a considerable anticancer effect on folate-positive MCF-7 cells by causing oxidative stress and inhibiting tumor growth superior in comparison to the free drug [201]. For anticancer drug release and real-time monitoring, mesoporous silica NPs containing carbon dots and the temperature-responsive polymer, poly(N-vinylcaprolactam) (PNVCL), were produced using pH-sensitive Schiff base bonds [202]. The temperature-sensitive polymer extended the drug release at 37 °C and the fluorescent feature of carbon dots allowed for monitoring the drug release in real-time. As a result, DOX is released through the pores of silica NPs due to the breakdown of pH-sensitive bonds in an acidic environment and internalized into lung cancer cells, resulting in cell death.

Carbon Nano Materials in Anticancer Drug Delivery

Carbon-based nanomaterials (CBNs), including fullerenes, graphene, carbon nanotubes, and their derivatives, have become widely utilized materials for a variety of applications because of their unique mechanical, thermal, electrical, chemical, and optical capabilities. Cancer therapy, targeted drug administration, bio-sensing, cell and tissue imaging, and regenerative medicine are only a few of the current applications of CBNs in biomedicine [203].

Quantum Dots

Quantum dots (QDs) are semiconductor nanometer-sized crystals that exhibit unique optical and electronic features. They have good chemical and photo-stability, as well as a high quantum yield and size-tunable light emission. With the same light wavelength, various types of QDs can be stimulated, and they can be monitored simultaneously due to their narrow emission bands [204].

QD-based multifunctional probes are promising vehicles for anticancer drug and siRNA delivery, in addition to their prospective application for molecular imaging. Their unique optical properties make them good candidates for luminous nano-probes and carriers in biological applications. In addition to the loading with drugs, QDs also can be labeled with specific ligands for tumor-targeting applications. In this way, cancer drugs can be produced specifically to attack the tumor region, whereas their distribution in healthy tissues is minimized [205].

Ruzycka-Ayoush *et al.* improved a formulation using Ag-In-Zn-S QDs. The QD nanocrystals were modified with 11-mercaptoundecanoic acid (MUA), L-cysteine (Cys), and lipoic acid that was adorned with FA for the targeted delivery of the DOX to folate receptors (FARs) of A549 cells. Results showed that QD-Cys and QD-MUA were bound to the maximum amount of FA and DOX structures, which caused them to bind more to (FARs) and release more DOX [206].

Paul *et al.* designed a new photoresponsive NP formulation that consists of photoluminescent silicon quantum dots (SiQDs); they used an o-nitrobenzyl (ONB) derivative as a photo trigger for controlled release of the anticancer agent chlorambucil (Cbl). *In vitro* biological experiments demonstrated that photoirradiation of ONBCbl-SiQDs resulted in efficient cellular internalization, and it showed an anticancer effect on HeLa (human cervix adenocarcinoma) cells [207].

Hettiarachchi *et al.* reported a triple conjugated system using carbon dots (C-dots) with a size of approximately 1.5-1.7 nm. To create the triple conjugated system, C-dots were conjugated with transferrin, a targeted ligand, and two anticancer medicines, including epirubicin and temozolomide. The maximum average particle size of this formulation was 3.5 nm. Results showed that the triple conjugated system (C-dots-trans-temo-epi (C-DT) was more cytotoxic to brain tumor cells than the dual conjugated systems (C-dots-trans-temo (C-TT) and C-dots-trans-epi (C-ET)) and free drugs [208].

Lin *et al.* produced nanocarriers based on cadmium sulphoselenide/zinc sulfide quantum dots (CdSSe/ZnS QDs). These CdSSe/ZnS QDs were functionalized with polyethyleneimine (PEI) to generate a stable complex (QD-PEI), which was then utilized to load siRNA that specifically targets human telomerase reverse transcriptase (TERT). *In vitro* studies were performed on U87 and U251 glioblastoma cell lines, and high gene transfection efficiency (>80%) was achieved. The gene and protein expression levels of TERT were found to be significantly lower 48 hours after transfection of tumor cells. More critically, inhibiting TERT gene expression drastically reduced glioblastoma cell proliferation [209].

Graphene and Graphene Oxide

The excellent mechanical, chemical, and biological capabilities of graphene family nanomaterials have increased the interest of researchers looking for novel materials for biomedical uses in the future [210]. Graphene and its derivatives have attracted the attention of researchers in the field of drug/gene delivery and tissue engineering due to their acceptable biocompatibility, very large surface areas, and excellent optical properties. Moreover, they can be easily functionalized [211] with biological molecules [212].

Among its outstanding electrical conductivity and thermal characteristics, graphene oxide has been investigated for drug delivery applications during the last ten years. There are also certain strategic aspects of using these materials in cancer treatment that are highlighted [213].

Xu *et al.* reported *in vitro* delivery of PTX to A549 and MCF-7 cells by using covalently bonded PEG-GO structures. Results showed that this PEG-GO nanocarrier was harmless to A549 and MCF-7 cells without being conjugated with PTX. In comparison to free PTX, GO-PEG-PTX displayed significantly strong cytotoxicity on A549 and MCF-7 cells throughout a wide concentration range of PTX [214].

Gong *et al.* reported a study on cancer chemo-photothermal therapy based on the interaction between fluorinated graphene oxide (FGO) and DOX. It was shown that by adding fluorine to the structure, more active sites were obtained for molecular interactions between DOX and FGO. The final formulation demonstrated great photothermal performance in the near-infrared region (NIR) region; it also exhibited a high drug loading capacity (more than 200%), pH-triggered drug release, and anticancer effect on HeLa cells [215].

Pei *et al.* developed a dual-drug delivery system based on PEGylated nano-graphene oxide (pGO) with cisplatin, *cis*-$[Pt(NH_3)_2Cl_2]$, and DOX to enable combination chemotherapy in one system. Results demonstrated that pGO-Pt/DOX NPs could be efficiently carried into the tumor cells, causing significant cell apoptosis and necrosis. Besides, cell growth of MCF-7 breast cancer and CAL-27 oral cancer cells was inhibited more effectively than a single drug delivery method or free drug application. Moreover, when compared to free drug applications, *in vivo* findings showed that the toxic effects of Pt and DOX on normal tissues were reduced by the pGO-Pt/DOX formulation. Tumor inhibition statistics, histology observations, and immunohistochemistry stains all demonstrated that the dual-drug delivery method was more effective than free-drug applications for cancer treatment [216].

Li *et al.* reported a study based on the nanoscale graphene oxide (NGO) as a drug carrier. Results showed that they achieved the oral squamous cell cancer (OSCC)-targeted DOX delivery by mediating the HN-1 (TSPLNIHNGQKL). In OSCC cells (CAL-27 and SCC-25), DOX@NGO-PEG-HN-1 demonstrated considerably higher cellular uptake and anticancer effect than free DOX. Furthermore, HN-1 (TSPLNIHNGQKL) demonstrated a significant tumor-targeting and competition-inhibition effect [217].

Carbon Nanotubes

Carbon nanotubes (CNTs) are carbon allotropes that consist of rolled graphene layer/layers and are shaped into nanometer-sized cylindrical tubes. Due to their small size, intense mechanical potency, and high electrical and thermal conductivity, they have amazing structural, mechanical, and electronic properties. Besides, thanks to their large surface area, CNTs have been successfully used in pharmacy and medicine to adsorb or conjugate a wide range of medicinal and diagnostic substances (drugs, genes, vaccines, antibodies, biosensors, *etc.*) [218].

Anticancer compounds (such as DTX, DOX, methotrexate (MTX), PTX, and GEM), anti-inflammatory substances, osteogenic dexamethasone (DEX) steroids, and others have been reported to be carried by CNTs [219].

The two main types of CNTs are single-walled CNTs (SWCNTs) and multi-walled CNTs (MWCNTs). MWCNTs consist of more than one graphene layer, whereas SWCNTs comprise of graphene monolayer. Carbon nanotubes, both single-walled and multi-walled, can be employed in anticancer drug delivery experiments [220].

Due to their unique mechanical properties, morphologies, and functionalizable properties, single-walled CNTs have been exploited as anticancer drug delivery devices [221]. SWCNTs have a more defined wall, while MWCNTs are more prone to structural faults, resulting in a less stable nanostructure. So, they continue to be highlighted in numerous studies. SWCNTs have no conclusive advantages over MWCNTs in terms of drug delivery; their defined smaller diameter may be helpful for quality control, whereas flaws and a less stable structure make modification easier [222]. Many single-walled CNT-based anticancer drug delivery platforms have been reported in the literature.

Tripisciano *et al.* reported a study based on cisplatin and SWCNTs combination. *In vitro* results showed that this formulation reduced the vitality of prostate cancer cells, PC-3, and DU145 [223].

Chen *et al.* developed ETO-loaded epidermal growth factor (EGF)-chitosan (CS)-SWCNTs (EGF/CS/SWCNT-COOHs/ETO). CNTs became more water-dispersible due to the addition of CS, which also functioned as a linker for conjugation with EGF. Results showed that EGF/CS/SWCNT-COOHs exhibited minor cytotoxicity, and the loading capacity of the ETO was found to be 25-27% (wt./wt.). Moreover, EGF/CS/SWCNT-COOHs/ ETO showed a 2.7-fold increased cytotoxic effect on human alveolar carcinoma epithelial cells compared to ETO alone. Their research demonstrated that this innovative drug delivery technology has the potential to improve ETO efficacy [224].

Feazell *et al.* reported that the platinum(IV) complex c,c,t-$[Pt(NH3)_2Cl_2(OEt)(O_2CCH_2CH_2CO_2H)$, a cisplatin derivative, which is linked to the surface of amine-functionalized soluble SWNTs, exhibited a dramatically increased cytotoxicity profile on NTERA-2 (human testicular carcinoma) cells [225].

MWCNTs consist of sp^2 carbons that are elongated cylindrical nano-objects. Their diameter ranges from 3 to 30 nm, and they can grow to be several centimeters long; therefore, their aspect ratio can range from 10 to 10×10^6 [226]. MWCNTs, in particular, have intrinsic anticancer capabilities, interfering with microtubule dynamics and triggering anti-proliferative, anti-migratory, and cytotoxic effects *in vitro*, which lead to tumor growth inhibition *in vivo* [227]. A few important studies in the field of anticancer drug delivery with MWCNTs are listed below.

Dipyridamole is a drug that can raise the concentration of anticancer medicines (5-Fluorouracil, methotrexate, piperidine, vincristine) in cancer cells, improving the efficacy of cancer treatment [228]. The application of MWCNTs loaded with dipyridamole, a poorly soluble medication, was investigated by Zhu *et al.* According to the results, MWCNTs are promising carriers for loading dipyridamole [229].

Neves *et al.* investigated the double-walled carbon nanotubes (DWNTs) as a new delivery vector for siRNA targeting in anticancer applications *via* inducing apoptosis. They loaded siRNA into the oxDWNT (oxidized double-walled CNTS) for blocking the expression of the anti-apoptotic protein survivin, which is over-expressed in most human tumors. It was successfully demonstrated that the level of apoptotic cells after transfection with oxDWNT-siRNAsurvivin complexes is similar to that obtained with siRNAsurvivin delivered by a standard siRNA transfection agent [230].

Ren *et al.* developed a dual-targeting drug delivery system. This formulation was based on PEGylated oxidized MWCNTs, and it was modified with angiopep-2 (O-MWNTs-PEG-ANG). Results showed that DOX-loaded O-MWNTs-P-G-ANG revealed a better anti-glioma effect than DOX. Finally, O-MWNTs-P-G-ANG was found to be a promising dual-targeting carrier for delivering DOX for brain tumor treatment [231].

Carbon Nanohorns

SWCNHs (single-walled carbon nanohorns) are graphene-based nanostructures, and their multifarious properties make them ideal nanosystems for drug delivery applications. They also have simple synthesis and functionalization procedures to obtain the appropriate physicochemical properties [232].

Ajima *et al.* utilized the oxidized SWNHs for cisplatin delivery, and they investigated the anticancer effect of this formulation on human lung cancer cells. They found that cisplatin was slowly released from the SWNHs in the aquatic environment, and this formulation could be used as a potential anticancer drug delivery vehicle [233].

Li *et al.* combined vincristine, an anticancer drug, with oxidized SWCNHs. Then, they functionalized its structure with the DSPE-PEG-IGF-IR monoclonal antibody. The results showed that the resulting formulation induced a more significant toxic effect on MCF-7 cells compared to the effect of free vincristine. *In vivo* experiments also revealed that this formulation showed a more efficient antitumor effect without any significant adverse effects on normal tissues [234].

Fullerenes

The fullerene family consists of many subspecies, such as C20, C28, C32, C44, C50, C58, C70, C76, C84, C240, C540, C960, and the most prevalent fullerene types are C60 and C70, whereas the others are extremely rare [235]. The fullerene family, particularly C60, offers impressive optical, electrochemical, and physical features that can be used in a variety of medical applications. Due to their distinctive carbon cage structure and extensive opportunity for derivatization, fullerenes have been employed as gene and drug delivery systems [236].

Misra *et al.* conjugated glycine-linked C60 fullerenes with *N*-desmethyltamoxifen and examined their anticancer activity on MCF-7 cells. The results revealed that the residence time of the drug in the biological system was increased as a result of this conjugation, and it was found to be more in the targeted cell, thereby increasing the effectiveness of the drug [237].

Magoulas *et al.* reported DOX-C60-PEG conjugates that were formed by anchoring DOX molecules on PEGylated C60 particles. Compared to the PBS medium, the drug was released more rapidly from the conjugate in the presence of tumor lysate. Also, this conjugate showed a considerably antiproliferative effect on MCF-7 cells compared to free DOX [238].

APPROVED NANODRUGS FOR CANCER THERAPY

The first clinically approved nano-drug was the PEGylated liposomal formulation of DOX, known as the brand name Doxil® or Caelyx®. It was approved by the FDA for the treatment of Kaposi's Sarcoma, ovarian cancer, and multiple myeloma in 1995, 2005, and 2008, respectively. The clinical use of albumin-based PTX formulation, under the brand name Abraxane®, was approved by the FDA against breast cancer in 2005, non-small-cell lung cancer (NSCLC) in 2012, and pancreatic cancer in 2013 [239]. A list of NPs approved by the FDA in the USA and/or the EMA in the EU and/or other countries against many types of cancer, is shown in Table **1**.

Table 1. List of the approved nanodrugs for anticancer therapy.

Nanostructure	Name	Active Molecule	Cancer Type(s)	Approval Year(s)	References
Polymeric NPs approved					
Albumin NP bound	Abraxane®	PTX	Breast cancer, pancreatic cancer, and NSCLC	2005 (FDA) 2008 (EMA)	[240, 241]
PEG	Oncaspar®/ Pegaspargase®	L-asparaginase	Acute lymphoblastic leukemia	1994 (FDA)	[242]
PLGH	Eligard®	Leuprolide acetate	Prostate cancer	2002 (FDA)	[243]
Polymeric micelle	Genexol-PM®	PTX	Breast cancer, gastic cancer, and NSCLC	2007 (Korea)	[244]
Polymeric micelle	Apealea®	PTX	Ovarian, fallopian tube, and peritoneal cancer	2018 (EMA)	[245]
Lipid-based NPs approved					
PEGylated liposome	Doxil®/ Caelyx®	DOX	HIV-related Kaposi sarcoma, ovarian cancer, and multiple myeloma	1995 (FDA) 1996 (EMA)	[246]
Liposome	DaunoXome®	Daunorubicin	HIV-related Kaposi sarcoma	1996 (FDA)	[88]

(Table 1) cont.....

Nanostructure	Name	Active Molecule	Cancer Type(s)	Approval Year(s)	References
Liposome	DepoCyt®	Cytarabin	Lymphomatous meningitis	1999 (FDA) 2001 (EMA)	[247]
Non-PEGylated liposome	Myocet®	DOX	Metastatic breast cancer	2000 (EMA) 2001 (Canada)	[248]
Liposome	Lipusu®	PTX	Breast cancer, gastric cancer, and NSCLC	2006 (China)	[249]
Liposome	Mepact®	Mifamurtide	Osteosarcoma	2009 (EMA)	[250]
Liposome	Marqibo®	Vincristine Sulfate	Acute lymphoblastic leukemia	2012 (FDA)	[251]
Liposome	LEP-ETU®	PTX	Ovarian cancer	2015 (FDA)	[252]
PEGylated liposome	Onivyde®/ MM-398®	Irinotecan	Metastatic pancreatic cancer	2015 (FDA)	[253]
Lipid NP	DHP107®	PTX	Gastric cancer	2016 (Korea)	[252]
Liposome	Vyxeos®	Daunorubicin and cytarabine	Acute myelogenous leukemia (AML)	2017 (FDA) 2018 (EMA)	[247]
Inorganic NPs approved					
Iron oxide NP	Nanotherm®	Aminosilane	Local ablation in glioblastoma, prostate, and pancreatic cancer	2010 (EMA) 2018 (FDA)	[187, 254]
Phosphate coating/ Hafnium oxide NP	NBTXR3®/ Hensify®	Radiotherapy	Locally-advanced soft tissue sarcoma	2019 (CE-MARK)	[247, 255]

Effect of the Size, Shape, and Surface Charge of the Nanocarriers on Anticancer Activities

Advances in nanotechnology allow the production of nanocarriers in many different shapes, sizes, and functional properties. The fact that nanocarriers can be produced and modified by many different procedures allows multiple characteristics, such as shape, size, and functionalization, which affect the anticancer activity.

Shape

Thanks to a wide variety of production methods, nanocarriers can be produced in many different shapes, such as spherical, rod [256], branched [257], snowflake [258], dendrite [259], and cone [260].

The shape of an NP affects its properties, such as blood circulation and binding affinity. Given this situation is expected that the shape of an NP also influences its tumor deposition rate and therapeutic efficacy. The shape of nanocarriers also changes the aspect ratio of these structures. Accordingly, since different tumors have different vascular wall porosity, this indicates that nanocarriers produced in different shapes will show extravasation on different tumors [261]. For instance, when comparing the uptake of cubic, spherical, and rod-like AuNPs, among them, spherical particles had the highest uptake with regards to weight, but rod-like NPs had the highest uptake in terms of quantity [262].

Size

Although particles up to 100 nm in size are accepted as NPs [263], NPs produced in larger sizes are encountered in the literature [264]. The sizes of nanocarriers also significantly affect the activities of these structures. Several studies have shown that 50 nm is the optimum size for the cellular uptake of NPs. Smaller particles (about 15-30 nm) and larger particles (about 70-240 nm) were shown to have lower NP uptake [265, 270]. Furthermore, NPs with sizes ranging from 30 to 50 nm efficiently interact with cell membrane receptors and are then internalized *via* receptor-mediated endocytosis [271].

The required NP size to cross physiological barriers is unclear and open to discussion. For example, NPs with hydrodynamic diameters of 100-400 nm were considered optimal for passive targeting due to their good EPR efficiency, while Cabral *et al.* stated that micelles with 30 nm size showed better tumor penetration than those of 100 nm size [272].

Surface Charge

The effect of the NP size and shape on biodistribution is being more investigated, but the charge of these structures also significantly affects their biodistribution and therapeutic efficacy.

Almost every part of the tumor microenvironment is electrostatically charged. The blood vessels are slightly negative because of glycocalyx; besides, HA in the interstitial space and collagen fibers are slightly positive. As a result, the electrostatic interaction between the tumor cells and NPs has a critical role in drug delivery applications [273].

Tumor cells are slightly negatively charged, so possibly positively charged NPs can be more easily taken [274] into the cell *via* electrostatic adhesion-mediated targeting [275]. On the other hand, in the tumor tissues, neutral as well as negatively charged NPs may travel longer distances. As a result, a "delayed charge reversal profile" could be a desirable alternative since tumor penetration could be increased without influencing cellular internalization [276].

TARGETED ANTICANCER DELIVERY

Effective cancer therapy depends on the ability of anticancer drugs to pass through the barriers in the body and reach the target tumor without losing their activity, and selectively kill cancer cells without damaging the surrounding cells. Various nanocarriers are widely employed for passive and active targeting strategies (Fig. **1**) in order to enhance the activity of chemotherapeutics by selective localization in the tumor tissues, minimize their adverse effect, and prevent drug resistance [26]. In this section, the mechanism of passive and active cancer targeting, and applications of targeted drug carriers will be summarized.

Passive Targeting

Passive targeting of anticancer drugs is provided by the small size of nanocarriers and the defected tumor vasculature. Briefly, the increasing food and oxygen demand causes an increase in the permeability of endothelium, thus enhancing vascular permeability in most tumor sites [277]. This process is called the EPR effect, and was first reported in 1986 by Matsumura and Maeda [278]. EPR effect leads to the accumulation of nanocarriers with diameters of 20 to 200 nm in the tumor tissue by 70-fold. Also, their escape from the tumor area becomes difficult because of the poor lymphatic drainage, resulting in the retention of NPs. These extraordinary characteristics hold great promise in the development of tumor-targeted chemotherapy [277, 279].

The passive targeting mechanism can also be achieved by circulating nanocarriers for a long time that allows them to continually travel through the tumor microenvironment. In addition, nanocarriers with cationic charge can be distributed in tumors because of electrostatic attractions between endothelial cells and nanocarriers [24]. However, some tumors may not be ideal for passive targeting. For example, EPR is not effective for every tumor tissue because the vascular permeability is different. Therefore, the active targeting of nanocarriers has emerged as an alternative strategy.

Active Targeting

Active targeting is achieved through interaction of nanocarriers modified with specific tumor tissue-specific targeting moiety and the cell surface receptors located on cancer cells. These targeting moieties can be antibodies, nucleic acids, peptides, glycoproteins, vitamins, *etc.*, that can bind to overexpressed receptors by tumor cells [280]. Active targeting of nanocarriers not only increases cellular uptake of anticancer agents through receptor-mediated endocytosis by accumulating at the target site but also reduces off-target delivery of anticancer agents, avoiding the limitations of passive tumor targeting, and contributes to the prevention of multidrug resistance [281]. Moreover, by developing multivalent active targeting strategies, many receptors located on cancer cells can be targeted to increase the specificity of the treatment [26].

Essentially, the targeting ligand(s) for active targeting-mediated cancer therapy should be specific for tumor-overexpressed target receptors, and these target receptor(s) should be expressed uniformly in all target cells, allowing them to distinguish cancer cells from healthy cells. Folate receptor, glycoproteins, and epidermal growth factor receptors are the most often targeted cell surface receptors in various types of malignancies [282].

CONCLUSION

Nanocarrier systems have a significant role in the delivery of anticancer agents. The use of nanocarriers in cancer therapy has contributed to improving the effectiveness and overcoming problems, such as toxicity, lack of specificity, and drug resistance. In recent years, the development of various nanocarrier formulations used as delivery vehicles for cancer therapeutics has received considerable attention. Many polymeric, lipid-based, and inorganic nanocarriers have been approved since the 1990s, and at present, more candidate compounds are under examination in clinical trials. Advanced delivery systems are also being developed to transport anticancer agents to the tumor site selectively and with a controlled delivery (because of the stimuli-responsive characteristics). In this chapter, various nanocarriers have been introduced as prospects to be used as anticancer delivery vehicles. The studies demonstrated that nanocarriers have the potential for the effective delivery of therapeutic agents. Although, some of these nanocarriers have limited clinical use due to their drawbacks, such as off-target effects, cancer heterogeneity, and the lack of standardized production methods. Currently, there are several novel anticancer delivery systems under the spotlight to find the final anticancer strategies.

CONSENT FOR PUBLICATION

Not applicable.

CONFLICT OF INTEREST

The author declares no conflict of interest, financial or otherwise.

ACKNOWLEDGEMENT

Declared none.

REFERENCES

[1]   Pérez-Herrero E, Fernández-Medarde A. Advanced targeted therapies in cancer: Drug nanocarriers, the future of chemotherapy. Eur J Pharm Biopharm 2015; 93: 52-79.
[http://dx.doi.org/10.1016/j.ejpb.2015.03.018] [PMID: 25813885]

[2]   Sung H, Ferlay J, Siegel RL, *et al.* Global cancer statistics 2020: GLOBOCAN estimates of incidence and mortality worldwide for 36 cancers in 185 countries. CA Cancer J Clin 2021; 71(3): 209-49.
[http://dx.doi.org/10.3322/caac.21660] [PMID: 33538338]

[3]   Peer D, Karp JM, Hong S, Farokhzad OC, Margalit R, Langer R. Nanocarriers as an Emerging Platform for Cancer Therapy Nano-Enabled Medical Applications. Jenny Stanford Publishing 2020; pp. 61-91.
[http://dx.doi.org/10.1201/9780429399039-2]

[4]   Kareliotis G, Tremi I, Kaitatzi M, *et al.* Combined radiation strategies for novel and enhanced cancer treatment. Int J Radiat Biol 2020; 96(9): 1087-103.
[http://dx.doi.org/10.1080/09553002.2020.1787544] [PMID: 32602416]

[5]   Barbari C, Fontaine T, Parajuli P, *et al.* Immunotherapies and combination strategies for immuno-oncology. Int J Mol Sci 2020; 21(14): 5009.
[http://dx.doi.org/10.3390/ijms21145009] [PMID: 32679922]

[6]   Delfi M, Sartorius R, Ashrafizadeh M, *et al.* Self-assembled peptide and protein nanostructures for anti-cancer therapy: Targeted delivery, stimuli-responsive devices and immunotherapy. Nano Today 2021; 38: 101119.
[http://dx.doi.org/10.1016/j.nantod.2021.101119] [PMID: 34267794]

[7]   Bolhassani A, Saleh T. Challenges in Advancing the Field of Cancer Gene Therapy: An Overview of the Multi-Functional Nanocarriers. Novel Gene Therapy Approaches 2013; pp. 197-259.

[8]   Xu H, Li Z, Si J. Nanocarriers in gene therapy: a review. J Biomed Nanotechnol 2014; 10(12): 3483-507.
[http://dx.doi.org/10.1166/jbn.2014.2044] [PMID: 26000367]

[9]   Hwang HS, Shin H, Han J, Na K. Combination of photodynamic therapy (PDT) and anti-tumor immunity in cancer therapy. J Pharm Investig 2018; 48(2): 143-51.
[http://dx.doi.org/10.1007/s40005-017-0377-x] [PMID: 30680248]

[10]   Ashrafizadeh M, Zarrabi A, Hashemi F, *et al.* Curcumin in cancer therapy: A novel adjunct for combination chemotherapy with paclitaxel and alleviation of its adverse effects. Life Sci 2020; 256: 117984.
[http://dx.doi.org/10.1016/j.lfs.2020.117984] [PMID: 32593707]

[11]   El-Hussein A, Manoto SL, Ombinda-Lemboumba S, Alrowaili ZA, Mthunzi-Kufa P. A review of chemotherapy and photodynamic therapy for lung cancer treatment. Anticancer Agents Med Chem 2021; 21(2): 149-61.
[http://dx.doi.org/10.2174/18715206MTA1uNjQp3] [PMID: 32242788]

[12]   Nikolaou M, Pavlopoulou A, Georgakilas AG, Kyrodimos E. The challenge of drug resistance in cancer treatment: a current overview. Clin Exp Metastasis 2018; 35(4): 309-18.
[http://dx.doi.org/10.1007/s10585-018-9903-0] [PMID: 29799080]

[13]   Da Silva CG, Peters GJ, Ossendorp F, Cruz LJ. The potential of multi-compound nanoparticles to bypass drug resistance in cancer. Cancer Chemother Pharmacol 2017; 80(5): 881-94.
[http://dx.doi.org/10.1007/s00280-017-3427-1] [PMID: 28887666]

[14] Slingerland M, Guchelaar HJ, Gelderblom H. Liposomal drug formulations in cancer therapy: 15 years along the road. Drug Discov Today 2012; 17(3-4): 160-6.
[http://dx.doi.org/10.1016/j.drudis.2011.09.015] [PMID: 21983329]

[15] Kumar M, Bishnoi RS, Shukla AK, Jain CP. Techniques for formulation of nanoemulsion drug delivery system: a review. Prev Nutr Food Sci 2019; 24(3): 225-34.
[http://dx.doi.org/10.3746/pnf.2019.24.3.225] [PMID: 31608247]

[16] Jeevanandam J, Barhoum A, Chan YS, Dufresne A, Danquah MK. Review on nanoparticles and nanostructured materials: history, sources, toxicity and regulations. Beilstein J Nanotechnol 2018; 9(1): 1050-74.
[http://dx.doi.org/10.3762/bjnano.9.98] [PMID: 29719757]

[17] DeFrates K, Markiewicz T, Gallo P, *et al.* Protein polymer-based nanoparticles: Fabrication and medical applications. Int J Mol Sci 2018; 19(6): 1717.
[http://dx.doi.org/10.3390/ijms19061717] [PMID: 29890756]

[18] Saad MZH, Jahan R, Bagul U. Nanopharmaceuticals: a new perspective of drug delivery system. Asian Journal of Biomedical & Pharmaceutical Sciences 2012; 2(14): 11-20.

[19] Singh S, Singh S, Lillard JW Jr, Singh R. Drug delivery approaches for breast cancer. Int J Nanomedicine 2017; 12: 6205-18.
[http://dx.doi.org/10.2147/IJN.S140325] [PMID: 28883730]

[20] Wang W, Hao Y, Liu Y, Li R, Huang DB, Pan YY. Nanomedicine in lung cancer: Current states of overcoming drug resistance and improving cancer immunotherapy. Wiley Interdiscip Rev Nanomed Nanobiotechnol 2021; 13(1): e1654.
[http://dx.doi.org/10.1002/wnan.1654] [PMID: 32700465]

[21] Tran P, Lee SE, Kim DH, Pyo YC, Park JS. Recent advances of nanotechnology for the delivery of anticancer drugs for breast cancer treatment. J Pharm Investig 2020; 50(3): 261-70.
[http://dx.doi.org/10.1007/s40005-019-00459-7]

[22] Ulldemolins A, Seras-Franzoso J, Andrade F, *et al.* Perspectives of nano-carrier drug delivery systems to overcome cancer drug resistance in the clinics. Cancer Drug Resist 2021; 4(1): 44-68.
[http://dx.doi.org/10.20517/cdr.2020.59] [PMID: 35582007]

[23] Mitchell MJ, Billingsley MM, Haley RM, Wechsler ME, Peppas NA, Langer R. Engineering precision nanoparticles for drug delivery. Nat Rev Drug Discov 2021; 20(2): 101-24.
[http://dx.doi.org/10.1038/s41573-020-0090-8] [PMID: 33277608]

[24] Jin KT, Lu ZB, Chen JY, *et al.* Recent trends in nanocarrier-based targeted chemotherapy: selective delivery of anticancer drugs for effective lung, colon, cervical, and breast cancer treatment. J Nanomater 2020; 2020: 1-14.
[http://dx.doi.org/10.1155/2020/9184284]

[25] Wei QY, Xu YM, Lau ATY. Recent progress of nanocarrier-based therapy for solid malignancies. Cancers (Basel) 2020; 12(10): 2783.
[http://dx.doi.org/10.3390/cancers12102783] [PMID: 32998391]

[26] Dadwal A, Baldi A, Kumar Narang R. Nanoparticles as carriers for drug delivery in cancer 2018; 46(sup2): 295-305.
[http://dx.doi.org/10.1080/21691401.2018.1457039]

[27] Navya PN, Kaphle A, Srinivas SP, Bhargava SK, Rotello VM, Daima HK. Current trends and challenges in cancer management and therapy using designer nanomaterials. Nano Converg 2019; 6(1): 23.
[http://dx.doi.org/10.1186/s40580-019-0193-2] [PMID: 31304563]

[28] Gagliardi A, Giuliano E, Venkateswararao E, *et al.* Biodegradable polymeric nanoparticles for drug delivery to solid tumors. Front Pharmacol 2021; 12: 601626.
[http://dx.doi.org/10.3389/fphar.2021.601626] [PMID: 33613290]

[29]   Taghipour-Sabzevar V, Sharifi T, Moghaddam MM. Polymeric nanoparticles as carrier for targeted and controlled delivery of anticancer agents. Ther Deliv 2019; 10(8): 527-50.
[http://dx.doi.org/10.4155/tde-2019-0044] [PMID: 31496433]

[30]   Kucuksayan E, Bozkurt F, Yilmaz MT, Sircan-Kucuksayan A, Hanikoglu A, Ozben T. A new combination strategy to enhance apoptosis in cancer cells by using nanoparticles as biocompatible drug delivery carriers. Sci Rep 2021; 11(1): 13027.
[http://dx.doi.org/10.1038/s41598-021-92447-x] [PMID: 34158544]

[31]   Wu J, Wang Q, Dong X, *et al.* Biocompatible AIEgen/p-glycoprotein siRNA@reduction-sensitive paclitaxel polymeric prodrug nanoparticles for overcoming chemotherapy resistance in ovarian cancer. Theranostics 2021; 11(8): 3710-24.
[http://dx.doi.org/10.7150/thno.53828] [PMID: 33664857]

[32]   Yallapu MM, Jaggi M, Chauhan SC. Scope of nanotechnology in ovarian cancer therapeutics. J Ovarian Res 2010; 3(1): 19.
[http://dx.doi.org/10.1186/1757-2215-3-19] [PMID: 20691083]

[33]   Senapati S, Mahanta AK, Kumar S, Maiti P. Controlled drug delivery vehicles for cancer treatment and their performance. Signal Transduct Target Ther 2018; 3(1): 7.
[http://dx.doi.org/10.1038/s41392-017-0004-3] [PMID: 29560283]

[34]   Kalyane D, Raval N, Maheshwari R, Tambe V, Kalia K, Tekade RK. Employment of enhanced permeability and retention effect (EPR): Nanoparticle-based precision tools for targeting of therapeutic and diagnostic agent in cancer. Mater Sci Eng C 2019; 98: 1252-76.
[http://dx.doi.org/10.1016/j.msec.2019.01.066] [PMID: 30813007]

[35]   González-Pastor R, Lancelot A, Morcuende-Ventura V, *et al.* Combination chemotherapy with cisplatin and chloroquine: effect of encapsulation in micelles formed by self-assembling hybrid dendritic-linear-dendritic block copolymers. Int J Mol Sci 2021; 22(10): 5223.
[http://dx.doi.org/10.3390/ijms22105223] [PMID: 34069278]

[36]   Patra JK, Das G, Fraceto LF, *et al.* Nano based drug delivery systems: recent developments and future prospects. J Nanobiotechnology 2018; 16(1): 71.
[http://dx.doi.org/10.1186/s12951-018-0392-8] [PMID: 30231877]

[37]   Gao S, Islam R, Fang J. Tumor environment-responsive hyaluronan conjugated zinc protoporphyrin for targeted anticancer photodynamic therapy. J Pers Med 2021; 11(2): 136.
[http://dx.doi.org/10.3390/jpm11020136] [PMID: 33671291]

[38]   Cavalcante CH, Fernandes RS, de Oliveira Silva J, *et al.* Doxorubicin-loaded pH-sensitive micelles: A promising alternative to enhance antitumor activity and reduce toxicity. Biomed Pharmacother 2021; 134: 111076.
[http://dx.doi.org/10.1016/j.biopha.2020.111076] [PMID: 33341054]

[39]   Huang Y, Yan J, Peng S, *et al.* pH/Reduction Dual-Stimuli-Responsive Cross-Linked Micelles Based on Multi-Functional Amphiphilic Star Copolymer: Synthesis and Controlled Anti-Cancer Drug Release. Polymers (Basel) 2020; 12(1): 82.
[http://dx.doi.org/10.3390/polym12010082] [PMID: 31947729]

[40]   Sun Z, Song C, Wang C, Hu Y, Wu J. Hydrogel-based controlled drug delivery for cancer treatment: a review. Mol. Pharma 2020; 17(2): 373–391.
[http://dx.doi.org/10.1021/acs.molpharmaceut.9b01020] [PMID: 31877054]

[41]   Maspes A, Pizzetti F, Rossetti A, Makvandi P, Sitia G, Rossi F. Advances in Bio-Based Polymers for Colorectal Cancer Treatment: Hydrogels and Nanoplatforms. Gels 2021; 7(1): 6.
[http://dx.doi.org/10.3390/gels7010006] [PMID: 33440908]

[42]   Omtvedt LA, Kristiansen KA, Strand WI, Aachmann FL, Strand BL, Zaytseva-Zotova DS. Alginate hydrogels functionalized with β-cyclodextrin as a local paclitaxel delivery system. J Biomed Mater Res A 2021; 109(12): 2625-39.

[http://dx.doi.org/10.1002/jbm.a.37255] [PMID: 34190416]

[43]   Ding F, Mou Q, Ma Y, *et al.* A crosslinked nucleic acid nanogel for effective siRNA delivery and antitumor therapy. Angew Chem Int Ed 2018; 57(12): 3064-8.
[http://dx.doi.org/10.1002/anie.201711242] [PMID: 29364558]

[44]   Ciolacu DE, Nicu R, Ciolacu F. Cellulose-based hydrogels as sustained drug-delivery systems. Materials (Basel) 2020; 13(22): 5270.
[http://dx.doi.org/10.3390/ma13225270] [PMID: 33233413]

[45]   Lu YJ, Lan YH, Chuang CC, *et al.* Injectable Thermo-Sensitive Chitosan Hydrogel Containing CPT-11-Loaded EGFR-Targeted Graphene Oxide and SLP2 shRNA for Localized Drug/Gene Delivery in Glioblastoma Therapy. Int J Mol Sci 2020; 21(19): 7111.
[http://dx.doi.org/10.3390/ijms21197111] [PMID: 32993166]

[46]   Xu Y, Sun L, Feng S, *et al.* Smart pH-Sensitive Nanogels for Enhancing Synergistic Anticancer Effects of Integrin $\alpha_v\beta_3$ Specific Apoptotic Peptide and Therapeutic Nitric Oxide. ACS Appl Mater Interfaces 2019; 11(38): 34663-75.
[http://dx.doi.org/10.1021/acsami.9b10830] [PMID: 31490654]

[47]   Tarach P, Janaszewska A. Recent Advances in Preclinical Research Using PAMAM Dendrimers for Cancer Gene Therapy. Int J Mol Sci 2021; 22(6): 2912.
[http://dx.doi.org/10.3390/ijms22062912] [PMID: 33805602]

[48]   Saluja V, Mishra Y, Mishra V, Giri N, Nayak P. Dendrimers based cancer nanotheranostics: An overview. Int J Pharm 2021; 600: 120485.
[http://dx.doi.org/10.1016/j.ijpharm.2021.120485] [PMID: 33744447]

[49]   Edis Z, Wang J, Waqas MK, Ijaz M, Ijaz M. Nanocarriers-mediated drug delivery systems for anticancer agents: an overview and perspectives. Int J Nanomedicine 2021; 16: 1313-30.
[http://dx.doi.org/10.2147/IJN.S289443] [PMID: 33628022]

[50]   Hu Q, Wang Y, Xu L, Chen D, Cheng L. Transferrin Conjugated pH- and Redox-Responsive Poly(Amidoamine) Dendrimer Conjugate as an Efficient Drug Delivery Carrier for Cancer Therapy. Int J Nanomedicine 2020; 15: 2751-64.
[http://dx.doi.org/10.2147/IJN.S238536] [PMID: 32368053]

[51]   Alfei S, Marengo B, Zuccari G, Turrini F, Domenicotti C. Dendrimer Nanodevices and Gallic Acid as Novel Strategies to Fight Chemoresistance in Neuroblastoma Cells. Nanomaterials (Basel) 2020; 10(6): 1243.
[http://dx.doi.org/10.3390/nano10061243] [PMID: 32604768]

[52]   Chang M, Zhang F, Wei T, *et al.* Smart linkers in polymer-drug conjugates for tumor-targeted delivery. J Drug Target 2016; 24(6): 475-91.
[http://dx.doi.org/10.3109/1061186X.2015.1108324] [PMID: 26560242]

[53]   Avramović N, Mandić B, Savić-Radojević A, Simić T. Polymeric nanocarriers of drug delivery systems in cancer therapy. Pharmaceutics 2020; 12(4): 298.
[http://dx.doi.org/10.3390/pharmaceutics12040298] [PMID: 32218326]

[54]   Arkaban H, Khajeh Ebrahimi A, Yarahmadi A, Zarrintaj P, Barani M. Development of a multifunctional system based on $CoFe_2O_4$ @polyacrylic acid NPs conjugated to folic acid and loaded with doxorubicin for cancer theranostics. Nanotechnology 2021; 32(30): 305101.
[http://dx.doi.org/10.1088/1361-6528/abf878] [PMID: 33857938]

[55]   Alven S, Nqoro X, Buyana B, Aderibigbe BA. Polymer-drug conjugate, a potential therapeutic to combat breast and lung cancer. Pharmaceutics 2020; 12(5): 406.
[http://dx.doi.org/10.3390/pharmaceutics12050406] [PMID: 32365495]

[56]   Sun H, Yan L, Zhang R, Lovell JF, Wu Y, Cheng C. A sulfobetaine zwitterionic polymer-drug conjugate for multivalent paclitaxel and gemcitabine co-delivery. Biomater Sci 2021; 9(14): 5000-10.
[http://dx.doi.org/10.1039/D1BM00393C] [PMID: 34105535]

[57]   Feng Q, Tong R. Anticancer nanoparticulate polymer-drug conjugate. Bioeng Transl Med 2016; 1(3): 277-96.
[http://dx.doi.org/10.1002/btm2.10033] [PMID: 29313017]

[58]   Du C, Ding Y, Qian J, Zhang R, Dong CM. Achieving traceless ablation of solid tumors without recurrence by mild photothermal-chemotherapy of triple stimuli-responsive polymer-drug conjugate nanoparticles. J Mater Chem B Mater Biol Med 2019; 7(3): 415-32.
[http://dx.doi.org/10.1039/C8TB02432D] [PMID: 32254729]

[59]   Wen J, Qiu N, Zhu Z, *et al.* A size-shrinkable matrix metallopeptidase-2-sensitive delivery nanosystem improves the penetration of human programmed death-ligand 1 siRNA into lung-tumor spheroids. Drug Deliv 2021; 28(1): 1055-66.
[http://dx.doi.org/10.1080/10717544.2021.1931560] [PMID: 34078185]

[60]   Grigoletto A, Martinez G, Gabbia D, *et al.* Folic acid-targeted paclitaxel-polymer conjugates exert selective cytotoxicity and modulate invasiveness of colon cancer cells. Pharmaceutics 2021; 13(7): 929.
[http://dx.doi.org/10.3390/pharmaceutics13070929] [PMID: 34201494]

[61]   Van de Sande L, Cosyns S, Willaert W, Ceelen W. Albumin-based cancer therapeutics for intraperitoneal drug delivery: a review. Drug Deliv 2020; 27(1): 40-53.
[http://dx.doi.org/10.1080/10717544.2019.1704945] [PMID: 31858848]

[62]   Lluch A, Álvarez I, Muñoz M, Seguí MÁ, Tusquets I, García-Estévez L. Treatment innovations for metastatic breast cancer: Nanoparticle albumin-bound (NAB) technology targeted to tumors. Crit Rev Oncol Hematol 2014; 89(1): 62-72.
[http://dx.doi.org/10.1016/j.critrevonc.2013.08.001] [PMID: 24071503]

[63]   Chen H, Huang S, Wang H, *et al.* Preparation and characterization of paclitaxel palmitate albumin nanoparticles with high loading efficacy: an *in vitro* and *in vivo* anti-tumor study in mouse models. Drug Deliv 2021; 28(1): 1067-79.
[http://dx.doi.org/10.1080/10717544.2021.1921078] [PMID: 34109887]

[64]   Kundranda M, Niu J. Albumin-bound paclitaxel in solid tumors: clinical development and future directions. Drug Des Devel Ther 2015; 9: 3767-77.
[http://dx.doi.org/10.2147/DDDT.S88023] [PMID: 26244011]

[65]   Xu J, Zhao Q, Jin Y, Qiu L. High loading of hydrophilic/hydrophobic doxorubicin into polyphosphazene polymersome for breast cancer therapy. Nanomedicine 2014; 10(2): 349-58.
[http://dx.doi.org/10.1016/j.nano.2013.08.004] [PMID: 23969103]

[66]   Vlakh E, Ananyan A, Zashikhina N, *et al.* Preparation, characterization, and biological evaluation of poly(glutamic acid)-b-polyphenylalanine polymersomes. Polymers (Basel) 2016; 8(6): 212.
[http://dx.doi.org/10.3390/polym8060212] [PMID: 30979309]

[67]   Rawal S, Patel MM. Threatening cancer with nanoparticle aided combination oncotherapy. J Control Release 2019; 301: 76-109.
[http://dx.doi.org/10.1016/j.jconrel.2019.03.015] [PMID: 30890445]

[68]   Ahmed F, Pakunlu RI, Brannan A, Bates F, Minko T, Discher DE. Biodegradable polymersomes loaded with both paclitaxel and doxorubicin permeate and shrink tumors, inducing apoptosis in proportion to accumulated drug. J Control Release 2006; 116(2): 150-8.
[http://dx.doi.org/10.1016/j.jconrel.2006.07.012] [PMID: 16942814]

[69]   Colley HE, Hearnden V, Avila-Olias M, *et al.* Polymersome-mediated delivery of combination anticancer therapy to head and neck cancer cells: 2D and 3D *in vitro* evaluation. Mol Pharm 2014; 11(4): 1176-88.
[http://dx.doi.org/10.1021/mp400610b] [PMID: 24533501]

[70]   Sanson C, Schatz C, Le Meins JF, *et al.* A simple method to achieve high doxorubicin loading in biodegradable polymersomes. J Control Release 2010; 147(3): 428-35.

[http://dx.doi.org/10.1016/j.jconrel.2010.07.123] [PMID: 20692308]

[71]   Rani S, Rana R, Saraogi GK, Kumar V, Gupta U. Self-emulsifying oral lipid drug delivery systems: advances and challenges. AAPS PharmSciTech 2019; 20(3): 129.
[http://dx.doi.org/10.1208/s12249-019-1335-x] [PMID: 30815765]

[72]   Jain AS, Dhawan VV, Sarmento B, Nagarsenker MS. *In vitro* and *ex vivo* evaluations of lipid anti-cancer nanoformulations: insights and assessment of bioavailability enhancement. AAPS PharmSciTech 2016; 17(3): 553-71.
[http://dx.doi.org/10.1208/s12249-016-0522-2] [PMID: 27068527]

[73]   Kalepu S, Manthina M, Padavala V. Oral lipid-based drug delivery systems - an overview. Acta Pharm Sin B 2013; 3(6): 361-72.
[http://dx.doi.org/10.1016/j.apsb.2013.10.001]

[74]   Shrestha H, Bala R, Arora S. Lipid-based drug delivery systems. Journal of pharmaceutics 2014.
[http://dx.doi.org/10.1155/2014/801820]

[75]   Chuang SY, Lin CH, Huang TH, Fang JY. Lipid-based nanoparticles as a potential delivery approach in the treatment of rheumatoid arthritis. Nanomaterials (Basel) 2018; 8(1): 42.
[http://dx.doi.org/10.3390/nano8010042] [PMID: 29342965]

[76]   Bayón-Cordero L, Alkorta I, Arana L. Application of solid lipid nanoparticles to improve the efficiency of anticancer drugs. Nanomaterials (Basel) 2019; 9(3): 474.
[http://dx.doi.org/10.3390/nano9030474] [PMID: 30909401]

[77]   Carvalho IPS, Miranda MA, Silva LB, *et al.* In vitro anticancer activity and physicochemical properties of Solanum lycocarpum alkaloidic extract loaded in natural lipid-based nanoparticles. Colloid Interface Sci Commun 2019; 28: 5-14.
[http://dx.doi.org/10.1016/j.colcom.2018.11.001]

[78]   García-Pinel B, Porras-Alcalá C, Ortega-Rodríguez A, *et al.* Lipid-based nanoparticles: application and recent advances in cancer treatment. Nanomaterials (Basel) 2019; 9(4): 638.
[http://dx.doi.org/10.3390/nano9040638] [PMID: 31010180]

[79]   Chiu G, Wong MY, Ling LU, *et al.* Lipid-based nanoparticulate systems for the delivery of anti-cancer drug cocktails: Implications on pharmacokinetics and drug toxicities. Curr Drug Metab 2009; 10(8): 861-74.
[http://dx.doi.org/10.2174/138920009790274531] [PMID: 20214582]

[80]   Mei L, Zhang Z, Zhao L, *et al.* Pharmaceutical nanotechnology for oral delivery of anticancer drugs. Adv Drug Deliv Rev 2013; 65(6): 880-90.
[http://dx.doi.org/10.1016/j.addr.2012.11.005] [PMID: 23220325]

[81]   Kang DI, Kang HK, Gwak HS, Han HK, Lim SJ. Liposome composition is important for retention of liposomal rhodamine in P-glycoprotein-overexpressing cancer cells. Drug Deliv 2009; 16(5): 261-7.
[http://dx.doi.org/10.1080/10717540902937562] [PMID: 19538007]

[82]   Olusanya T, Haj Ahmad R, Ibegbu D, Smith J, Elkordy A. Liposomal drug delivery systems and anticancer drugs. Molecules 2018; 23(4): 907.
[http://dx.doi.org/10.3390/molecules23040907] [PMID: 29662019]

[83]   Antimisiaris S, Kallinteri P, Fatouros D. Liposomes and Drug Delivery. Pharmaceutical Sciences Encyclopedia 2010.
[http://dx.doi.org/10.1002/9780470571224.pse352]

[84]   Kraft JC, Freeling JP, Wang Z, Ho RJY. Emerging research and clinical development trends of liposome and lipid nanoparticle drug delivery systems. J Pharm Sci 2014; 103(1): 29-52.
[http://dx.doi.org/10.1002/jps.23773] [PMID: 24338748]

[85]   Kaddah S, Khreich N, Kaddah F, Charcosset C, Greige-Gerges H. Cholesterol modulates the liposome membrane fluidity and permeability for a hydrophilic molecule. Food Chem Toxicol 2018; 113: 40-8.
[http://dx.doi.org/10.1016/j.fct.2018.01.017] [PMID: 29337230]

[86]   Yingchoncharoen P, Kalinowski DS, Richardson DR. Lipid-based drug delivery systems in cancer therapy: what is available and what is yet to come. Pharmacol Rev 2016; 68(3): 701-87.
[http://dx.doi.org/10.1124/pr.115.012070] [PMID: 27363439]

[87]   Feng L, Mumper RJ. A critical review of lipid-based nanoparticles for taxane delivery. Cancer Lett 2013; 334(2): 157-75.
[http://dx.doi.org/10.1016/j.canlet.2012.07.006] [PMID: 22796606]

[88]   Dawidczyk CM, Kim C, Park JH, *et al.* State-of-the-art in design rules for drug delivery platforms: Lessons learned from FDA-approved nanomedicines. J Control Release 2014; 187: 133-44.
[http://dx.doi.org/10.1016/j.jconrel.2014.05.036] [PMID: 24874289]

[89]   Saraf S, Jain A, Tiwari A, Verma A, Panda PK, Jain SK. Advances in liposomal drug delivery to cancer: An overview. J Drug Deliv Sci Technol 2020; 56: 101549.
[http://dx.doi.org/10.1016/j.jddst.2020.101549]

[90]   Fouladi F, Steffen KJ, Mallik S. Enzyme-responsive liposomes for the delivery of anticancer drugs. Bioconjug Chem 2017; 28(4): 857-68.
[http://dx.doi.org/10.1021/acs.bioconjchem.6b00736] [PMID: 28201868]

[91]   Fonseca C, Moreira J, Ciudad C, Pedrosodelima M, Simões S. Targeting of sterically stabilised pH-sensitive liposomes to human T-leukaemia cells. Eur J Pharm Biopharm 2005; 59(2): 359-66.
[http://dx.doi.org/10.1016/j.ejpb.2004.08.012] [PMID: 15661509]

[92]   Yang MM, Wilson WR, Wu Z. pH-Sensitive PEGylated liposomes for delivery of an acidic dinitrobenzamide mustard prodrug: Pathways of internalization, cellular trafficking and cytotoxicity to cancer cells. Int J Pharm 2017; 516(1-2): 323-33.
[http://dx.doi.org/10.1016/j.ijpharm.2016.11.041] [PMID: 27871834]

[93]   Pourhassan H, Clergeaud G, Hansen AE, *et al.* Revisiting the use of sPLA$_2$-sensitive liposomes in cancer therapy. J Control Release 2017; 261: 163-73.
[http://dx.doi.org/10.1016/j.jconrel.2017.06.024] [PMID: 28662900]

[94]   Yang Y, Zhao Z, Xie C, Zhao Y. Dual-targeting liposome modified by glutamic hexapeptide and folic acid for bone metastatic breast cancer. Chem Phys Lipids 2020; 228: 104882.
[http://dx.doi.org/10.1016/j.chemphyslip.2020.104882] [PMID: 32017901]

[95]   Jose A, Labala S, Venuganti VVK. Co-delivery of curcumin and STAT3 siRNA using deformable cationic liposomes to treat skin cancer. J Drug Target 2017; 25(4): 330-41.
[http://dx.doi.org/10.1080/1061186X.2016.1258567] [PMID: 27819148]

[96]   Kumar R. Lipid-based nanoparticles for drug-delivery systems Nanocarriers for Drug Delivery. Elsevier 2019; pp. 249-84.

[97]   Tavano L, Aiello R, Ioele G, Picci N, Muzzalupo R. Niosomes from glucuronic acid-based surfactant as new carriers for cancer therapy: Preparation, characterization and biological properties. Colloids Surf B Biointerfaces 2014; 118: 7-13.
[http://dx.doi.org/10.1016/j.colsurfb.2014.03.016] [PMID: 24709252]

[98]   Salem HF, Kharshoum RM, Abo El-Ela FI, F AG, Abdellatif KRA. Evaluation and optimization of pH-responsive niosomes as a carrier for efficient treatment of breast cancer. Drug Deliv Transl Res 2018; 8(3): 633-44.
[http://dx.doi.org/10.1007/s13346-018-0499-3] [PMID: 29488171]

[99]   Ag Seleci D, Seleci M, Stahl F, Scheper T. Tumor homing and penetrating peptide-conjugated niosomes as multi-drug carriers for tumor-targeted drug delivery. RSC Advances 2017; 7(53): 33378-84.
[http://dx.doi.org/10.1039/C7RA05071B]

[100]   Barani M, Mirzaei M, Torkzadeh-Mahani M, Adeli-sardou M. Evaluation of carum-loaded niosomes on breast cancer cells: Physicochemical properties, *in vitro* cytotoxicity, flow cytometric, DNA fragmentation and cell migration assay. Sci Rep 2019; 9(1): 7139.

[http://dx.doi.org/10.1038/s41598-019-43755-w] [PMID: 31073144]

[101]   Rajput S, Puvvada N, Kumar BNP, *et al.* Overcoming Akt induced therapeutic resistance in breast cancer through siRNA and thymoquinone encapsulated multilamellar gold niosomes. Mol Pharm 2015; 12(12): 4214-25.
[http://dx.doi.org/10.1021/acs.molpharmaceut.5b00692] [PMID: 26505213]

[102]   Akbarzadeh I, Tavakkoli Yaraki M, Ahmadi S, Chiani M, Nourouzian D. Folic acid-functionalized niosomal nanoparticles for selective dual-drug delivery into breast cancer cells: An in-vitro investigation. Adv Powder Technol 2020; 31(9): 4064-71.
[http://dx.doi.org/10.1016/j.apt.2020.08.011]

[103]   de Mendoza AE-H, Campanero MA, Mollinedo F, Blanco-Prieto MJ. Lipid nanomedicines for anticancer drug therapy. J Biomed Nanotechnol 2009; 5(4): 323-43.
[http://dx.doi.org/10.1166/jbn.2009.1042] [PMID: 20055079]

[104]   Manjunath K, Reddy JS, Venkateswarlu V. Solid lipid nanoparticles as drug delivery systems. Methods Find Exp Clin Pharmacol 2005; 27(2): 127-44.
[http://dx.doi.org/10.1358/mf.2005.27.2.876286] [PMID: 15834465]

[105]   Subedi RK, Kang KW, Choi HK. Preparation and characterization of solid lipid nanoparticles loaded with doxorubicin. Eur J Pharm Sci 2009; 37(3-4): 508-13.
[http://dx.doi.org/10.1016/j.ejps.2009.04.008] [PMID: 19406231]

[106]   Lingayat VJ, Zarekar NS, Shendge RS. Solid lipid nanoparticles: a review. Nanoscience Nanotechnology Research 2017; 2: 67-72.

[107]   Güney G, Kutlu H. Importance of solid lipid nanoparticles in cancer therapy. Nanotechnology 2011; 3: 400-3.

[108]   Wong H, Bendayan R, Rauth A, Li Y, Wu X. Chemotherapy with anticancer drugs encapsulated in solid lipid nanoparticles. Adv Drug Deliv Rev 2007; 59(6): 491-504.
[http://dx.doi.org/10.1016/j.addr.2007.04.008] [PMID: 17532091]

[109]   Severino P, Andreani T, Macedo AS, Fangueiro JF, Santana MHA, Silva AM, *et al.* Current state-of-art and new trends on lipid nanoparticles (SLN and NLC) for oral drug delivery. Journal of Drug Delivery 2012.

[110]   Geszke-Moritz M, Moritz M. Solid lipid nanoparticles as attractive drug vehicles: Composition, properties and therapeutic strategies. Mater Sci Eng C 2016; 68: 982-94.
[http://dx.doi.org/10.1016/j.msec.2016.05.119] [PMID: 27524099]

[111]   Xu Z, Chen L, Gu W, *et al.* The performance of docetaxel-loaded solid lipid nanoparticles targeted to hepatocellular carcinoma. Biomaterials 2009; 30(2): 226-32.
[http://dx.doi.org/10.1016/j.biomaterials.2008.09.014] [PMID: 18851881]

[112]   Miglietta A, Cavalli R, Bocca C, Gabriel L, Rosa Gasco M. Cellular uptake and cytotoxicity of solid lipid nanospheres (SLN) incorporating doxorubicin or paclitaxel. Int J Pharm 2000; 210(1-2): 61-7.
[http://dx.doi.org/10.1016/S0378-5173(00)00562-7] [PMID: 11163988]

[113]   Serpe L, Catalano MG, Cavalli R, *et al.* Cytotoxicity of anticancer drugs incorporated in solid lipid nanoparticles on HT-29 colorectal cancer cell line. Eur J Pharm Biopharm 2004; 58(3): 673-80.
[http://dx.doi.org/10.1016/j.ejpb.2004.03.026] [PMID: 15451544]

[114]   Wang W, Zhang L, Chen T, *et al.* Anticancer effects of resveratrol-loaded solid lipid nanoparticles on human breast cancer cells. Molecules 2017; 22(11): 1814.
[http://dx.doi.org/10.3390/molecules22111814] [PMID: 29068422]

[115]   Wang W, Chen T, Xu H, *et al.* Curcumin-loaded solid lipid nanoparticles enhanced anticancer efficiency in breast cancer. Molecules 2018; 23(7): 1578.
[http://dx.doi.org/10.3390/molecules23071578] [PMID: 29966245]

[116]   Han Y, Zhang P, Chen Y, Sun J, Kong F. Co-delivery of plasmid DNA and doxorubicin by solid lipid

   nanoparticles for lung cancer therapy. Int J Mol Med 2014; 34(1): 191-6.
   [http://dx.doi.org/10.3892/ijmm.2014.1770] [PMID: 24804644]

[117] Naseri N, Valizadeh H, Zakeri-Milani P. Solid lipid nanoparticles and nanostructured lipid carriers: structure, preparation and application. Adv Pharm Bull 2015; 5(3): 305-13.
   [http://dx.doi.org/10.15171/apb.2015.043] [PMID: 26504751]

[118] Beloqui A, Solinís MÁ, Rodríguez-Gascón A, Almeida AJ, Préat V. Nanostructured lipid carriers: Promising drug delivery systems for future clinics. Nanomedicine 2016; 12(1): 143-61.
   [http://dx.doi.org/10.1016/j.nano.2015.09.004] [PMID: 26410277]

[119] Nasirizadeh S, Malaekeh-Nikouei B. Solid lipid nanoparticles and nanostructured lipid carriers in oral cancer drug delivery. J Drug Deliv Sci Technol 2020; 55: 101458.
   [http://dx.doi.org/10.1016/j.jddst.2019.101458]

[120] Liu D, Liu Z, Wang L, Zhang C, Zhang N. Nanostructured lipid carriers as novel carrier for parenteral delivery of docetaxel. Colloids Surf B Biointerfaces 2011; 85(2): 262-9.
   [http://dx.doi.org/10.1016/j.colsurfb.2011.02.038] [PMID: 21435845]

[121] Selvamuthukumar S, Velmurugan R. Nanostructured Lipid Carriers: A potential drug carrier for cancer chemotherapy. Lipids Health Dis 2012; 11(1): 159.
   [http://dx.doi.org/10.1186/1476-511X-11-159] [PMID: 23167765]

[122] Bhise K, Kashaw SK, Sau S, Iyer AK. Nanostructured lipid carriers employing polyphenols as promising anticancer agents: Quality by design (QbD) approach. Int J Pharm 2017; 526(1-2): 506-15.
   [http://dx.doi.org/10.1016/j.ijpharm.2017.04.078] [PMID: 28502895]

[123] Salvi VR, Pawar P. Nanostructured lipid carriers (NLC) system: A novel drug targeting carrier. J Drug Deliv Sci Technol 2019; 51: 255-67.
   [http://dx.doi.org/10.1016/j.jddst.2019.02.017]

[124] Haider M, Abdin SM, Kamal L, Orive G. Nanostructured lipid carriers for delivery of chemotherapeutics: A review. Pharmaceutics 2020; 12(3): 288.
   [http://dx.doi.org/10.3390/pharmaceutics12030288] [PMID: 32210127]

[125] Yang X, Li Y, Li M, Zhang L, Feng L, Zhang N. Hyaluronic acid-coated nanostructured lipid carriers for targeting paclitaxel to cancer. Cancer Lett 2013; 334(2): 338-45.
   [http://dx.doi.org/10.1016/j.canlet.2012.07.002] [PMID: 22776563]

[126] Poonia N, Kaur Narang J, Lather V, *et al.* Resveratrol loaded functionalized nanostructured lipid carriers for breast cancer targeting: Systematic development, characterization and pharmacokinetic evaluation. Colloids Surf B Biointerfaces 2019; 181: 756-66.
   [http://dx.doi.org/10.1016/j.colsurfb.2019.06.004] [PMID: 31234063]

[127] Makeen HA, Mohan S, Al-Kasim MA, *et al.* Gefitinib loaded nanostructured lipid carriers: characterization, evaluation and anti-human colon cancer activity *in vitro*. Drug Deliv 2020; 27(1): 622-31.
   [http://dx.doi.org/10.1080/10717544.2020.1754526] [PMID: 32329374]

[128] Taratula O, Kuzmov A, Shah M, Garbuzenko OB, Minko T. Nanostructured lipid carriers as multifunctional nanomedicine platform for pulmonary co-delivery of anticancer drugs and siRNA. J Control Release 2013; 171(3): 349-57.
   [http://dx.doi.org/10.1016/j.jconrel.2013.04.018] [PMID: 23648833]

[129] Chaudhari VS, Murty US, Banerjee S. Lipidic nanomaterials to deliver natural compounds against cancer: a review. Environ Chem Lett 2020; 18(6): 1803-12.
   [http://dx.doi.org/10.1007/s10311-020-01042-5]

[130] Lim SB, Banerjee A, Önyüksel H. Improvement of drug safety by the use of lipid-based nanocarriers. J Control Release 2012; 163(1): 34-45.
   [http://dx.doi.org/10.1016/j.jconrel.2012.06.002] [PMID: 22698939]

[131] Palazzolo S, Bayda S, Hadla M, *et al.* The clinical translation of organic nanomaterials for cancer

therapy: a focus on polymeric nanoparticles, micelles, liposomes and exosomes. Curr Med Chem 2018; 25(34): 4224-68.
[http://dx.doi.org/10.2174/0929867324666170830113755] [PMID: 28875844]

[132] Sawant RR, Torchilin VP. Multifunctionality of lipid-core micelles for drug delivery and tumour targeting. Mol Membr Biol 2010; 27(7): 232-46.
[http://dx.doi.org/10.3109/09687688.2010.516276] [PMID: 20929339]

[133] Lukyanov AN, Torchilin VP. Micelles from lipid derivatives of water-soluble polymers as delivery systems for poorly soluble drugs. Adv Drug Deliv Rev 2004; 56(9): 1273-89.
[http://dx.doi.org/10.1016/j.addr.2003.12.004] [PMID: 15109769]

[134] Gill KK, Kaddoumi A, Nazzal S. PEG-lipid micelles as drug carriers: physiochemical attributes, formulation principles and biological implication. J Drug Target 2015; 23(3): 222-31.
[http://dx.doi.org/10.3109/1061186X.2014.997735] [PMID: 25547369]

[135] Lovelyn C, Attama AA. Current state of nanoemulsions in drug delivery. J Biomater Nanobiotechnol 2011; 2(5): 626-39.
[http://dx.doi.org/10.4236/jbnb.2011.225075]

[136] McClements DJ. Nanoemulsions *versus* microemulsions: terminology, differences, and similarities. Soft Matter 2012; 8(6): 1719-29.
[http://dx.doi.org/10.1039/C2SM06903B]

[137] Jaiswal M, Dudhe R, Sharma P. Nanoemulsion: an advanced mode of drug delivery system. 3 Biotech 5(2): 123-7.2015;

[138] Singh Y, Meher JG, Raval K, *et al.* Nanoemulsion: Concepts, development and applications in drug delivery. J Control Release 2017; 252: 28-49.
[http://dx.doi.org/10.1016/j.jconrel.2017.03.008] [PMID: 28279798]

[139] Shakeel F, Ramadan W. Transdermal delivery of anticancer drug caffeine from water-in-oil nanoemulsions. Colloids Surf B Biointerfaces 2010; 75(1): 356-62.
[http://dx.doi.org/10.1016/j.colsurfb.2009.09.010] [PMID: 19783127]

[140] Mendes Miranda SE, Alcântara Lemos J, Fernandes RS, *et al.* Enhanced antitumor efficacy of lapachol-loaded nanoemulsion in breast cancer tumor model. Biomed Pharmacother 2021; 133: 110936.
[http://dx.doi.org/10.1016/j.biopha.2020.110936] [PMID: 33254016]

[141] Alkhatib M, Binsiddiq B, Backer W. In Vivo Evaluation of the Anticancer Activity of a Water-in-Garlic Oil Nanoemulsion Loaded with Docetaxel. Int J Pharm Sci Res 2017; 8: 5373-9.

[142] Raviadaran R, Ng MH, Chandran D, Ooi KK, Manickam S. Stable W/O/W multiple nanoemulsion encapsulating natural tocotrienols and caffeic acid with cisplatin synergistically treated cancer cell lines (A549 and HEP G2) and reduced toxicity on normal cell line (HEK 293). Mater Sci Eng C 2021; 121: 111808.
[http://dx.doi.org/10.1016/j.msec.2020.111808] [PMID: 33579452]

[143] Wang F, Li C, Cheng J, Yuan Z. Recent advances on inorganic nanoparticle-based cancer therapeutic agents. Int J Environ Res Public Health 2016; 13(12): 1182.
[http://dx.doi.org/10.3390/ijerph13121182] [PMID: 27898016]

[144] Juarranz Á, Jaén P, Sanz-Rodríguez F, Cuevas J, González S. Photodynamic therapy of cancer. Basic principles and applications. Clin Transl Oncol 2008; 10(3): 148-54.
[http://dx.doi.org/10.1007/s12094-008-0172-2] [PMID: 18321817]

[145] Chatterjee DK, Diagaradjane P, Krishnan S. Nanoparticle-mediated hyperthermia in cancer therapy. Ther Deliv 2011; 2(8): 1001-14.
[http://dx.doi.org/10.4155/tde.11.72] [PMID: 22506095]

[146] Chandran PR, Thomas RT. Gold Nanoparticles in Cancer Drug Delivery.Nanotechnology Applications for Tissue Engineering. Oxford: William Andrew Publishing 2015; pp. 221-37.

[http://dx.doi.org/10.1016/B978-0-323-32889-0.00014-5]

[147] Mei W, Wu Q. Applications of metal nanoparticles in medicine/metal nanoparticles as anticancer agents. Metal Nanoparticles 2018; pp. 169-90.
[http://dx.doi.org/10.1002/9783527807093.ch7]

[148] Yafout M, Ousaid A, Khayati Y, El Otmani IS. Gold nanoparticles as a drug delivery system for standard chemotherapeutics: A new lead for targeted pharmacological cancer treatments. Sci Am 2021; 11: e00685.

[149] Ahmad T, Sarwar R, Iqbal A, *et al.* Recent advances in combinatorial cancer therapy *via* multifunctionalized gold nanoparticles. Nanomedicine (Lond) 2020; 15(12): 1221-37.
[http://dx.doi.org/10.2217/nnm-2020-0051] [PMID: 32370608]

[150] Jain S, Hirst DG, O'Sullivan JM. Gold nanoparticles as novel agents for cancer therapy. Br J Radiol 2012; 85(1010): 101-13.
[http://dx.doi.org/10.1259/bjr/59448833] [PMID: 22010024]

[151] Kim H, Lee D. Near-infrared-responsive cancer photothermal and photodynamic therapy using gold nanoparticles. Polymers (Basel) 2018; 10(9): 961.
[http://dx.doi.org/10.3390/polym10090961] [PMID: 30960886]

[152] Safwat MA, Soliman GM, Sayed D, Attia MA. Gold nanoparticles enhance 5-fluorouracil anticancer efficacy against colorectal cancer cells. Int J Pharm 2016; 513(1-2): 648-58.
[http://dx.doi.org/10.1016/j.ijpharm.2016.09.076] [PMID: 27693737]

[153] Yahyaei B, Pourali P. One step conjugation of some chemotherapeutic drugs to the biologically produced gold nanoparticles and assessment of their anticancer effects. Sci Rep 2019; 9(1): 10242.
[http://dx.doi.org/10.1038/s41598-019-46602-0] [PMID: 31308430]

[154] Taghdisi SM, Danesh NM, Lavaee P, *et al.* Double targeting, controlled release and reversible delivery of daunorubicin to cancer cells by polyvalent aptamers-modified gold nanoparticles. Mater Sci Eng C 2016; 61: 753-61.
[http://dx.doi.org/10.1016/j.msec.2016.01.009] [PMID: 26838906]

[155] Zhang L, Shen S, Cheng L, *et al.* Mesoporous gold nanoparticles for photothermal controlled anticancer drug delivery. Nanomedicine (Lond) 2019; 14(11): 1443-54.
[http://dx.doi.org/10.2217/nnm-2018-0242] [PMID: 31169451]

[156] Zhu DM, Xie W, Xiao YS, *et al.* Erythrocyte membrane-coated gold nanocages for targeted photothermal and chemical cancer therapy. Nanotechnology 2018; 29(8): 084002.
[http://dx.doi.org/10.1088/1361-6528/aa9ca1] [PMID: 29339567]

[157] Saeed BA, Lim V, Yusof NA, Khor KZ, Rahman HS, Abdul Samad N. Antiangiogenic properties of nanoparticles: a systematic review. Int J Nanomedicine 2019; 14: 5135-46.
[http://dx.doi.org/10.2147/IJN.S199974] [PMID: 31371952]

[158] You YH, Lin YF, Nirosha B, Chang HT, Huang YF. Polydopamine-coated gold nanostar for combined antitumor and antiangiogenic therapy in multidrug-resistant breast cancer. Nanotheranostics 2019; 3(3): 266-83.
[http://dx.doi.org/10.7150/ntno.36842] [PMID: 31263658]

[159] Gomes HIO, Martins CSM, Prior JAV. Silver nanoparticles as carriers of anticancer drugs for efficient target treatment of cancer cells. Nanomaterials (Basel) 2021; 11(4): 964.
[http://dx.doi.org/10.3390/nano11040964] [PMID: 33918740]

[160] Zhang XF, Liu ZG, Shen W, Gurunathan S. Silver nanoparticles: synthesis, characterization, properties, applications, and therapeutic approaches. Int J Mol Sci 2016; 17(9): 1534.
[http://dx.doi.org/10.3390/ijms17091534] [PMID: 27649147]

[161] Ivanova N, Gugleva V, Dobreva M. IvayloPehlivanov. Silver nanoparticles as multi-functional drug delivery systems 2018.

[162] Almalki MA, Khalifa AYZ. Silver nanoparticles synthesis from Bacillus sp KFU36 and its anticancer effect in breast cancer MCF-7 cells *via* induction of apoptotic mechanism. J Photochem Photobiol B 2020; 204: 111786.
[http://dx.doi.org/10.1016/j.jphotobiol.2020.111786] [PMID: 31982671]

[163] Majeed S, Aripin FHB, Shoeb NSB, Danish M, Ibrahim MNM, Hashim R. Bioengineered silver nanoparticles capped with bovine serum albumin and its anticancer and apoptotic activity against breast, bone and intestinal colon cancer cell lines. Mater Sci Eng C 2019; 102: 254-63.
[http://dx.doi.org/10.1016/j.msec.2019.04.041] [PMID: 31146998]

[164] Akther T, Vabeiryureilai Mathipi , Davoodbasha M, Srinivasan H, Srinivasan H. Nachimuthu Senthil Kumar. Fungal-mediated synthesis of pharmaceutically active silver nanoparticles and anticancer property against A549 cells through apoptosis. Environ Sci Pollut Res Int 2019; 26(13): 13649-57.
[http://dx.doi.org/10.1007/s11356-019-04718-w] [PMID: 30919178]

[165] Inbakandan D, Kumar C, Bavanilatha M, Ravindra DN, Kirubagaran R, Khan SA. Ultrasonic-assisted green synthesis of flower like silver nanocolloids using marine sponge extract and its effect on oral biofilm bacteria and oral cancer cell lines. Microb Pathog 2016; 99: 135-41.
[http://dx.doi.org/10.1016/j.micpath.2016.08.018] [PMID: 27554277]

[166] Sarkar S, Kotteeswaran V. Green synthesis of silver nanoparticles from aqueous leaf extract of Pomegranate (*Punica granatum*) and their anticancer activity on human cervical cancer cells. 2018.

[167] Mousavi B, Tafvizi F, Zaker Bostanabad S. Green synthesis of silver nanoparticles using Artemisia turcomanica leaf extract and the study of anti-cancer effect and apoptosis induction on gastric cancer cell line (AGS). Artificial Cells, Nanomedicine, and Biotechnology 46((sup 1)): 499-510.2018;

[168] Datkhile KD, Patil SR, Durgawale PP, *et al.* Biogenic synthesis of gold nanoparticles using Argemone mexicana L. and their cytotoxic and genotoxic effects on human colon cancer cell line (HCT-15). J Genet Eng Biotechnol 2021; 19(1): 9.
[http://dx.doi.org/10.1186/s43141-020-00113-y] [PMID: 33443619]

[169] Dadashpour M, Firouzi-Amandi A, Pourhassan-Moghaddam M, *et al.* Biomimetic synthesis of silver nanoparticles using Matricaria chamomilla extract and their potential anticancer activity against human lung cancer cells. Mater Sci Eng C 2018; 92: 902-12.
[http://dx.doi.org/10.1016/j.msec.2018.07.053] [PMID: 30184820]

[170] Mofolo MJ, Kadhila P, Chinsembu KC, Mashele S, Sekhoacha M. Green synthesis of silver nanoparticles from extracts of *Pechuel-loeschea leubnitziae* : their anti-proliferative activity against the U87 cell line. Inorganic and Nano-Metal Chemistry 2020; 50(10): 949-55.
[http://dx.doi.org/10.1080/24701556.2020.1729191]

[171] Singh SP, Mishra A, Shyanti RK, Singh RP, Acharya A. Silver Nanoparticles Synthesized Using Carica papaya Leaf Extract (AgNPs-PLE) Causes Cell Cycle Arrest and Apoptosis in Human Prostate (DU145) Cancer Cells. Biol Trace Elem Res 2021; 199(4): 1316-31.
[http://dx.doi.org/10.1007/s12011-020-02255-z] [PMID: 32557113]

[172] Pei J, Fu B, Jiang L, Sun T. Biosynthesis, characterization, and anticancer effect of plant-mediated silver nanoparticles using *Coptis chinensis*. Int J Nanomedicine 2019; 14: 1969-78.
[http://dx.doi.org/10.2147/IJN.S188235] [PMID: 30936697]

[173] Erdogan O, Abbak M, Demirbolat GM, *et al.* Green synthesis of silver nanoparticles *via* Cynara scolymus leaf extracts: The characterization, anticancer potential with photodynamic therapy in MCF7 cells. PLoS One 2019; 14(6): e0216496.
[http://dx.doi.org/10.1371/journal.pone.0216496] [PMID: 31220110]

[174] Cheng Q, Liu Y. Multifunctional platinum-based nanoparticles for biomedical applications. Wiley Interdiscip Rev Nanomed Nanobiotechnol 2017; 9(2): e1410.
[http://dx.doi.org/10.1002/wnan.1410] [PMID: 27094725]

[175] Shoshan MS, Vonderach T, Hattendorf B, Wennemers H. Peptide-coated platinum nanoparticles with

selective toxicity against liver cancer cells. Angew Chem Int Ed 2019; 58(15): 4901-5.
[http://dx.doi.org/10.1002/anie.201813149] [PMID: 30561882]

[176] Manikandan M, Hasan N, Wu HF. Platinum nanoparticles for the photothermal treatment of Neuro 2A cancer cells. Biomaterials 2013; 34(23): 5833-42.
[http://dx.doi.org/10.1016/j.biomaterials.2013.03.077] [PMID: 23642996]

[177] Daneshvar F, Salehi F, Karimi M, Vais RD, Mosleh-Shirazi MA, Sattarahmady N. Combined X-ray radiotherapy and laser photothermal therapy of melanoma cancer cells using dual-sensitization of platinum nanoparticles. J Photochem Photobiol B 2020; 203: 111737.
[http://dx.doi.org/10.1016/j.jphotobiol.2019.111737] [PMID: 31862636]

[178] Gu T, Wang Y, Lu Y, *et al.* Platinum nanoparticles to enable electrodynamic therapy for effective cancer treatment. Adv Mater 2019; 31(14): 1806803.
[http://dx.doi.org/10.1002/adma.201806803] [PMID: 30734370]

[179] Fu B, Dang M, Tao J, Li Y, Tang Y. Mesoporous platinum nanoparticle-based nanoplatforms for combined chemo-photothermal breast cancer therapy. J Colloid Interface Sci 2020; 570: 197-204.
[http://dx.doi.org/10.1016/j.jcis.2020.02.051] [PMID: 32151829]

[180] Ma Z, Zhang Y, Zhang J, *et al.* Ultrasmall Peptide-Coated Platinum Nanoparticles for Precise NIR-II Photothermal Therapy by Mitochondrial Targeting. ACS Appl Mater Interfaces 2020; 12(35): 39434-43.
[http://dx.doi.org/10.1021/acsami.0c11469] [PMID: 32805937]

[181] Ullah S, Ahmad A, Wang A, *et al.* Bio-fabrication of catalytic platinum nanoparticles and their *in vitro* efficacy against lungs cancer cells line (A549). J Photochem Photobiol B 2017; 173: 368-75.
[http://dx.doi.org/10.1016/j.jphotobiol.2017.06.018] [PMID: 28646755]

[182] Alshatwi AA, Athinarayanan J, Vaiyapuri Subbarayan P. Green synthesis of platinum nanoparticles that induce cell death and G2/M-phase cell cycle arrest in human cervical cancer cells. J Mater Sci Mater Med 2015; 26(1): 7.
[http://dx.doi.org/10.1007/s10856-014-5330-1] [PMID: 25577212]

[183] Aygun A, Gülbagca F, Ozer LY, *et al.* Biogenic platinum nanoparticles using black cumin seed and their potential usage as antimicrobial and anticancer agent. J Pharm Biomed Anal 2020; 179: 112961.
[http://dx.doi.org/10.1016/j.jpba.2019.112961] [PMID: 31732404]

[184] Baskaran B, Muthukumarasamy A, Chidambaram S, Sugumaran A, Ramachandran K, Rasu Manimuthu T. Cytotoxic potentials of biologically fabricated platinum nanoparticles from *Streptomyces sp.* on MCF-7 breast cancer cells. IET Nanobiotechnol 2017; 11(3): 241-6.
[http://dx.doi.org/10.1049/iet-nbt.2016.0040] [PMID: 28476980]

[185] Martinkova P, Brtnicky M, Kynicky J, Pohanka M. Iron oxide nanoparticles: innovative tool in cancer diagnosis and therapy. Adv Healthc Mater 2018; 7(5): 1700932.
[http://dx.doi.org/10.1002/adhm.201700932] [PMID: 29205944]

[186] Norouzi M, Yathindranath V, Thliveris JA, Miller DW. Salinomycin-loaded iron oxide nanoparticles for glioblastoma therapy. Nanomaterials (Basel) 2020; 10(3): 477.
[http://dx.doi.org/10.3390/nano10030477] [PMID: 32155938]

[187] Soetaert F, Korangath P, Serantes D, Fiering S, Ivkov R. Cancer therapy with iron oxide nanoparticles: Agents of thermal and immune therapies. Adv Drug Deliv Rev 2020; 163-164: 65-83.
[http://dx.doi.org/10.1016/j.addr.2020.06.025] [PMID: 32603814]

[188] Khan S, Setua S, Kumari S, *et al.* Superparamagnetic iron oxide nanoparticles of curcumin enhance gemcitabine therapeutic response in pancreatic cancer. Biomaterials 2019; 208: 83-97.
[http://dx.doi.org/10.1016/j.biomaterials.2019.04.005] [PMID: 30999154]

[189] Fang Z, Li X, Xu Z, *et al.* Hyaluronic acid-modified mesoporous silica-coated superparamagnetic Fe_3O_4 nanoparticles for targeted drug delivery. Int J Nanomedicine 2019; 14: 5785-97.
[http://dx.doi.org/10.2147/IJN.S213974] [PMID: 31440047]

[190]   Jordan A, Scholz R, Maier-Hauff K, *et al.* The effect of thermotherapy using magnetic nanoparticles on rat malignant glioma. J Neurooncol 2006; 78(1): 7-14.
[http://dx.doi.org/10.1007/s11060-005-9059-z] [PMID: 16314937]

[191]   Maier-Hauff K, Ulrich F, Nestler D, *et al.* Efficacy and safety of intratumoral thermotherapy using magnetic iron-oxide nanoparticles combined with external beam radiotherapy on patients with recurrent glioblastoma multiforme. J Neurooncol 2011; 103(2): 317-24.
[http://dx.doi.org/10.1007/s11060-010-0389-0] [PMID: 20845061]

[192]   Khaledian M, Nourbakhsh MS, Saber R, Hashemzadeh H, Darvishi MH. Preparation and Evaluation of Doxorubicin-Loaded PLA-PEG-FA Copolymer Containing Superparamagnetic Iron Oxide Nanoparticles (SPIONs) for Cancer Treatment: Combination Therapy with Hyperthermia and Chemotherapy. Int J Nanomedicine 2020; 15: 6167-82.
[http://dx.doi.org/10.2147/IJN.S261638] [PMID: 32922000]

[193]   Quinto CA, Mohindra P, Tong S, Bao G. Multifunctional superparamagnetic iron oxide nanoparticles for combined chemotherapy and hyperthermia cancer treatment. Nanoscale 2015; 7(29): 12728-36.
[http://dx.doi.org/10.1039/C5NR02718G] [PMID: 26154916]

[194]   Yang Z, Duan J, Wang J, *et al.* Superparamagnetic iron oxide nanoparticles modified with polyethylenimine and galactose for siRNA targeted delivery in hepatocellular carcinoma therapy. Int J Nanomedicine 2018; 13: 1851-65.
[http://dx.doi.org/10.2147/IJN.S155537] [PMID: 29618926]

[195]   Alyassin Y, Sayed EG, Mehta P, *et al.* Application of mesoporous silica nanoparticles as drug delivery carriers for chemotherapeutic agents. Drug Discov Today 2020; 25(8): 1513-20.
[http://dx.doi.org/10.1016/j.drudis.2020.06.006] [PMID: 32561300]

[196]   Zhou Y, Quan G, Wu Q, *et al.* Mesoporous silica nanoparticles for drug and gene delivery. Acta Pharm Sin B 2018; 8(2): 165-77.
[http://dx.doi.org/10.1016/j.apsb.2018.01.007] [PMID: 29719777]

[197]   Ansari L, Jaafari MR, Bastami TR, Malaekeh-Nikouei B. Improved anticancer efficacy of epirubicin by magnetic mesoporous silica nanoparticles: *in vitro* and *in vivo* studies. 2018; 46(sup2): 606-6.

[198]   Chang B, Sha X, Guo J, Jiao Y, Wang C, Yang W. Thermo and pH dual responsive, polymer shell coated, magnetic mesoporous silica nanoparticles for controlled drug release. J Mater Chem 2011; 21(25): 9239-47.
[http://dx.doi.org/10.1039/c1jm10631g]

[199]   Naz S, Wang M, Han Y, *et al.* Enzyme-responsive mesoporous silica nanoparticles for tumor cells and mitochondria multistage-targeted drug delivery. Int J Nanomedicine 2019; 14: 2533-42.
[http://dx.doi.org/10.2147/IJN.S202210] [PMID: 31114189]

[200]   Paris JL, Manzano M, Cabañas MV, Vallet-Regí M. Mesoporous silica nanoparticles engineered for ultrasound-induced uptake by cancer cells. Nanoscale 2018; 10(14): 6402-8.
[http://dx.doi.org/10.1039/C8NR00693H] [PMID: 29561558]

[201]   Kundu M, Chatterjee S, Ghosh N, Manna P, Das J, Sil PC. Tumor targeted delivery of umbelliferone *via* a smart mesoporous silica nanoparticles controlled-release drug delivery system for increased anticancer efficiency. Mater Sci Eng C 2020; 116: 111239.
[http://dx.doi.org/10.1016/j.msec.2020.111239] [PMID: 32806268]

[202]   Li X, Hu S, Lin Z, *et al.* Dual-responsive mesoporous silica nanoparticles coated with carbon dots and polymers for drug encapsulation and delivery. Nanomedicine (Lond) 2020; 15(25): 2447-58.
[http://dx.doi.org/10.2217/nnm-2019-0440] [PMID: 32945224]

[203]   Gupta TK, Budarapu PR, Chappidi SR, y B SS, Paggi M, Bordas SP. YB SS, Paggi M, Bordas SP. Advances in carbon based nanomaterials for bio-medical applications. Curr Med Chem 2019; 26(38): 6851-77.
[http://dx.doi.org/10.2174/0929867326666181126113605] [PMID: 30474523]

[204] Matea C, Mocan T, Tabaran F, *et al.* Quantum dots in imaging, drug delivery and sensor applications. Int J Nanomedicine 2017; 12: 5421-31.
[http://dx.doi.org/10.2147/IJN.S138624] [PMID: 28814860]

[205] Iannazzo D, Ziccarelli I, Pistone A. Graphene quantum dots: multifunctional nanoplatforms for anticancer therapy. J Mater Chem B Mater Biol Med 2017; 5(32): 6471-89.
[http://dx.doi.org/10.1039/C7TB00747G] [PMID: 32264412]

[206] Ruzycka-Ayoush M, Kowalik P, Kowalczyk A, Bujak P, Nowicka AM, Wojewodzka M, *et al.* Quantum dots as targeted doxorubicin drug delivery nanosystems. Cancer Nanotechnol 2021; 12(1): 1-27.
[PMID: 33456622]

[207] Paul A, Jana A, Karthik S, Bera M, Zhao Y, Singh NDP. Photoresponsive real time monitoring silicon quantum dots for regulated delivery of anticancer drugs. J Mater Chem B Mater Biol Med 2016; 4(3): 521-8.
[http://dx.doi.org/10.1039/C5TB02045J] [PMID: 32263215]

[208] Hettiarachchi SD, Graham RM, Mintz KJ, *et al.* Triple conjugated carbon dots as a nano-drug delivery model for glioblastoma brain tumors. Nanoscale 2019; 11(13): 6192-205.
[http://dx.doi.org/10.1039/C8NR08970A] [PMID: 30874284]

[209] Lin G, Chen T, Zou J, *et al.* Quantum dots-siRNA nanoplexes for gene silencing in central nervous system tumor cells. Front Pharmacol 2017; 8: 182.
[http://dx.doi.org/10.3389/fphar.2017.00182] [PMID: 28420995]

[210] Ge Z, Yang L, Xiao F, Wu Y, Yu T, Chen J, *et al.* Graphene family nanomaterials: properties and potential applications in dentistry 2018.
[http://dx.doi.org/10.1155/2018/1539678]

[211] Jiang J-H, Pi J, Jin H, Cai J-Y. Functional graphene oxide as cancer-targeted drug delivery system to selectively induce oesophageal cancer cell apoptosis 2018; 46((sup3)): S297-46(sup3):S297-S307..
[http://dx.doi.org/10.1080/21691401.2018.1492418]

[212] Muñoz R, Singh DP, Kumar R, Matsuda A. Graphene oxide for drug delivery and cancer therapy. Nanostructured polymer composites for biomedical applications 447-88.2019;

[213] Durán N, Martinez D, Silveira C, *et al.* Graphene oxide: a carrier for pharmaceuticals and a scaffold for cell interactions. Curr Top Med Chem 2015; 15(4): 309-27.
[http://dx.doi.org/10.2174/1568026615666150108144217] [PMID: 25579346]

[214] Xu Z, Zhu S, Wang M, Li Y, Shi P, Huang X. Delivery of paclitaxel using PEGylated graphene oxide as a nanocarrier. ACS Appl Mater Interfaces 2015; 7(2): 1355-63.
[http://dx.doi.org/10.1021/am507798d] [PMID: 25546399]

[215] Gong P, Du J, Wang D, *et al.* Fluorinated graphene as an anticancer nanocarrier: an experimental and DFT study. J Mater Chem B Mater Biol Med 2018; 6(18): 2769-77.
[http://dx.doi.org/10.1039/C8TB00102B] [PMID: 32254229]

[216] Pei X, Zhu Z, Gan Z, *et al.* PEGylated nano-graphene oxide as a nanocarrier for delivering mixed anticancer drugs to improve anticancer activity. Sci Rep 2020; 10(1): 2717.
[http://dx.doi.org/10.1038/s41598-020-59624-w] [PMID: 32066812]

[217] Li R, Wang Y, Du J, *et al.* Graphene oxide loaded with tumor-targeted peptide and anti-cancer drugs for cancer target therapy. Sci Rep 2021; 11(1): 1725.
[http://dx.doi.org/10.1038/s41598-021-81218-3] [PMID: 33462277]

[218] He H, Pham-Huy LA, Dramou P, Xiao D, Zuo P, Pham-Huy C. Carbon nanotubes: applications in pharmacy and medicine 2013.
[http://dx.doi.org/10.1155/2013/578290]

[219] Zare H, Ahmadi S, Ghasemi A, *et al.* Carbon nanotubes: smart drug/gene delivery carriers. Int J

Nanomedicine 2021; 16: 1681-706.
[http://dx.doi.org/10.2147/IJN.S299448] [PMID: 33688185]

[220] Nahle S, Safar R, Grandemange S, *et al.* Single wall and multiwall carbon nanotubes induce different toxicological responses in rat alveolar macrophages. J Appl Toxicol 2019; 39(5): 764-72.
[http://dx.doi.org/10.1002/jat.3765] [PMID: 30605223]

[221] Dineshkumar B, Krishnakumar K, Bhatt AR, *et al.* Single-walled and multi-walled carbon nanotubes based drug delivery system: Cancer therapy: A review. Indian J Cancer 2015; 52(3): 262-4.
[http://dx.doi.org/10.4103/0019-509X.176720] [PMID: 26905103]

[222] Zhang W, Zhang Z, Zhang Y. The application of carbon nanotubes in target drug delivery systems for cancer therapies. Nanoscale Res Lett 2011; 6(1): 555.
[http://dx.doi.org/10.1186/1556-276X-6-555] [PMID: 21995320]

[223] Tripisciano C, Kraemer K, Taylor A, Borowiak-Palen E. Single-wall carbon nanotubes based anticancer drug delivery system. Chem Phys Lett 2009; 478(4-6): 200-5.
[http://dx.doi.org/10.1016/j.cplett.2009.07.071]

[224] Chen C, Xie XX, Zhou Q, *et al.* EGF-functionalized single-walled carbon nanotubes for targeting delivery of etoposide. Nanotechnology 2012; 23(4): 045104.
[http://dx.doi.org/10.1088/0957-4484/23/4/045104] [PMID: 22222202]

[225] Feazell RP, Nakayama-Ratchford N, Dai H, Lippard SJ. Soluble single-walled carbon nanotubes as longboat delivery systems for platinum(IV) anticancer drug design. J Am Chem Soc 2007; 129(27): 8438-9.
[http://dx.doi.org/10.1021/ja073231f] [PMID: 17569542]

[226] Kukovecz Á, Kozma G, Kónya Z. Multi-walled carbon nanotubes Springer handbook of nanomaterials. Springer 2013; pp. 147-88.
[http://dx.doi.org/10.1007/978-3-642-20595-8_5]

[227] González-Lavado E, Valdivia L, García-Castaño A, *et al.* Multi-walled carbon nanotubes complement the anti-tumoral effect of 5-Fluorouracil. Oncotarget 2019; 10(21): 2022-9.
[http://dx.doi.org/10.18632/oncotarget.26770] [PMID: 31007845]

[228] Ge S-M, Zhan D-L, Zhang S-H, Song L-Q, Han W-W. Reverse screening approach to identify potential anti-cancer targets of dipyridamole. Am J Transl Res 2016; 8(12): 5187-98.
[PMID: 28077994]

[229] Zhu W, Huang H, Dong Y, Han C, Sui X, Jian B. Multi-walled carbon nanotube-based systems for improving the controlled release of insoluble drug dipyridamole. Exp Ther Med 2019; 17(6): 4610-6.
[http://dx.doi.org/10.3892/etm.2019.7510] [PMID: 31105789]

[230] Neves V, Heister E, Costa S, *et al.* Design of double-walled carbon nanotubes for biomedical applications. Nanotechnology 2012; 23(36): 365102.
[http://dx.doi.org/10.1088/0957-4484/23/36/365102] [PMID: 22914449]

[231] Ren J, Shen S, Wang D, *et al.* The targeted delivery of anticancer drugs to brain glioma by PEGylated oxidized multi-walled carbon nanotubes modified with angiopep-2. Biomaterials 2012; 33(11): 3324-33.
[http://dx.doi.org/10.1016/j.biomaterials.2012.01.025] [PMID: 22281423]

[232] Moreno-Lanceta A, Medrano-Bosch M, Melgar-Lesmes P. Single-walled carbon nanohorns as promising nanotube-derived delivery systems to treat cancer. Pharmaceutics 2020; 12(9): 850.
[http://dx.doi.org/10.3390/pharmaceutics12090850] [PMID: 32906852]

[233] Ajima K, Yudasaka M, Murakami T, Maigné A, Shiba K, Iijima S. Carbon nanohorns as anticancer drug carriers. Mol Pharm 2005; 2(6): 475-80.
[http://dx.doi.org/10.1021/mp0500566] [PMID: 16323954]

[234] Li N, Zhao Q, Shu C, *et al.* Targeted killing of cancer cells *in vivo* and *in vitro* with IGF-IR antibody-directed carbon nanohorns based drug delivery. Int J Pharm 2015; 478(2): 644-54.

[http://dx.doi.org/10.1016/j.ijpharm.2014.12.015] [PMID: 25510600]

[235] Yan QL, Gozin M, Zhao FQ, Cohen A, Pang SP. Highly energetic compositions based on functionalized carbon nanomaterials. Nanoscale 2016; 8(9): 4799-851.
[http://dx.doi.org/10.1039/C5NR07855E] [PMID: 26880518]

[236] Bakry R, Vallant RM, Najam-ul-Haq M, *et al.* Medicinal applications of fullerenes. Int J Nanomedicine 2007; 2(4): 639-49.
[PMID: 18203430]

[237] Misra C, Kumar M, Sharma G, *et al.* Glycinated fullerenes for tamoxifen intracellular delivery with improved anticancer activity and pharmacokinetics. Nanomedicine (Lond) 2017; 12(9): 1011-23.
[http://dx.doi.org/10.2217/nnm-2016-0432] [PMID: 28440713]

[238] Magoulas GE, Bantzi M, Messari D, *et al.* Synthesis and evaluation of anticancer activity in cells of novel stoichiometric pegylated fullerene-doxorubicin conjugates. Pharm Res 2015; 32(5): 1676-93.
[http://dx.doi.org/10.1007/s11095-014-1566-1] [PMID: 25380982]

[239] Salvioni L, Rizzuto MA, Bertolini JA, Pandolfi L, Colombo M, Prosperi D. Thirty years of cancer nanomedicine: success, frustration, and hope. Cancers (Basel) 2019; 11(12): 1855.
[http://dx.doi.org/10.3390/cancers11121855] [PMID: 31769416]

[240] Altundag K, Bulut N, Dizdar O, Harputluoglu H. Albumin-bound paclitaxel, ABI-007 may show better efficacy than paclitaxel in basal-like breast cancers: association between caveolin-1 expression and ABI-007. Breast Cancer Res Treat 2006; 100(3): 329-30.
[http://dx.doi.org/10.1007/s10549-006-9250-8] [PMID: 16897435]

[241] Miele E, Spinelli GP, Miele E, Tomao F, Tomao S. Albumin-bound formulation of paclitaxel (Abraxane ABI-007) in the treatment of breast cancer. Int J Nanomedicine 2009; 4: 99-105.
[PMID: 19516888]

[242] Dinndorf PA, Gootenberg J, Cohen MH, Keegan P, Pazdur R. FDA drug approval summary: pegaspargase (oncaspar) for the first-line treatment of children with acute lymphoblastic leukemia (ALL). Oncologist 2007; 12(8): 991-8.
[http://dx.doi.org/10.1634/theoncologist.12-8-991] [PMID: 17766659]

[243] Sartor O. Eligard: leuprolide acetate in a novel sustained-release delivery system. Urology 2003; 61(2) (Suppl. 1): 25-31.
[http://dx.doi.org/10.1016/S0090-4295(02)02396-8] [PMID: 12667884]

[244] Werner ME, Cummings ND, Sethi M, *et al.* Preclinical evaluation of Genexol-PM, a nanoparticle formulation of paclitaxel, as a novel radiosensitizer for the treatment of non-small cell lung cancer. Int J Radiat Oncol Biol Phys 2013; 86(3): 463-8.
[http://dx.doi.org/10.1016/j.ijrobp.2013.02.009] [PMID: 23708084]

[245] He H, Liu L, Morin EE, Liu M, Schwendeman A. Survey of clinical translation of cancer nanomedicines—lessons learned from successes and failures. Acc Chem Res 2019; 52(9): 2445-61.
[http://dx.doi.org/10.1021/acs.accounts.9b00228] [PMID: 31424909]

[246] Barenholz YC. Doxil®-the first FDA-approved Nano-drug: from basics *via* CMC, cell culture and animal studies to clinical use. Nanomedicines: Design. Delivery Detection 2016; 51: 315-45.

[247] Pérez-López A, Martín-Sabroso C, Torres-Suárez AI, Aparicio-Blanco J. Timeline of translational formulation technologies for cancer therapy: Successes, failures, and lessons learned therefrom. Pharmaceutics 2020; 12(11): 1028.
[http://dx.doi.org/10.3390/pharmaceutics12111028] [PMID: 33126622]

[248] Ponce A, Wright A, Dewhirst M, Needham D. Targeted bioavailability of drugs by triggered release from liposomes. Future Lipidol 2006; 1(1): 25-34.
[http://dx.doi.org/10.2217/17460875.1.1.25]

[249] Bernabeu E, Cagel M, Lagomarsino E, Moretton M, Chiappetta DA. Paclitaxel: What has been done and the challenges remain ahead. Int J Pharm 2017; 526(1-2): 474-95.

[http://dx.doi.org/10.1016/j.ijpharm.2017.05.016] [PMID: 28501439]

[250] Meyers PA. Muramyl tripeptide (mifamurtide) for the treatment of osteosarcoma. Expert Rev Anticancer Ther 2009; 9(8): 1035-49.
[http://dx.doi.org/10.1586/era.09.69] [PMID: 19671023]

[251] Silverman JA, Deitcher SR. Marqibo® (vincristine sulfate liposome injection) improves the pharmacokinetics and pharmacodynamics of vincristine. Cancer Chemother Pharmacol 2013; 71(3): 555-64.
[http://dx.doi.org/10.1007/s00280-012-2042-4] [PMID: 23212117]

[252] Sofias AM, Dunne M, Storm G, Allen C. The battle of "nano" paclitaxel. Adv Drug Deliv Rev 2017; 122: 20-30.
[http://dx.doi.org/10.1016/j.addr.2017.02.003] [PMID: 28257998]

[253] Passero FC Jr, Grapsa D, Syrigos KN, Saif MW. The safety and efficacy of Onivyde (irinotecan liposome injection) for the treatment of metastatic pancreatic cancer following gemcitabine-based therapy. Expert Rev Anticancer Ther 2016; 16(7): 697-703.
[http://dx.doi.org/10.1080/14737140.2016.1192471] [PMID: 27219482]

[254] Huang H, Feng W, Chen Y, Shi J. Inorganic nanoparticles in clinical trials and translations. Nano Today 2020; 35: 100972.
[http://dx.doi.org/10.1016/j.nantod.2020.100972]

[255] Ragelle H, Danhier F, Préat V, Langer R, Anderson DG. Nanoparticle-based drug delivery systems: a commercial and regulatory outlook as the field matures. Expert Opin Drug Deliv 2017; 14(7): 851-64.
[http://dx.doi.org/10.1080/17425247.2016.1244187] [PMID: 27730820]

[256] Nima ZA, Alwbari AM, Dantuluri V, *et al.* Targeting nano drug delivery to cancer cells using tunable, multi-layer, silver-decorated gold nanorods. J Appl Toxicol 2017; 37(12): 1370-8.
[http://dx.doi.org/10.1002/jat.3495] [PMID: 28730725]

[257] Tarhan T, Tural B, Tural S. Synthesis and characterization of new branched magnetic nanocomposite for loading and release of topotecan anti-cancer drug. J Anal Sci Technol 2019; 10(1): 30.
[http://dx.doi.org/10.1186/s40543-019-0189-x]

[258] Huang D, Zhao J, Wang M, Zhu S. Snowflake-like gold nanoparticles as SERS substrates for the sensitive detection of organophosphorus pesticide residues. Food Control 2020; 108: 106835.
[http://dx.doi.org/10.1016/j.foodcont.2019.106835]

[259] Luo J, Zhang M, Yang Y, Yu C. Synthesis of dendritic mesoporous organosilica nanoparticles under a mild acidic condition with homogeneous wall structure and near-neutral surface. Chem Commun (Camb) 2021; 57(36): 4416-9.
[http://dx.doi.org/10.1039/D0CC08017A] [PMID: 33949408]

[260] Dong HAN, Shu-Chao ZHANG. Synthesis of hexagonal cone-shaped ZnO nanoparticles using solvothermal reaction and its photoluminescence property. Wuli Huaxue Xuebao 2008; 24(3): 539-42.
[http://dx.doi.org/10.3866/PKU.WHXB20080333]

[261] Toy R, Peiris PM, Ghaghada KB, Karathanasis E. Shaping cancer nanomedicine: the effect of particle shape on the *in vivo* journey of nanoparticles. Nanomedicine (Lond) 2014; 9(1): 121-34.
[http://dx.doi.org/10.2217/nnm.13.191] [PMID: 24354814]

[262] Hoshyar N, Gray S, Han H, Bao G. The effect of nanoparticle size on *in vivo* pharmacokinetics and cellular interaction. Nanomedicine (Lond) 2016; 11(6): 673-92.
[http://dx.doi.org/10.2217/nnm.16.5] [PMID: 27003448]

[263] Soares S, Sousa J, Pais A, Vitorino C. Nanomedicine: principles, properties, and regulatory issues. Front Chem 2018; 6: 360.
[http://dx.doi.org/10.3389/fchem.2018.00360] [PMID: 30177965]

[264] Medina C, Santos-Martinez MJ, Radomski A, Corrigan OI, Radomski MW. Nanoparticles: pharmacological and toxicological significance. Br J Pharmacol 2007; 150(5): 552-8.

[http://dx.doi.org/10.1038/sj.bjp.0707130] [PMID: 17245366]

[265] Chithrani BD, Chan WCW. Elucidating the mechanism of cellular uptake and removal of protein-coated gold nanoparticles of different sizes and shapes. Nano Lett 2007; 7(6): 1542-50.
[http://dx.doi.org/10.1021/nl070363y] [PMID: 17465586]

[266] Geiser M, Rothen-Rutishauser B, Kapp N, *et al.* Ultrafine particles cross cellular membranes by nonphagocytic mechanisms in lungs and in cultured cells. Environ Health Perspect 2005; 113(11): 1555-60.
[http://dx.doi.org/10.1289/ehp.8006] [PMID: 16263511]

[267] Jin H, Heller DA, Sharma R, Strano MS. Size-dependent cellular uptake and expulsion of single-walled carbon nanotubes: single particle tracking and a generic uptake model for nanoparticles. ACS Nano 2009; 3(1): 149-58.
[http://dx.doi.org/10.1021/nn800532m] [PMID: 19206261]

[268] Lu F, Wu SH, Hung Y, Mou CY. Size effect on cell uptake in well-suspended, uniform mesoporous silica nanoparticles. Small 2009; 5(12): 1408-13.
[http://dx.doi.org/10.1002/smll.200900005] [PMID: 19296554]

[269] Osaki F, Kanamori T, Sando S, Sera T, Aoyama Y. A quantum dot conjugated sugar ball and its cellular uptake. On the size effects of endocytosis in the subviral region. J Am Chem Soc 2004; 126(21): 6520-1.
[http://dx.doi.org/10.1021/ja048792a] [PMID: 15161257]

[270] Wang SH, Lee CW, Chiou A, Wei PK. Size-dependent endocytosis of gold nanoparticles studied by three-dimensional mapping of plasmonic scattering images. J Nanobiotechnology 2010; 8(1): 33.
[http://dx.doi.org/10.1186/1477-3155-8-33] [PMID: 21167077]

[271] Foroozandeh P, Aziz AA. Insight into cellular uptake and intracellular trafficking of nanoparticles. Nanoscale Res Lett 2018; 13(1): 339.
[http://dx.doi.org/10.1186/s11671-018-2728-6] [PMID: 30361809]

[272] Kang H, Rho S, Stiles WR, *et al.* Size-dependent EPR effect of polymeric nanoparticles on tumor targeting. Adv Healthc Mater 2020; 9(1): 1901223.
[http://dx.doi.org/10.1002/adhm.201901223] [PMID: 31794153]

[273] Lahir Y. Understanding the basic role of glycocalyx during cancer. Journal of Radiation and Cancer Research 2016; 7(3): 79.
[http://dx.doi.org/10.4103/0973-0168.197974]

[274] Zhang H, Kong X, Tang Y, Lin W. Hydrogen sulfide triggered charge-reversal micelles for cancer-targeted drug delivery and imaging. ACS Appl Mater Interfaces 2016; 8(25): 16227-39.
[http://dx.doi.org/10.1021/acsami.6b03254] [PMID: 27280335]

[275] Ji T, Ding Y, Zhao Y, *et al.* Peptide assembly integration of fibroblast-targeting and cell-penetration features for enhanced antitumor drug delivery. Adv Mater 2015; 27(11): 1865-73.
[http://dx.doi.org/10.1002/adma.201404715] [PMID: 25651789]

[276] Zein R, Sharrouf W, Selting K. Physical properties of nanoparticles that result in improved cancer targeting 2020.
[http://dx.doi.org/10.1155/2020/5194780]

[277] Attia MF, Anton N, Wallyn J, Omran Z, Vandamme TF. An overview of active and passive targeting strategies to improve the nanocarriers efficiency to tumour sites. J Pharm Pharmacol 2019; 71(8): 1185-98.
[http://dx.doi.org/10.1111/jphp.13098] [PMID: 31049986]

[278] Matsumura Y, Maeda H. A new concept for macromolecular therapeutics in cancer chemotherapy: mechanism of tumoritropic accumulation of proteins and the antitumor agent smancs. Cancer Res 1986; 46(12 Pt 1): 6387-92.
[PMID: 2946403]

[279] Wakaskar R. Passive and active targeting in tumor microenvironment. Int J Drug Dev Res 2017; 9(2).

[280] Din F, Aman W, Ullah I, *et al.* Effective use of nanocarriers as drug delivery systems for the treatment of selected tumors. Int J Nanomedicine 2017; 12: 7291-309.
[http://dx.doi.org/10.2147/IJN.S146315] [PMID: 29042776]

[281] Muhamad N, Plengsuriyakarn T, Na-Bangchang K. Application of active targeting nanoparticle delivery system for chemotherapeutic drugs and traditional/herbal medicines in cancer therapy: a systematic review. Int J Nanomedicine 2018; 13: 3921-35.
[http://dx.doi.org/10.2147/IJN.S165210] [PMID: 30013345]

[282] Yao Y, Zhou Y, Liu L, *et al.* Nanoparticle-based drug delivery in cancer therapy and its role in overcoming drug resistance. Front Mol Biosci 2020; 7: 193.
[http://dx.doi.org/10.3389/fmolb.2020.00193] [PMID: 32974385]

CHAPTER 6

Application of Bioceramics to Cancer Therapy

Shirin B. Hanaei[1] and **Yvonne Reinwald[2,*]**

[1] *College of Engineering and Physical Sciences, Aston University, Birmingham, United Kingdom*

[2] *Department of Engineering, School of Science and Technology, Nottingham Trent University, Nottingham, United Kingdom*

Abstract: Despite the great medical developments, cancer remains the main cause of death amongst individuals under 85 years. Novel therapeutic approaches for cancer therapy are constantly being developed, and bioactive ceramics show great promise in this respect. Bioceramics contain inorganic components, which help in the repair, replacement, and regeneration of human cells; for that reason, their use is growing in scope. Bioceramics have a flexible nature and can be modified with biologically active substances for a particular treatment or improvement of tissue or organ functionality. Materials, including glass-ceramics and calcium phosphate, can be loaded with specific drugs, growth factors, peptides, and hormones in a particular fashion. Also, for the elimination of infections and inflammations after surgery, the surface of bioceramics can be modified, and antibiotics can be introduced to prevent bacterial biofilm formation. In the context of bone cancer diagnosis and treatment, mesoporous bioceramics have demonstrated excellent properties not only for being osteoinductive and osteoconductive but also for drug delivery, therefore, being rendered as a remarkable platform for the creation of bone tissue engineering scaffolds for the purpose of bone cancer treatment. Furthermore, the creation of ceramic magnetic nanoparticles as thermoseeds for hyperthermia exhibits promising development for cancer treatment. The conjugation of ceramic nanoparticles with therapeutic agents and heat treatment *via* different magnetic fields improve the efficacy of hyperthermia to the extent that it makes them an alternative to chemotherapy. This chapter discusses the therapeutic value of bioceramics.

Keywords: Bioactive, Bioceramics, Cancer therapy, Clinical trials, Drug release, Glass-ceramic, Hyperthermia, Magnetic nanoparticles.

INTRODUCTION

Cancer is the second most common cause of death in the world. In the United States, around 1.9 million new cases of cancer are estimated, while 608,570 died from cancer in 2020 [1]. The cost of cancer treatment for patients directly was

---

* **Corresponding author Yvonne Reinwald:** Department of Engineering, School of Science and Technology, Nottingham Trent University, Nottingham, United Kingdom; E-mail: yvonne.reinwald@ntu.ac.uk

Felipe López-Saucedo (Ed.)

about $5.6 billion, which included radiation treatment, surgical procedures, and chemotherapy. In 2015, the United States spent $183 billion on cancer-related health care, which is expected to rise to $246 billion by 2030, amounting to a 34% increase [2]. In the United Kingdom, approximately 250,000 individuals are diagnosed with cancer each year, with over 130,000 resulting in death. The annual cost of cancer care for the NHS is £5 billion, but the total cost for society, including lost productivity, is around £18 billion [3].

Cancer can be divided into five categories, namely carcinomas, sarcomas, leukaemia, lymphoma, and myeloma. Carcinomas are one of the most common types of cancer, which originate in epithelial tissue that covers the surface of internal organs and glands. They mostly occur in skin, breasts, lungs, and pancreas, and usually cause solid tumours. Different types of carcinomas include basal cell carcinoma, melanoma, and Merkle cell carcinoma [4]. Sarcomas are malignant cancers that occur in connective tissues, including muscles, fat, cartilage, tendons, and bone. Chondrosarcoma, Ewing's sarcoma, osteosarcoma, and soft tissue sarcoma are a few examples [5]. Leukaemia is a blood cancer originating in the bone marrow. It prevents the marrow from producing normal white and red blood cells. Consequently, blood cells grow and divide uncontrollably. Acute lymphocytic leukaemia, chronic myeloid leukaemia, chronic lymphocytic leukaemia, and acute myeloid leukaemia are the main types of leukaemia [6]. Lymphoma is a cancer that starts in the lymphatic nodes or glands as well as in organs, like the breast and brain. Therefore, lymphoma is considered a cancer of the immune system [7]. Myeloma is a cancer type that develops in the plasma cells of the bone marrow. Myeloma cells may also congregate in the bone to form a single tumour known as plasmacytoma. In certain cases, myeloma cells accumulate in several bones, resulting in several bone tumours or multiple myeloma [4].

The most common cancer types include breast, lung, and colorectal cancer, which accounted for approximately 50% of all new cancer diagnoses in women in 2020. In men, prostate, lung, and colorectal cancers accounted for an estimated 43% of all cancers diagnosed in 2020. In the history of medicine, cancer treatment has been a long-standing issue. Every year, millions of cases are identified, along with millions of deaths. It is assumed that approximately, 90-95% of cases are caused by environmental factors, while the remaining 5-10% are due to genetic factors [8]. Generally, all cancer-causing factors can be categorised into six groups, namely chemical carcinogens, radiation, viral and bacterial infection, heredity, hormones, and immune system dysfunction [8, 9].

Novel cancer treatment strategies have been introduced thanks to the advancement of new technologies. Amongst these, bioceramics offer great hope and promises.

They are considered an important material class for biomedical engineering applications [10]. During the last few decades, bioceramics have been applied in direct contact with living tissues mainly for the repair and regeneration of tissues affected by trauma or disease. The aim of this chapter is to introduce the current state-of-the-art applications of bioceramics in research and clinic with a focus on cancer therapy.

Extracellular Matrix and Tumour Microenvironment

Healthy cells follow a normal cycle of growth, apoptosis, and cell regeneration. However, cancer cells divide uncontrollably, spread *via* the bloodstream and the lymphatic system, and invade other parts of the body, so-called metastasis [11 - 13]. The extracellular matrix (ECM) plays an important role in the progression and spread of cancer cells in the tumour microenvironment (TME). The ECM provides mechanical and physical support for the cells in the TME, and contains essential chemokines and angiogenic factors, which provide compressive and tensile strength *via* cell surface receptors [14].

Remodelling mechanisms that occur in ECM are categorised into four key processes (Fig. **1**). First, ECM is deposited, which alters the concentration of ECM components. Post-translational modifications (PTM) then occur where the biochemical properties and structural aspects of the ECM are altered (Fig. **1a**). Next, proteolytic degradation occurs, aiding the bioactive ECM fragments and ECM-bound factors (Fig. **1b**). The final step is a force-mediated physical remodelling process (Fig. **1c**), which modifies the ECM arrangement by aligning ECM fibres and open channels for cell migration [15].

Tissue homoeostasis is based on accurate ECM remodelling leading to tissue-specific biochemical and biophysical ECM characteristics. ECM components serve as ligands for diverse cell surface receptors, such as integrins, receptor tyrosine kinases, and syndecans. Hence, changes to the delicate balance in the ECM remodelling process impact complicated cellular signalling networks.

Therefore, it is expected that malignant and tumour-associated stromal cells alter the ECM remodelling processes, resulting in a cancer-supporting matrix, the TEM, that actively supports the tumour's pathogenesis [16]. The TEM consists of malignant cells and a variety of cell types, including immune cells, vascular endothelium cells, adipocytes, lymphocytes, fibroblasts and pericytes. There is also a complex network of cytokines, chemokines, GFs, and multiple subtypes of interleukins that mediate intracellular communication. Table **1** presents different cell lines and their relevant functions in the TME.

MECHANISMS OF ECM REMODELLING

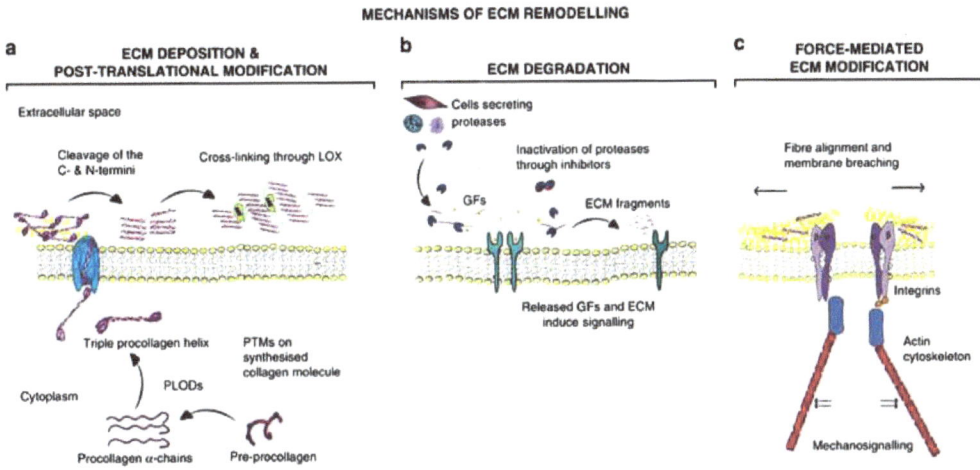

Fig. (1). Remodelling of the ECM. **a**) Pre-collagen is translocated to the Golgi apparatus and changed into procollagen α-chain. PTMs occur on procollagen molecules. PLODs induce lysine hydroxylation of procollagen chains, resulting in the formation of a triple procollagen helix. Collagen fibrils are created through proteases on the C- and N-terminal of the pro-peptides and cross-linking of collagens fibrils. **b**) In the ECM degradation process, MMPs cleave the ECM, resulting in the release of cytokines, matrix-bound growth factors (GFs) and ECM fragments, facilitating cell migration. **c**) ECM remodelling includes conformational modification through the binding of integrin to ECM molecules. Therefore, binding sites are exposed to provide self-assembly for fibrils, inducing fibre alignment [15].

Intercellular communication occurs through cell surface molecules, which are cell-type-specific markers [17, 18]. The TME's lysyl oxidase and transglutaminase induce the reorientation of collagen and elastin fibres, which results in more rigid and/or larger fibrils, leading to increased stiffness of tumour tissue compared to adjacent normal tissues [19]. Cathepsins, a large family of cysteine proteases, are upregulated in TME's cells. For example, cathepsin L expresses and activates heparanase, promoting inflammation, angiogenesis, and metastasis [20, 21]. Fig. (2) represents the ECM remodelling in the primary tumour.

Table 1. Cells and their functions in the tumour microenvironment

Cell	Anti-tumour effects	Pro-tumour effects	Impact from the tumour environment	References
T-lymphocytes	CD8+ T, CD4+ TH1, and IFN-γ exhibit inhibitory impact.	CD4+, TH2, TH17, FOXP3+ are associated with poor prognosis.	Located in the surrounding and the tumour mass. Depending on the type of cancer and the stage, they can be either pro-tumour or anti-tumour.	[22]

(Table 1) cont.....

Cell	Anti-tumour effects	Pro-tumour effects	Impact from the tumour environment	References
B lymphocytes	Anti-tumour effects in some breast and ovarian cancers.	Bregs or B10 inhibits tumour-specific immune responses in mouse models. Bregs also prevents tumour clearance, which is induced by anti-CD20 antibodies in lymphoma mouse model.	Occasionally located within the invasive margin of tumours, but mainly in draining lymph nodes and lymphoid structures next to the TME. Pro-tumour effects are not related to Bregs infiltration of the TME, but it seems they impact other immune cells in the neighbouring lymphoid tissue.	[44 - 46]
NK and NKT cells	Anti-tumour effects in colorectal, gastric, lung, renal and liver cancer.		Located outside the tumour area.	[47]
Tumour-associated macrophages (TAM)		Angiogenic markers, including VEGF and EMAP2, attract TAM to hypoxia and necrosis area of the tumour.	TMA accumulation in necrotic and/or hypoxic parts of tumour is associated with poor prognosis and plays a key role in cell migration, invasion, and metastases.	[48, 49]
Myeloid-derived suppressor cells (MDSCs)		They inhibit CD8+ T cells and polarize macrophages to a tumour-promoting phenotype. Also, they play an immune suppressive role by producing IL-10.	Inhibitory effects of the immune cells.	[50, 51]
Dendritic cells (DCs)		The inflammation and hypoxic condition of TME compromise the DC functionality for immune system activation. Also, T-cell responses at the tumour site have been shown to be suppressed by certain DCs.	DCs play a role in antigen processing and presentation. However, in TME, they are unable to properly induce an immune response to antigens found in tumours.	[52, 53]

(Table 1) cont.....

Cell	Anti-tumour effects	Pro-tumour effects	Impact from the tumour environment	References
Tumour-associated neutrophils (TANs)	The anti-tumour effects have been reported *via* immunological or cytokine activation, where neutrophils can destroy disseminated tumour cells and inhibit TGF-b.	Pro-tumour properties *via* enhancing angiogenesis, increasing degradation of the extracellular matrix and immune suppression.	TANs play a role in both pro- and anti- tumour activities.	[54, 55]
Cancer-associated fibroblasts (CAFs)		CAFs cause the secretion of GFs, such as HGF, FGFs and IGF1, which are mitogenic for malignant cells. Fibroblast-derived CXCL12 chemokine induces cancer cell growth and promotes the migration of progenitor of other cell types into the TME due to chemoattractant properties.	Myofibroblasts, also known as cancer-associated fibroblasts, are abundant in many TMEs. CAFs may originate from a variety of precursors, including smooth muscle cells, myoepithelial cells, endothelial cells, as well as mesenchymal stem cells.	[56 - 61]
Adipocytes		Adipocytes effectively support the development of malignant cells in certain cancers, such as intra-abdominal cancers, that metastasize to the momentum by secreting adipokines.	By supplying fatty acids as food for cancer cells, adipocytes promote the growth of malignant cells.	[62]
Vascular endothelial cells		The branching patterns of blood vessels are chaotic, and the vessel lumen is irregular. The interstitial fluid pressure rises as the channels leak, creating irregular blood supply, oxygenation, nutrient and drug delivery in the TME. As a result, hypoxia rises, making metastasis easier.	Angiogenic factors produced by malignant cells, myeloid cells or CAFs, and chemokines in the TME stimulate the sprouting of endothelial cells, which are needed for cancer growth and survival.	[63]

(Table 1) cont.....

Cell	Anti-tumour effects	Pro-tumour effects	Impact from the tumour environment	References
Pericytes		Pro-tumour and higher metastases in the bladder and colorectal cancer are linked to low pericyte protection of the vasculature. They also induce activation of the MET receptor, resulting in a poor prognosis in malignant breast cancer.	Pericytes, or perivascular stromal cells, are a key component of the vascularisation in the tumour, providing structural stability to blood vessels in TME.	[64, 65]
Lymphatic endothelial cells		Through secretion of factors, like VEGFC and VEGFD, tumour cells can invade lymphatics or induce lymphatic vessel sprouting in TME.	Lymphatic endothelial cells in the TME play a critical role in the spread of cancerous cells. They also influence cancer development by modifying the immune response to cancer and physically altering the TME.	[66, 67]

CURRENT METHODS FOR CANCER TREATMENT

There are numerous methods for cancer treatment. Their choice and success rate depend on the cancer type, stage, severity, tumour mass, the location of the tumour and patient's background and medical history [7]. This section provides a brief overview of current treatment strategies for different types of cancer.

Surgery

The most common treatment for many cancer types is surgery, which involves the removal of abnormal tissue. Surgery is also employed for diagnostic purposes to determine the stage of cancer, reconstruction surgery to restore a body part, or for cancer prevention.

Chemotherapy

Chemotherapy is the administration of cytotoxic antineoplastic drugs to destroy cancer cells by disrupting their growth and division. The drugs reach the target organs through the vascular system. For leukaemia and lymphoma, chemotherapy is the only treatment option. However, chemotherapy drugs may have other uses, such as shrinking the tumour prior to surgery and/or radiotherapy, reducing the risk of cancer recurrence, chemoradiation and palliative chemotherapy.

Combinations of surgery and chemotherapy can cure cancers, such as leukaemia, colorectal cancer, breast cancer, pancreatic cancer, and ovarian cancer [23, 24]. Although chemotherapy is not a permanent treatment, it can also be beneficial to decrease pain and size of tumour mass, especially inoperable tumours. However, chemotherapeutic agents also destroy healthy cells, such as bone marrow cells, lining cells of the gastrointestinal tract, and hair follicle cells. Most patients experience anaemia due to suppression of bone marrow and reduced generation of red blood cells, white blood cells, and platelets. Other side effects can include diarrhea, mouth sores, nausea, and vomiting [25].

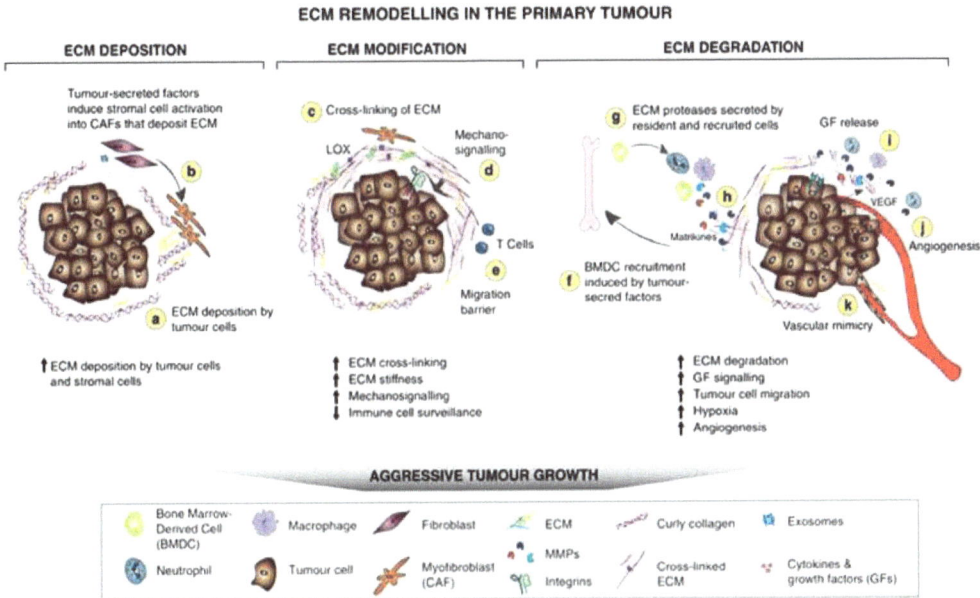

Fig. (2). Remodelling of the ECM in the primary tumour. a,b) Cancer-derived factors stimulate stromal cells to secrete and deposit large quantities of ECM alongside cancer cells; c) cross-linking of collagen fibres results in increased matrix stiffness surrounding the tumour, and e) barrier formation to elude T-cell immune surveillance; d) the matrix stiffness stimulates interactions between the ECM and tumour cell receptors, triggering mechano-signalling *via* integrins; f) cancerous and immune cells produce chemokines, cytokines, and GFs, which aid in the maintenance of the TEM; g) ECM-degrading proteases are released, and tumour cells are either cell surface-bound or secreted; h,i) matrices are generated, and matrix-bound GFs promoting tumour proliferation, migration, invasion and angiogenesis are released; j) a hypoxic environment is formed due to the ECM alteration. Secretion of MMP-9 results in ECM degradation and releases matrix-bound vascular endothelial growth factor (VEGF), forming a concentration gradient for angiogenesis branching; k) tumour cells may develop endothelial-like properties and replicate the vasculature that connects blood vessels [15].

Radiotherapy

Radiotherapy involves the application of ionising radiation with high energy to kill cancer cells in the targeted region by destroying their DNA. Damaged healthy cells constitute one of the common side effects of radiotherapy. This damage is reversible since healthy cells can repair themselves. Multiple radiation rays are released from different exposure directions to converge the treatment area, delivering a much higher dosage than in the surrounding healthy tissue [26]. Radiotherapy can be performed either *via* external beam therapy, where the radiation is given from outside the body, or internal radiotherapy, also called brachytherapy or radioisotope therapy. Some patients can undergo both internal and external radiotherapy. Radiotherapy may be used on its own to treat, for example, head and neck cancer or combined with other therapies, such as chemotherapy (chemoradiation) or surgery [27]. It can also be used to reduce cancer recurrence post-surgery. Some patients receive radiotherapy (neo-adjuvant radiotherapy) prior to surgery to shrink the tumour size for ease of tumour removal [27].

Immunotherapy

Immunotherapy harnesses the patient's immune system to attack cancer cells. It includes monoclonal antibodies, checkpoint inhibitors (CIs), vaccine therapy, and adoptive cell transfer (ACT). Monoclonal antibodies act by binding to specific cancer cell receptors. This allows the immune system to recognise these cells, causing an immune response in the body which then destroys the cancer cells [28]. CIs are an important part of the immune system as they can affect lymphocytes. When lymphocytes are active, they attack and destroy cancer cells. By binding to lymphocytes, CIs prevent them from turning off by blocking the signals that cause them to do so. Hence, lymphocytes keep their efficacy and capability of attacking cancer cells [29]. Currently, vaccine therapy is one of the most important areas of cancer research. Vaccines are designed to recognize receptors on specific cancer cells in the same manner as they do against diseases. They enhance the immune system in recognising and attacking the cancer cells [30]. ACT stimulates T-cells to destroy cancer cells. Usually, cells with the highest efficacy at recognising cancer cells are isolated from the patient and expanded on a large scale *in vitro* before being re-injected into the bloodstream. Gene modification can also be performed on these cells to improve the cells' ability to recognise and destroy cancer cells [31].

Hormone Therapy

Hormone therapy functions by modifying the hormones released or active in the body to reduce or stop tumour growth either by inhibiting hormone production or by suppressing cancer cell growth and spread. It is mainly used for breast cancer and prostate cancer. Hormone treatment may have unwanted side effects, including diarrhea, nausea, hot flashes, weakened bones, and fatigue [32].

Targeted Therapeutics

Targeted therapeutics are the group of drugs that work through 'targeting' those distinctions that help cancer cells to grow and divide uncontrollably. Targeted therapies include angiogenesis inhibitors, which block the cell signalling pathways for blood vessels to grow, resulting in cancer cell death due to blood and oxygen deprivation. Furthermore, cancer growth inhibitors compromise chemical signals responsible for cancer growth and development, which ultimately slows the growth of cancer. Finally, monoclonal antibodies bind to a target cell receptor to inhibit cancer growth and angiogenesis, improving the immune system by distinguishing cancer cells and delivering a chemotherapy drug directly to a target cell. One of the major side effects of targeted therapies is cancer cell resistance to that specific therapeutic; hence, combination therapy, including surgery or chemotherapy, is recommended [33].

BIOACTIVE MATERIALS AND BIOCERAMICS FOR MEDICAL THERAPIES

During the last century, many different ceramics have been developed for different biomedical purposes, including diagnostic instruments, fibre optics for endoscopy, gold porcelain crowns, carriers for antibodies and enzyme deliveries. Ceramics are brittle inorganic non-metallic materials. They are biologically and chemically non-reactive and have wide applications, ranging from tiles and bricks to fibre optics and space shuttles. Bioceramics, on the other hand, refer to a category of materials with high biocompatibility. Bioceramics consist of inorganic compounds with a mixture of ionic and covalent bonds, which have often been utilised for tissue regeneration. Biomaterial-based tissue regeneration includes a biomaterial platform, also called scaffold, which acts as a temporary matrix guiding cell proliferation and differentiation. To facilitate the modulation of cell functions, GFs and other biomolecules may be introduced into the scaffolds alongside the cells. Important scaffold characteristics for biomedical application have been widely described and should include biocompatibility, appropriate mechanical properties, and porosity [34, 35].

Development of Bioceramics

Bioceramics can be divided into first, second and third-generation bioceramics. The main goal of the first generation in the 1960s was to replace damaged tissue with the lowest possible toxic reaction for the patient. Biological "inertness" was considered the most important feature of the materials [36], and common examples included zirconia (ZrO_2) and alumina (Al_2O_3), which were used clinically due to their high strength, exceptional corrosion/wear resistance, and non-toxicity *in vivo* [37].

In the 1980s, the characteristics of bioceramics underwent significant modifications, while the interaction with living tissue was integrated as an essential aspect. The focus of biomaterials moved away from achieving solely bioinert tissue responses toward developing bioactive components capable of eliciting a controlled action and reaction *in vivo*, a so-called bioactive response. Known examples of second-generation bioceramics are calcium phosphates or sulphate compositions and glass-ceramics employed for bone tissue engineering, such as bone cement or coatings for metallic implants [38].

Third-generation bioceramics were developed for the stimulation of a specific biological and cellular response with the aim of tissue regeneration. The molecular modification of biodegradable bioceramics stimulates cellular interactions, cell proliferation and differentiation, as well as ECM production [39]. These glass-ceramics and porous bioceramics can also stimulate the expression of genes that play a role in the regeneration of living tissues. They were developed by doping porous second-generation bioceramics with biologically active compounds, as well as novel bioceramics, such as organic-inorganic hybrids, mesoporous ordered glasses, or mesoporous silica materials. Currently, different types of bioceramics are available from a variety of manufacturers and are employed by orthopaedics and maxillofacial surgeons [40]. Table **2** represents the different biomedical applications of bioceramics and human tissue responses following their implantation.

Reactivity of Bioceramics

Many different factors affect bioceramics' reactivity and hence reaction kinetics. One of the most important properties of synthetic bioceramics is their biocompatibility while containing specified therapeutic agents aiding tissue regeneration. The reactivity of ceramics helps to prevent immunological responses and the formation of foreign body giant cell post-implantation [41]. Bioceramics are mainly biocompatible and functional due to their chemical stability and ionic bonds [34].

The development of a functional interface with the living host tissue is a necessity for bioceramics *in vivo*. The type of tissue behaviour at the implant site has a strong correlation with the mechanism of action for tissue attachment [38, 42]. Implant failure over the last 20 years has commonly been traced back to the biomaterial–tissue interface. The inert material without chemical or biological interfacial integration with the host tissue induces the formation of a non-adherent fibrous capsule over time, which is the result of micromovement between the implant interface and the host tissue that can occur in both hard and soft tissue. The fibrous tissue thickness varies depending on the material type and relative micromovement [36]. Biodegradable materials degrade over time and can be substituted by native host tissue, which results in almost no interfacial fibrotic tissue. However, the most important changes to bioresorbable materials include i) appropriate strength at the interface area throughout the degradation stage and ii) equivalent degradation and body tissue repair rates. Additionally, bioresorbable materials must therefore be made up of only biologically and metabolically safe and appropriate substances to avoid chronic inflammation and discomfort for the patients [43].

Table 2. Therapeutic applications of bioceramics and corresponding tissue responses.

Medical Field	Therapeutic/Medical Device	Example of Implant/Material	Cellular Response	Implant Type	References
Orthopaedic applications	Arthroplasty, orthopaedic fixation devices for arthrodesis and spinal fusion.	Alumina, yettrium-stabilized, zirconia, silicon, nitride.	Fibrous tissue formation.	Inert	[68 - 70]
Coatings for chemical bonding	Hydroxyapatite (HA) coatings, diamond-like carbon coatings, calcium phosphates coatings, and nitride coatings.	Calcium phosphate-based bioceramic, diamond-like carbon films, HA/tricalcium phosphate (TCP), titanium nitride.	Interfacial bond formation.	Bioactive	[69, 71 - 73]
Bone tissue engineering	Bioceramic bone fillers, tissue-engineered bone constructs.	HA/collagen nanocomposite scaffold, Bioglass®, 3D porous graphene/nano-bioglass 58S composite scaffold.	Implant integration with the surrounding tissue.	Bioactive/ biodegradable	[74, 75]
Dental applications	Dental prosthesis, laminate veneers, dental implants, restorative dentistry, preventive dentistry, endodontics.	Alumina, spinel, lithium disilicate reinforced ceramics, and zirconia, porcelain veneers. Zirconia dental implant, fluoroalumino silicate glass powder, β-TCP, amorphous calcium phosphate, and HA, calcium-enriched mixture cement.	Fibrous tissue formation.	Inert	[76 - 81]
Ocular prosthesis	Glass ocular implants, silicon ocular implants, porous HA ocular implants, Al_2O_3 ocular implants.	Mules implant, solid or porous silicon episcleral implants, porous coralline HA sphere, bioceramic orbital implant.	Implant replacement with the surrounding tissue.	Bioactive/ biodegradable	[82 - 84]

Cellular Uptake and Therapeutic Dosage of Bioceramics

The conjugation of different bioactive ions with a specific medicinal benefit is a unique strategy to improve bioceramics' medicinal performance. The therapeutic dose can differ depending on the cell types and bioceramic compositional elements due to their different dissolution constants and degradation rates.

The effects of the ionic compound cuprorivaite ($CaCuSi_4O_{10}$) and calcium silicate (Ca_2SiO_4) bioceramics were investigated on human umbilical vein endothelial cells (HUVECs). Cytocompatibility was observed with concentrations between 0.195–50 mg mL^{-1} post. Also, a concentration of 25 mg mL^{-1} of Ca_2SiO_4 stimulated the proliferation of HUVECs, while Ca_2SiO_4 extracts at 0.78 mg mL^{-1} improved alkaline phosphatase (ALP) activity. Non-toxic concentrations for $CaCuSi_4O_{10}$ bioceramics ranged between 0.195–0.78 mg mL^{-1} for HUVECs. Furthermore, the expression of VEGF was observed with both Ca_2SiO_4 and $CaCuSi_4O_{10}$ bioceramic extracts [85] (Fig. **3**).

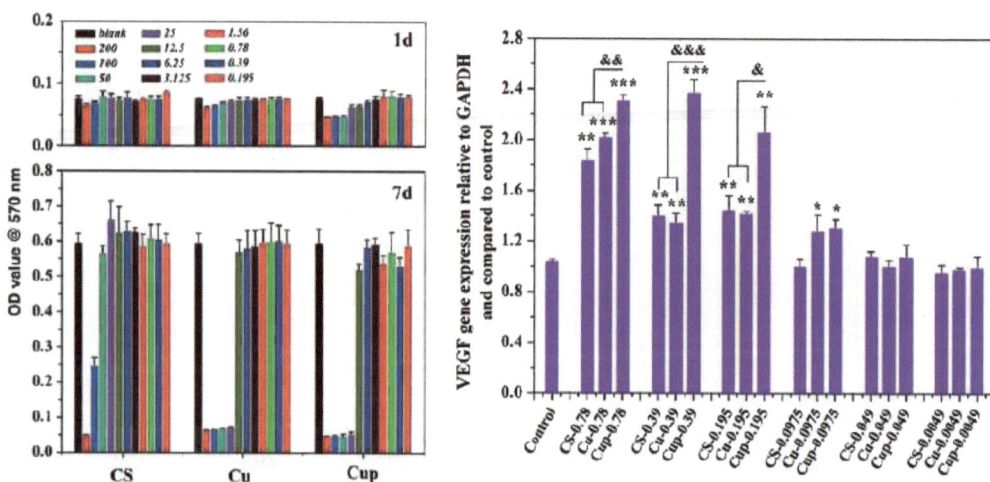

Fig. (3). The effect of $CaCuSi_4O_{10}$ (labelled as "Cup") and Ca_2SiO_4 (labelled as CS) bioceramics on HUVECs cells *in vitro*. **Left:** MTT viability assay demonstrated HUVECs proliferation when cultured in contact with the ionic dissolution product of the bioceramics in the form of conditioned media on day 1 and day 7. **Right:** VEGF was expressed at varying levels in HUVECs exposed to the conditioned media containing $CaCuSi_4O_{10}$, Ca_2SiO_4 extracts, and Cu ions for 72 h [85].

Another study investigated the effect of different concentrations of ionic dissolution products on cellular responses. The study showed that the growth of human periodontal ligament-derived cells (hPDLCs) was promoted by Li$^+$ ions released from Li-doped mesoporous bioactive glass (MBG) scaffolds at a dosage of 5 mM. Furthermore, upregulation of osteogenic markers, osteopontin, osteocalcin and ALP, in hPDLCs on day 3 and day 7 has been reported, while cell proliferation was suppressed at a higher concentration of Li^+, such as 10 mM and 20 mM [86]. Between 13–22 mg L^{-1}, Sr^{2+} from Sr-MBG scaffolds induced ALP activity and osteogenesis-related gene expression of hPDLCs [87].

Wu *et al.* (2013) studied the effect of Cu^{2+} ion products from 5Cu-MBG particles on the viability of human bone marrow stromal cells (hBMSCs). No cytotoxic effects were observed at 14.2 mg L^{-1}, but cell growth was impaired at 56.6 mg L^{-1} of Cu^{2+} ions. When hBMSCs were cultured at Cu^{2+} concentrations of 14.2 mg L^{-1} and 56.6 mg L^{-1}, their ALP activity and VEGF expression were significantly increased compared to MBG extracts without Cu^{2+} ions [88].

Furthermore, a study on the effect of Co^{2+} release from 2Co-MBG scaffolds reported that Co^{2+} aids BMSCs differentiation and enhances VEGF secretion from BMSCs compared to the MBG scaffolds without Co^{2+} [89]. Excessive amounts of ionic dissolution products can harm cells, although low ion concentrations may have little to no impact. Therefore, to successfully balance ion release and cytotoxicity, the administration of appropriate biocompatible bioceramics matrix is essential.

BIOACTIVE CERAMICS FOR CANCER THERAPY

Conventional cancer therapies after operation include chemotherapy and radiotherapy, which can cause severe side effects on the nearby healthy tissues [90]. Improvements are needed to supplement existing medical protocols for the elimination of this fatal disease. Hyperthermia, brachytherapy, and mesoporous bioactive glasses are the new state of art applications for cancer therapy. Biomaterial's research has grown in recent years because of the therapeutic need for tissue repair and regeneration. In addition, the concept of cancer nanomedicine has emerged as a promising method for combining nanotechnology with cancer therapy. The tuneable nature of bioceramics with the ability to release, for example, chemotherapy drugs to the target organ, makes them a perfect platform for cancer treatment. The following sections discuss the most recent studies on bioactive glass ceramics and their applications to cancer therapy.

Bifunctional Bioceramics for Cancer Therapy

Bioceramics have been used for both tissue regeneration and cancer therapy. This bifunctionality of bioceramics can be achieved through two main approaches.

First is the addition of photothermal elements to silicate bioceramics during the fabrication process. Biomaterials containing Cu-ions possess effective heat generation properties due to the d–d electron transition of Cu-ions provoked by near-infrared (NIR) irradiation [91, 92].

The Cu-ions were incorporated into scaffolds by spin coating $CaCuSi_4O_{10}$-nanoparticles (NPs) on the surface of poly(ε-caprolactone) and poly(D,L-lactic acid) (PLA) fibres for the treatment of postoperative wounds and melanoma skin cancer [93]. Significant cancer cell death was observed both *in vitro* and *in vivo* due to the excellent photothermal property of $CaCuSi_4O_{10}$ bioceramics. In addition, $CaCuSi_4O_{10}$ represented an effective healing of the wound of diabetic mice and tumour-bearing mice (Fig. **4**).

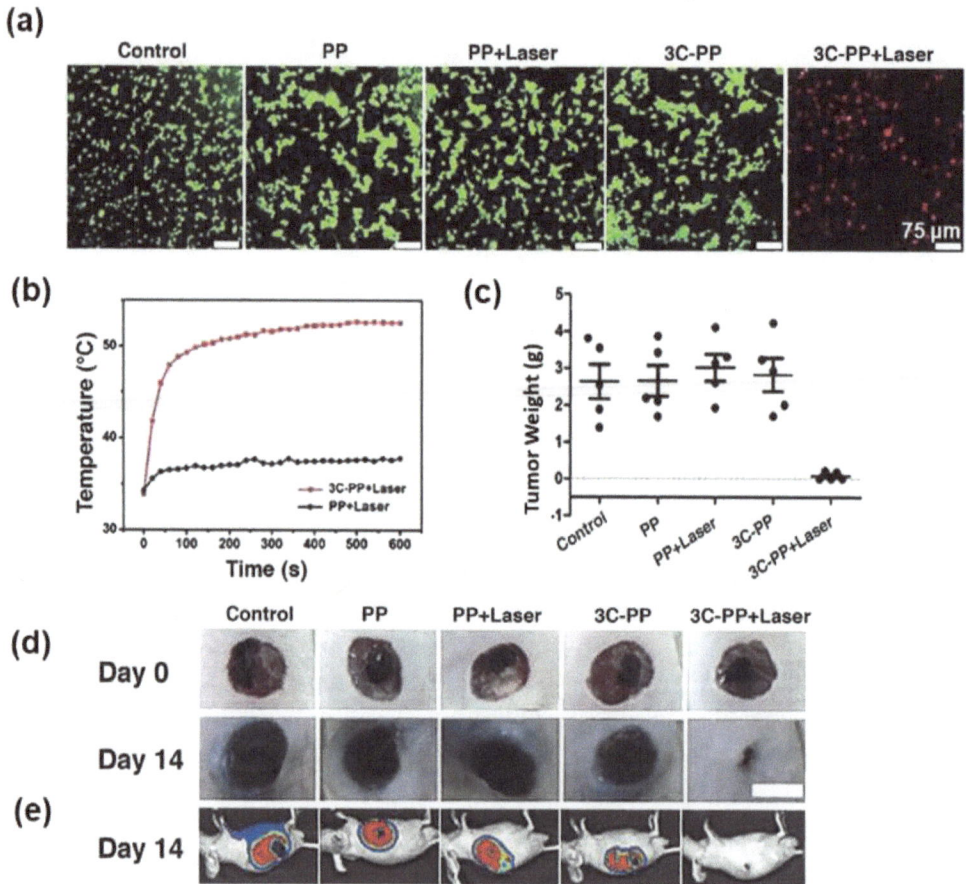

Fig. (4). Antitumour analysis of laser-irradiated PLA and (triple coated)-PLA scaffolds *in vitro* and *in vivo*. **a)** Live/dead staining (green: live cells; red: dead cells) of B16F10 cells treated with PLA fibre or triple coated-PLA scaffolds with or without irradiation (808 nm, 0.85 W cm^{-2}, 15 min). **b)** Temperature heating curve of 3C-PLA scaffolds under wet conditions with different power densities. **c)** Excised tumour weight at five different conditions (day 14). **d)** The effect of the scaffolds on wound healing on day 0 and day 14 in B16F10 melanoma-bearing mice (scale bar: 10 mm). **e)** Whole body and tumour fluorescence images at day 14 [93].

Second, biofunctionality can be achieved through the modification of silicate bioceramics with photothermal reagents. A Ca–P/polydopamine nanolayer was self-assembled on 3D-printed Ca–Si-based nagelschmidtite (Nagel, $Ca_7P_2Si_2O_{16}$) bioceramic scaffolds. The exceptional photothermal effect of polydopamine alongside the repair property of the Nagel bioceramic resulted in the destruction of Saos-2 (osteosarcoma) and MDA-MB-231 cells (breast cancer) in rodent *in vivo* models. Experiments showed that tumour cell growth was inhibited significantly due to the hyperthermia effect (Fig. **5**) [94].

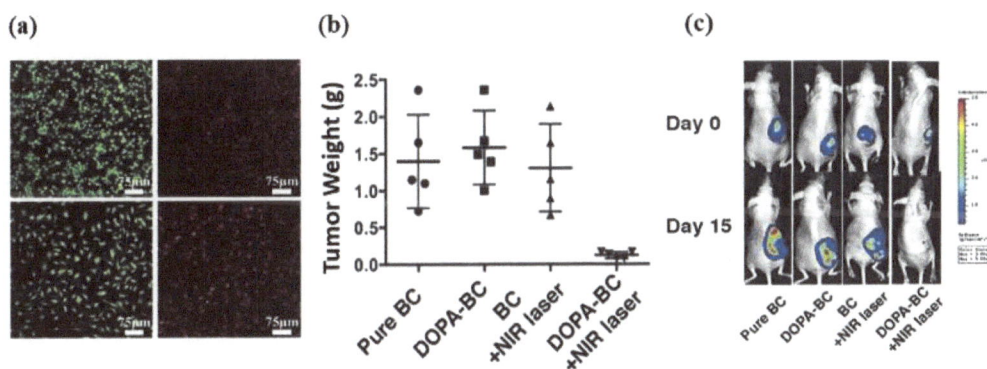

Fig. (5). Antitumour efficacy of pure bioceramic and polydopamine (DOPA-BC) scaffolds *in vitro* and *in vivo*. **a**) Live (green)/dead (red) staining of Saos-2 cells (top) and MDA-MB-231 cells (bottom) on scaffolds with or without 808 nm NIR laser irradiation. **b**) Tumour weight at four different conditions on day 15, and **c**) bioluminescence whole-body images of tumour bearing mice at day 0 and 15 indicate reduced tumour size [94].

Magnetic Bioceramics for Hypothermia Therapy

Implantable magnetic bioceramics have been introduced for the treatment of localized tumours using hyperthermia. Hyperthermia treatment is used for tumours in deep regions of the body. The tumour and surrounding tissues are heated to 43-47 °C, resulting in the loss of membrane integrity, reduction of cancer blood flow and blood vessel destruction, tissue necrosis, and ultimately the destruction of the cancer cells [95 - 97]. The treatment's efficacy can be enhanced when combined with chemotherapy and radiotherapy [98]. However, there are several drawbacks of hyperthermia therapy, including reaching the target temperature and target organs and the potential for overheating of the surrounding healthy tissues. Therefore, implantable magnetic mediators, such as magnetic bioceramics, may be suitable for tumours that are difficult to reach, such as bone tumours [99]. Magnetite-Wollastonite-based thermoseeds are sol-gel glass containing SiO_2, P_2O_5, and CaO at different proportions. These bioactive composites generate heat through the alteration of magnetic fields [100].

Ferromagnetic glass-ceramics have also attracted significant interest for the treatment of deep regional bone cancer. As they are biologically active, ceramic NPs form bone-like apatite, which results in the generation of new bone tissue [101]. Introducing a magnetic phase, such as a ferromagnetic particle (Fe_3O_4), can modify the glass properties so that it can be used as a hyperthermia thermo-seed [102]. Heat treatment of the ceramics *via* magnetic phase crystallization or incorporating magnetic NPs into the glass-ceramic network can also induce the magnetic phase in the glass ceramics [103]. An ongoing challenge for producing magnetic bioceramics is the balance between adequate magnetisation and required bioactivity since an increasing concentration of magnetic phase reduces the ceramics' bioactivity [104].

One of the challenges with hyperthermia therapy is tumour recurrence. Therefore, heat treatment *via* implanted materials can help destroy the remaining malignant cells around the tumour site [105]. One of the earliest ferromagnetic ceramics was introduced by Luderer *et al.* (1983). It has been shown that non-bioactive ferrimagnetic ceramics containing lithium ferrite $LiFe_5O_8$ induced a delay of tumour regrowth and permanently controlled breast adenocarcinoma in mice [106].

Exposure of iron oxide NPs (IONPs) to alternating magnetic fields resulted in thermal energy production by both Brownian fluctuations, the rapid rotation of the particles, and Néel fluctuations, the fluctuation of the magnetic moment within the particle [107]. IONPs can be used for localised magnetic hyperthermia treatment *via* two different approaches. First involves the use of magnetic hyperthermia in combination with conventional methods, such as radiotherapy or chemotherapy, where a resulting synergistic effect enhances the efficacy of other therapies [108]. Second includes magnetic thermoablation, which involves a higher range of temperatures (43-52 °C), to control the tumour growth in glioblastoma [109].

Different coating techniques, including polymers [110], silica [102, 111], and gold [112], have been used to improve the biodistribution, biocompatibility, and toxicity of IONPs. Coated NPs can also act as a micro-carrier for transporting different cytotoxic compounds, such as DOX bound to PEGylated or polyamidoamine-coated IONPs [113, 114] for localised drug delivery. In this strategy, the polymer layer maintains pH 11 for DOX, while it allows its release at pH 5-7, demonstrating pH-sensitive drug release action. Another advantage of IONPs in cancer therapy is their application as magnetic resonance imaging contrast agents, making them theragnostic (therapeutics and diagnostics) agents by utilising their ability to transport and release various therapeutic compounds to a target tissue [115].

Mesoporous Bioceramics as Cancer Drug Delivery Systems

Silica-based mesoporous bioceramics have gained attention due to their usability as drug delivery systems for tissue regeneration [116, 117]. Mesoporous bioceramics are fabricated using a surfactant as a mesostructure template to aid the inorganic precursors' assembly and condensation. A pore network is formed within the silica matrix after the surfactant is removed, which defines their physicochemical characteristics. Mesoporous bioceramics are characterised by high surface area ($\sim$ 1000 m^2 g^{-1}), large pore volume (ca. 1 cm^3 g^{-1}), modifiable mesopore size (2-50 nm), and the formation of homogenic 3D pore structure.

The interaction between these materials and the living tissue is mediated through chemical compounds located on the materials' surface. Silica-based mesoporous bioceramics consist of an ordered silica network with silanol groups occupying the internal pore surface and the outer surface of each particle. Therefore, they have similar physicochemical properties as bioactive glasses and produce nanoapatite coatings with a crystallinity comparable to biological apatite [117].

One benefit of organized mesoporous silicas is their ability to harbour osteoinductive agents to facilitate and accelerate bone formation. The fabrication of mesoporous bioceramics is straightforward, cheap, scalable, and manageable. Furthermore, due to the high proportion of silanol groups on their surface, they offer possibilities for developing multifunctional materials *via* covalent grafting using silane chemistry. The application of mesoporous ceramic NPs for drug delivery purposes in cancer therapy necessitates their functional suspension in biological solutions containing high levels of salt and protein. Pure mesoporous silica particles usually accumulate, especially after exposure to biological fluids, due to the formation of interparticle hydrogen bonds. This can be avoided by incorporating hydrophilic moieties, including phosphonates or poly-ethylen--glycol-derived phospholipids (PEG-lipid), on the particle surface [118].

Mesoporous bioceramics particles coated with polyethyleneimine polyethylene glycol copolymer (PEI-PEG) were investigated in a murine xenograft model to treat human squamous carcinoma. The particles demonstrated strong passive aggregation (12%) inside the tumoral mass and enhanced therapeutic effectiveness compared to free medication [119]. PEGylated silica nano-rattles containing docetaxel have shown 15% more tumour growth suppression compared to Taxotere, the therapeutic version of docetaxel, while still causing less systemic toxicity [120]. Also, mesoporous bioceramics can be used to trap and capture the essential biomolecules in tumour cells, which ultimately interfere with their cellular development. Vivero-Escoto *et al.* (2010) coated the MSC surface with phenanthridinium molecules that bind to cytoplasmic oligonucleotides, such as messenger RNA, inhibiting cell growth and development [121]. Additionally, Ravanbakhsh *et al.* produced mesoporous bioceramics submicron particles through sol-gel synthesis for regional alendronate delivery. Alendronate has vast applications in bone-related diseases, specifically osteosarcoma. The benefit of this method is high encapsulation of alendronate, which results in controlled release of alendronate and exhibits a lower risk of toxicity.

Overall, mesoporous bioceramics provide a dual treatment strategy for bone cancer through the ionic dissolution products, such as Si^{4+}, Ca^{2+}, and PO_4^{3-}, from bioactive glass for bone regeneration, while also allowing local delivery of antitumor reagents [122].

Bioactive Glass and Calcium Phosphates for Cancer Drug Delivery

Due to their mechanical properties and chemical resistance, calcium phosphates have traditionally been used as coatings and fillers for orthopaedic and dental implants, maxillofacial surgery, alveolar ridge augmentation, and powders in total knee and hip surgery [36, 123, 124]. Their biocompatibility, bioactivity, biodegradability, osteoconductivity and osteoinductivity also make them suitable materials for bone regeneration [125, 126].

More recently, due to their pH-dependent solubility, ease of preparation and functionalization and predictable pharmacokinetic pathways, biodegradable calcium phosphate NPs have been identified as reliable cancer therapy candidates [127, 128]. Physicochemical properties of calcium phosphates NPs, such as size, morphology, charge, structure, and surface chemistry, have also been shown to be important factors in determining the path of internalization [129 - 132].

A potential application for bioactive glasses is brachytherapy through the delivery of radioactive glass particles/microspheres to a target organ using either the bloodstream or by injection directly into the tumour. Bioactive glasses are highly biocompatible, biodegradable, and possess the capability to increase the growth of new healthy tissues post-radionuclide therapy [133]. Compared to conventional radiotherapy, brachytherapy allows the use of high doses of β and/or γ rays from the radioactive source in the tumour due to the proximity of the seed to the cancer tissue without damaging the healthy tissue [97]. For example, TheraSphere™ is a commercially available targeted liver cancer therapy using Yttrium-90 (90Y) biocompatible glass microspheres based on brachytherapy. It works through both 90Y radiation and embolization of capillaries by glass microspheres in tumours, which results in tumour shrinkage and destruction [134, 135].

For the insertion of desired genes into target cancer cells, different nano-delivery approaches have been introduced. *In vitro* and *in vivo* studies have shown that calcium phosphate-medicated gene delivery is one of the most promising approaches for targeted cancer therapy [136 - 138].

The precipitation of nucleic acids (*e.g.,* DNA and siRNA) with calcium phosphate is a simple and well-understood procedure. The electrostatic affinity between positively charged calcium ions and the negatively charged phosphate groups in the nucleic acids is believed to be the basis for the strong integration of nucleic acids in calcium phosphate NPs [139, 140]. Furthermore, adding positively charged moieties to the outer layer of calcium phosphate NPs, like amino groups, has been shown to increase nucleic acid loading and improve transfection [141, 142]. The development of nucleic acid-loaded calcium phosphates can be adjusted in a way to use not only nucleic acids but also drug molecules in calcium phosphates for parallel transfection and chemotherapy treatment [143, 144].

However, the production of calcium phosphate-based nucleic acid nanocarriers lacks reproducibility. The control and monitoring of transfection parameters are difficult due to the NP morphology, composition, and crystallinity. Moreover, the transfection efficiency is limited by the ability to efficiently encapsulate and shield nucleic acids from enzymatic attacks within the circulatory system or cells [89, 145].

Additionally, several other techniques for the delivery of cancer drugs into calcium phosphates have been reported, dependent on the size of calcium phosphate particles, the hydrophilicity level of the drug and the type of cancer. One of the most widely used methods involves drugs loading into different kinds of micelles that serve as a scaffold for the mineralization of calcium phosphate nano-shells. The combination of calcium phosphates with drug solutions, such as DOX, α-tocopheryl succinate (α-TOS), and 2-deoxy-D-glucose, has also been identified as a viable drug-loading strategy for *in vivo* studies [146 - 148]. The mechanism applied for drug loading is mainly based on the physical integration of drug molecules to the surface of calcium phosphates through electrostatic or hydrogen bonding. However, one of the major drawbacks of this mechanism is the detachment of weakly bound drugs post-exposure to the body circulation before being taken up by tumour cells. Therefore, it is critical to monitor the distribution of drug-loaded calcium phosphates within target cells and surrounding normal cells to improve their efficacy [36, 123, 124].

Silicate Nano-bioceramics

Halloysite ($Al_2Si_2O_5(OH)_4 \cdot nH_2O$) is tubular aluminosilicate clay that is cheap and naturally available in comparison to other inorganic nanotubes. The internal and external diameter of halloysite nanotubes (HNT) is 10 to 15 nm and 50 to 70 nm, respectively [149]. The cytocompatibility and the intracellular uptake of HNTs have been investigated [150]. In a previous study, a concentration of 75 μg mL^{-1} HNT was non-toxic for cervical cancer cells (HeLa) and breast cancer cells (MCF-7). In addition, fluorescently labelled HNTs were found in the nuclear surrounding following cellular uptake. The concentration of 0.5 mg mL^{-1} HNTs was safe for cells. Due to their hollow nanostructure and slow release, HNTs are considered effective nanocarriers for the delivery of a wide range of biologically active molecules, specifically antineoplastic drugs [151 - 153]. In a previous study, paclitaxel was encapsulated in HNTs for the treatment of colon cancer. Therefore, compounds were integrated into tablet form to inhibit paclitaxel release in the acidic environment of the stomach and to enhance the efficacy of paclitaxel by releasing it into the intestine. The HNTs surface was covered with pH-responsive polymer to monitor drug release in various pH conditions [154].

Laponites (LAP) consist of silicate clay that can be broken into non-toxic components, such as Na$^+$, Li$^+$, Mg^{2+}, and Si(OH)$_4$. LAPs are non-toxic at a concentration lower than 1 mg mL^{-1}. In addition, in the absence of any external osteoinductive stimuli, LAP may actively stimulate the osteogenic differentiation of human mesenchymal stem cells (hMSCs) *in vitro*. LAP bioceramics have excellent properties for tissue engineering applications, including surface hydrophilicity, exceptional serum absorption capacity and outstanding cytocompatibility and hemocompatibility. Another important characteristic is the formation of the HA layer following exposure to simulated body fluids, making them suitable for bone tissue regeneration [155]. Drug encapsulation for cancer therapy is one of the most common applications of LAP. Xu *et al.* (2018) developed an LAP-based nanoplatform that combines photodynamic and photothermal therapy for cancer treatment through overexpression of integrin $\alpha v \beta 3$ [156]. Indocyanine green (ICG) was first integrated with LAP nanodiscs, followed by coverage with polydopamine (PDA), and ultimately conjugation as target agent with polyethylene glycol–arginine–glycine–aspartic acid (PEG–RGD). The ICG/LAP–PDA–PEG–RGD (ILPR) NPs produced reactive oxygen species (ROS) through NIR light irradiation, and also improved photothermal conversion performance. Furthermore, *via* RGD-mediated targeting, ILPR NPs can directly target cancer cells, thus improving the cellular uptake. In another study, a LAP-based nanoplatform was designed incorporating AuNPs for theranostic purposes [157]. LAP-AuNPS were conjugated with hyaluronic acid for tumour-targeted CT imaging and DOX antineoplastic drugs for chemotherapy. After targeting, cancer cells overexpressed CD44, which then induced the release of the DOX, leading to the inhibition of cancer cell proliferation.

Mg ions play a critical role in human metabolism. Because of their low level of cytotoxicity, magnesium silicate hollow materials have been developed for drug delivery for cancer treatment [157, 158]. Glycol-modified hollow magnesium silicate NPs (HMMSNs) were designed and loaded with the drug DOX [159]. A Mg^{2+}-dependent DNAzyme was integrated into the HMMSNs, which trapped DOX within the HMMSNs. The acidic condition of tumour cells induced the biodegradation of HMMSNs, followed by which the presence of Mg^{2+} stimulated the Mg^{2+}-dependent DNAzyme, and ultimately, DOX was released. Another magnesium silicate platform for drug delivery has been designed by Sun *et al.*, named iron oxide@magnesium silicate (SPIO@MS) nanospheres. The unique yolk-shell porous structure of the nanospheres and the high surface area allowed the loading of DOX, which was released through pH-responsive degradation of the SPIO@MS nanosphere [160].

BIOCERAMICS IN CLINICAL TRIALS

The use of bioceramics for cancer therapy is a rapidly expanding field, with numerous clinical trials being conducted. Fig. (6) demonstrates different stages identifying the clinical need for fabrication and clinical application of the novel bioceramic platform. Information about past, current and upcoming clinical trials can be found on repositories, such as clinicaltrials.gov, ciscrp.org, or isrctn.com.

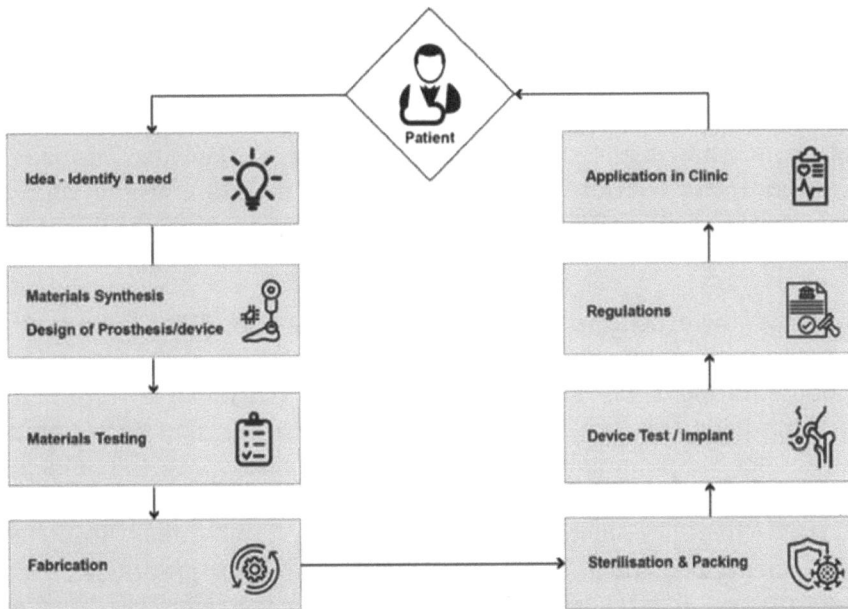

Fig. (6). Schematic view of different stages for the development of a novel bioceramic from generating the idea to the point where said material can be used in the patient; adapted from [161].

One of the most common applications for bioceramics in clinical trials is the treatment of bone defects, which result from benign bone cancers post-surgery. Traditionally, these defects are treated using allografts and autografts due to properties that are essential for a functional and optimal bone graft. However, limited supply, morbidity at the donor site, and potential disease transmission are some of the major drawbacks. Synthetic bone grafts are an excellent alternative for both autografts and allografts, and can eliminate their disadvantages. Often these synthetic bone substitutes are based on calcium phosphates because they mimic the structure and composition of bone minerals and play an important role in bone remodelling and regeneration.

Two ceramic bone graft substitutes, namely bioactive glass (BG) under the name Bonalive®, and β-TCP named ChronOs, were compared for filling bone defects after surgical removal of benign bone tumour (NCT00841152) between March 2009 and December 2018. According to the preclinical studies, BG was expected to be more effective than TCP in promoting defect repair and functional recovery following surgery. In this study, 120 participants were enrolled (> 18 years, both male and female) who were diagnosed with a primary or recurring benign bone tumour that needed surgical removal and subsequent filling of the defects. The patients were categorised into two strata. Stratum I investigated hand lesions, comparing BG and TCP with autologous bone grafts from the iliac crest as the standard of care. Stratum II investigated large long-bone lesions comparing the clinical performance of BG, TCP, and allogeneic bone graft (frozen femoral head) as control. After the treatment, hand-grip strength (Stratum I only), healing of cortical bone window with computed tomography (CT) scan (Stratum II only), biomaterial incorporation through radiographs, pain intensity, surgical wound healing, bone defect healing, and soft tissue complications in the surgical area were evaluated. The clinical trial had not revealed the ultimate results of the study [162].

In another trial, the efficacy of bioactive glass (BG)-S53P4 had been investigated and compared with autogenic bone (AB) grafts in patients diagnosed with benign bone tumours during 1993–1997. In this study, 25 participants were enrolled (> 18 years, both male and female) where a primary tumour had been found *via* X-ray due to either the patients' pain or pathologic fracture. All fractures had been treated before the participants were categorised into two groups. One group received BonAlive® biomaterials (BG group) and the other group received autogenic graft transplanted from iliac crest bone (AB group). Based on the results, significant differences in bone remodelling were observed between BG and AB groups, where AB outperformed BG. However, 36 months post-implantation, no difference was reported in the defect volume between the two groups. Results from CT scan and X-ray showed a high bone density in the BG group. Magnetic resonance imaging also demonstrated mainly or partly fatty bone marrow; in addition, the residue of BG granules had been visualised in groups with large bone tumours. Therefore, it was concluded that BG-S53P4 is a promising material with appropriate safety, osteoinductivity, bone bonding and antimicrobial properties for the treatment of benign bone tumours [163].

Furthermore, injectable bi-phasic calcium phosphate (BCP) ceramic bone substitute (CERAMENT™ |Bone void filler) was investigated in patients with benign bone tumours for bone remodelling and regeneration (NCT02567084). The substitute consists of 60% (wt./wt.) synthetic calcium sulphate ($CaSO_4$) and 40% (wt./wt.) HA. CERAMENT™ was also assessed for its ability to transform bone cysts into the bone and potentially stimulate bone growth in cystic areas not filled with the substitute. Here, 14 participants with benign bone tumour took part in the study, which started in February 2011 and was completed in December 2014. Percutaneous injection into cysts or mini-invasive surgery to remove solid tumours were performed. Subsequently, the defects were filled with CERAMENT™. The filler was mixed with a water-soluble radio-contrast agent iohexol (180 mg mL^{-1}) to make the material radiopaque. Participants were permitted to bear full weight following the procedure and were monitored clinically and radiologically for one year. After 12 months, bone healing and remodelling were observed through CT-scan and X-ray using Modified Neer classification of radiological results. Also, the cyst and bone formation volume were measured [164].

In a more recent study, bioceramic microspheres were utilised for the administration of anticancer drugs. Here, TheraSphere® (NCT03295006) researchers investigated the combination of radioembolization with sorafenib, the standard treatment for hepatocellular carcinoma (HCC). TheraSphere® consists of millions of microscopic glass beads containing radioactive yttrium (Y-90). Directly injected live, the beads emit radiation over several weeks damaging malignant cancer cells as well as the cancer's blood supply. It is assumed that radioembolisation before commencing sorafenib treatment may be more effective in extending life expectancy than sorafenib alone. In this trial, half of the participants underwent radioembolisation before starting sorafenib, whereas the other half only received sorafenib. The main objective is to reduce or stop the progression of liver tumours without significantly increasing the number or severity of side effects. In total, 526 participants (male and female, >18 years) were enrolled in this trial. So far, positive results have been reported for the use of TheraSphere® Y-90 glass microspheres. TheraSphere® therapy was evaluated in patients with HCC, which is the most common type of primary liver cancer, using a dosing method known as multicompartment dosimetry, which maximizes the dose of Y-90 targeting the tumour while minimizing the radiation dose reaching normal liver tissue. The dose administered for each patient's liver tissue was calculated retroactively using imaging software. The medication was shown to be safe and well tolerated, with just 4.8% of patients reporting adverse effects, which were characterized as grade 3 hyperbilirubinemia. Data also revealed a link between the amount of radiation absorbed by the tumour and a three-year increase in survival probability, with a median overall survival of 20.3 months [165, 166].

A further clinical trial investigated newly designed intratumoral thermotherapy employing magnetic NPs in combination with percutaneous irradiation for the treatment of recurrent glioblastoma multiforme (GBM) [109]. Magnetic NPs called nano-cancer therapy were composed of biocompatible IONPs, which were injected into the tumour mass before being exposed to an alternating magnetic field to produce heat. In total, 59 patients with GBM took part in this study, which was performed in the timeframe between April 2005 and September 2009. Participants received neuronavigational controlled intratumoral instillation of an aqueous dispersion of magnetic IONPs. Fractionated stereotactic radiation was used in combination with the treatment. Overall survival after the diagnosis of a first tumour recurrence was the primary research outcome, whereas overall survival after the diagnosis of an initial tumour was the secondary outcome. From the point of the first tumour recurrence diagnosis, the overall survival was 13.4 months whilst the interval of the media time was 8.0 months between the first diagnosis and the primary tumour recurrence. The tumour volume significantly correlated with the survival rate. The new therapeutic strategy had very minor adverse effects and no major side effects. Overall, this study indicated that thermotherapy employing magnetic NPs in combination with a low radiation dosage for the treatment of GBM is effective and safe. A significant increase in the survival rate and longer times between treatment and first tumour recurrence were observed compared to conventional therapeutic approaches [109].

CONCLUSION AND FUTURE PERSPECTIVE

The discovery and development of novel bioceramics for a wide range of therapeutic applications offer great promise for health care and the medical field. Bioceramics in conjugation with other therapeutic approaches for cancer therapy have been introduced in many different studies. Amongst these, MBGs have demonstrated promising routes for cancer therapy due to their excellent bioactivity and drug delivery [167]. However, there is a lack of large-size MBGs on the market, and the very first MBGs scaffold with appropriate pores was introduced only recently and is still in the early stages of research, especially for clinical trials. One of the biggest challenges with glass ceramics is the formation of the crystal during fabrication, specifically for producing porous and multi-doped glass ceramic [168]. Crystallisation can compromise the bioactivity of the glasses, which can diminish the ion exchange between the glass and the aqueous solution. The temperatures for glass transition (T_g) and melting temperature (T_m) play an important role in the characteristics of the glass and its fabrication process [169]. Therefore, more research into MBGs fabrication conditions is needed. Another challenge is understanding *in vivo* drug release profiles, degradation after implantation and cell metabolism before clinical trials could commence. Bioactive glass ceramics have been introduced with high *in vivo* degradation rates to eliminate immune reactions or the requirement for revision surgery. MBGs have the potential to be successful cancer therapy platforms if the mechanisms of their *in vivo* performance and related biological activities can be better understood [111, 170, 171].

Furthermore, multi-doping techniques for fabricating bioceramics have been promising for cancer therapy. However, it is important to investigate the bioceramics' compositional characteristics before introducing additional elements for cancer therapy as it can affect properties, such as ion release profile, mechanical strength, crystallisation, the accuracy of the porosity, and the overall efficacy for the therapeutic goal [35].

Also, currently, there are many studies on the application of bioactive ceramics for cancer in hard tissue, while not much consideration has been given to the treatment of cancer in soft tissues. Therefore, further research is required to open avenues toward management and treatment of soft tissue cancers, such as leiomyosarcoma, gastrointestinal stromal tumours, liposarcoma, fibrosarcoma and Ewing's sarcoma [172].

Many novel bioceramics have been proposed for cancer therapy, and further research, including *in vitro* and *in vivo* studies, would be extremely beneficial to gain a better understanding of the practical factors that lead to effective therapies. Simulation approaches where an analytical model can be created *via* computational modelling software are effective methods for evaluating the material's compositional properties in advance and eliminating time-consuming experimental trials [173]. Currently, the functionality of new compounds for cancer therapy is tested based on trial-and-error methods. Due to their non-crystalline and multicomponent characteristics, rationalising the behaviour of these materials is a difficult endeavour. This will be a significant step forward in patient health care, and it will be one of the most significant contributions of bioactive glass ceramic research to humanity. The future of cancer therapy is to provide patients with even more personalised therapy options. Depending on the genetic alterations that occur in cancer cells, a variety of therapeutic approaches could be achieved using bioceramics. However, many challenges remain unmet, including mechanical strength, controlled ion release, and accurate porosity without crystallisation, as well as safety and efficacy of bioceramics for cancer therapy.

CONSENT FOR PUBLICATION

Not applicable.

CONFLICT OF INTEREST

The author declares no conflict of interest, financial or otherwise.

ACKNOWLEDGEMENT

Authors would like to thank Nottingham Trent University, Aston University and Sarcoma UK for the financial support.

REFERENCES

[1]   Cancer Statistics National Cancer Institute 2021. https://www.cancer.gov/about-cancer/understanding/statistics

[2]   Available from: https://www.fightcancer.org/sites/default/files/National%20Documents/Costs-of-Cancer-2020-10222020.pdf

[3]   https://www.gov.uk/government/publications/2010-to-2015-government-policy-cancer-research-and-treatment/2010-to-2015-government-policy-cancer-research-and-treatment

[4]   DeVita VT, Lawrence TS, Rosenberg SA. Lawrence TS, Rosenberg SA. Cancer: principles & practice of oncology: primer of the molecular biology of cancer. Lippincott Williams & Wilkins 2012.

[5]   Skubitz KM, D'Adamo DR. Sarcoma. Mayo Clin Proc 2007; 82(11): 1409-32.
        [http://dx.doi.org/10.4065/82.11.1409] [PMID: 17976362]

[6]   Leukemia SS. Stem Cell Rev 2005; 1(3): 197-205.
        [http://dx.doi.org/10.1385/SCR:1:3:197] [PMID: 17142856]

[7]   Weinberg RA. The biology of cancer. Garland Science; 2013.
        [http://dx.doi.org/10.1201/9780429258794]

[8]   Anand P, Kunnumakara AB, Sundaram C, *et al.* Cancer is a preventable disease that requires major lifestyle changes. Pharm Res 2008; 25(9): 2097-116.
        [http://dx.doi.org/10.1007/s11095-008-9661-9] [PMID: 18626751]

[9]   Parsa N. Environmental factors inducing human cancers. Iran J Public Health 2012; 41(11): 1-9.
        [PMID: 23304670]

[10]   Vallet-Regí M, Ruiz-Hernández E. Bioceramics: from bone regeneration to cancer nanomedicine. Adv Mater 2011; 23(44): 5177-218.
        [http://dx.doi.org/10.1002/adma.201101586] [PMID: 22009627]

[11]   Wang JJ, Lei KF, Han F. Tumor microenvironment: recent advances in various cancer treatments. Eur Rev Med Pharmacol Sci 2018; 22(12): 3855-64.
        [PMID: 29949179]

[12]   Visvader JE. Cells of origin in cancer. Nature 2011; 469(7330): 314-22.
        [http://dx.doi.org/10.1038/nature09781] [PMID: 21248838]

[13]   Patel A. Benign *vs.* Malignant Tumors. JAMA Oncol 2020; 6(9): 1488-8.

[http://dx.doi.org/10.1001/jamaoncol.2020.2592] [PMID: 32729930]

[14] Frantz C, Stewart KM, Weaver VM. The extracellular matrix at a glance. J Cell Sci 2010; 123(24): 4195-200.
[http://dx.doi.org/10.1242/jcs.023820] [PMID: 21123617]

[15] Winkler J, Abisoye-Ogunniyan A, Metcalf KJ, Werb Z. Concepts of extracellular matrix remodelling in tumour progression and metastasis. Nat Commun 2020; 11(1): 5120.
[http://dx.doi.org/10.1038/s41467-020-18794-x] [PMID: 33037194]

[16] Kai F, Drain AP, Weaver VM. The extracellular matrix modulates the metastatic journey. Dev Cell 2019; 49(3): 332-46.
[http://dx.doi.org/10.1016/j.devcel.2019.03.026] [PMID: 31063753]

[17] Wang M, Zhao J, Zhang L, *et al.* Role of tumor microenvironment in tumorigenesis. J Cancer 2017; 8(5): 761-73.
[http://dx.doi.org/10.7150/jca.17648] [PMID: 28382138]

[18] Hanahan D, Weinberg RA. Hallmarks of cancer: the next generation 144(5): 646-74. cell. 2011 Mar 4;144(5):646-674. 2011;

[19] Levental KR, Yu H, Kass L, *et al.* Matrix crosslinking forces tumor progression by enhancing integrin signaling. Cell 2009; 139(5): 891-906.
[http://dx.doi.org/10.1016/j.cell.2009.10.027] [PMID: 19931152]

[20] Edovitsky E, Elkin M, Zcharia E, Peretz T, Vlodavsky I. Heparanase gene silencing, tumor invasiveness, angiogenesis, and metastasis. J Natl Cancer Inst 2004; 96(16): 1219-30.
[http://dx.doi.org/10.1093/jnci/djh230] [PMID: 15316057]

[21] Lerner I, Hermano E, Zcharia E, *et al.* Heparanase powers a chronic inflammatory circuit that promotes colitis-associated tumorigenesis in mice. J Clin Invest 2011; 121(5): 1709-21.
[http://dx.doi.org/10.1172/JCI43792] [PMID: 21490396]

[22] Fridman WH, Pagès F, Sautès-Fridman C, Galon J. The immune contexture in human tumours: impact on clinical outcome. Nat Rev Cancer 2012; 12(4): 298-306.
[http://dx.doi.org/10.1038/nrc3245] [PMID: 22419253]

[23] Sullivan R, Alatise OI, Anderson BO, *et al.* Global cancer surgery: delivering safe, affordable, and timely cancer surgery. Lancet Oncol 2015; 16(11): 1193-224.
[http://dx.doi.org/10.1016/S1470-2045(15)00223-5] [PMID: 26427363]

[24] Chabner BA, Roberts TG Jr. Chemotherapy and the war on cancer. Nat Rev Cancer 2005; 5(1): 65-72.
[http://dx.doi.org/10.1038/nrc1529] [PMID: 15630416]

[25] Ma J, Waxman DJ. Combination of antiangiogenesis with chemotherapy for more effective cancer treatment. Mol Cancer Ther 2008; 7(12): 3670-84.
[http://dx.doi.org/10.1158/1535-7163.MCT-08-0715] [PMID: 19074844]

[26] Baumann M, Krause M, Overgaard J, *et al.* Radiation oncology in the era of precision medicine. Nat Rev Cancer 2016; 16(4): 234-49.
[http://dx.doi.org/10.1038/nrc.2016.18] [PMID: 27009394]

[27] Chen HHW, Kuo MT. Improving radiotherapy in cancer treatment: Promises and challenges. Oncotarget 2017; 8(37): 62742-58.
[http://dx.doi.org/10.18632/oncotarget.18409] [PMID: 28977985]

[28] Schuster M, Nechansky A, Kircheis R. Cancer immunotherapy. Biotechnol J 2006; 1(2): 138-47.
[http://dx.doi.org/10.1002/biot.200500044] [PMID: 16892244]

[29] Pardoll DM. The blockade of immune checkpoints in cancer immunotherapy. Nat Rev Cancer 2012; 12(4): 252-64.
[http://dx.doi.org/10.1038/nrc3239] [PMID: 22437870]

[30] Rosenberg SA, Yang JC, Restifo NP. Cancer immunotherapy: moving beyond current vaccines. Nat

Med 2004; 10(9): 909-15.
[http://dx.doi.org/10.1038/nm1100] [PMID: 15340416]

[31]   Riley RS, June CH, Langer R, Mitchell MJ. Delivery technologies for cancer immunotherapy. Nat Rev Drug Discov 2019; 18(3): 175-96.
[http://dx.doi.org/10.1038/s41573-018-0006-z] [PMID: 30622344]

[32]   Deli T, Orosz M, Jakab A. Hormone replacement therapy in cancer survivors–review of the literature. Pathol Oncol Res 2020; 26(1): 63-78.
[http://dx.doi.org/10.1007/s12253-018-00569-x] [PMID: 30617760]

[33]   Pérez-Herrero E, Fernández-Medarde A. Advanced targeted therapies in cancer: Drug nanocarriers, the future of chemotherapy. Eur J Pharm Biopharm 2015; 93: 52-79.
[http://dx.doi.org/10.1016/j.ejpb.2015.03.018] [PMID: 25813885]

[34]   Gul H, Khan M, Khan AS. Bioceramics: Types and clinical applications. Handbook of Ionic Substituted Hydroxyapatites 2020; 53-83.

[35]   Jones JR, Gibson IR. Ceramics, glasses, and glass-ceramics: Basic principles. Biomater Sci 2020; 289-305.

[36]   Hench LL. Bioceramics, a clinical success. Am Ceram Soc Bull 1998; 77(7): 67-74.

[37]   Hench LL, Best SM. Ceramics, glasses, and glass-ceramics: basic principles. Biomater Sci 2013; 128-51.

[38]   Hench LL, Wilson J. Surface-Active Biomaterials. Science 1984; 226(4675): 630-6.
[http://dx.doi.org/10.1126/science.6093253] [PMID: 6093253]

[39]   Hench LL, Polak JM. Third-generation biomedical materials. Science 2002; 295(5557): 1014-7.
[http://dx.doi.org/10.1126/science.1067404] [PMID: 11834817]

[40]   Vallet-Regi M. Bio-ceramics with clinical applications. John Wiley & Sons; 2014.
[http://dx.doi.org/10.1002/9781118406748]

[41]   Goncalves AD, Balestri W, Reinwald Y. Biomedical Implants for Regenerative Therapies. Biomaterials 2020.

[42]   Gross U, Kinne R, Schmitz HJ, Strunz V. The Response of bone to surface-active glasses glass-ceramics. CRC critical reviews in biocompatibility 1988; 4(2): 155-79.

[43]   de Groot K. Bioceramics Calcium Phosphate. CRC Press; 2018.
[http://dx.doi.org/10.1201/9781351070133]

[44]   Coronella JA, Telleman P, Kingsbury GA, Truong TD, Hays S, Junghans RP. Evidence for an antigen-driven humoral immune response in medullary ductal breast cancer. Cancer Res 2001; 61(21): 7889-99.
[PMID: 11691809]

[45]   Schioppa T, Moore R, Thompson RG, *et al.* B regulatory cells and the tumor-promoting actions of TNF-α during squamous carcinogenesis. Proc Natl Acad Sci USA 2011; 108(26): 10662-7.
[http://dx.doi.org/10.1073/pnas.1100994108] [PMID: 21670304]

[46]   Horikawa M, Minard-Colin V, Matsushita T, Tedder TF. Regulatory B cell production of IL-10 inhibits lymphoma depletion during CD20 immunotherapy in mice. J Clin Invest 2011; 121(11): 4268-80.
[http://dx.doi.org/10.1172/JCI59266] [PMID: 22019587]

[47]   Tachibana T, Onodera H, Tsuruyama T, *et al.* Increased intratumor Valpha24-positive natural killer T cells: a prognostic factor for primary colorectal carcinomas. Clin Cancer Res 2005; 11(20): 7322-7.
[http://dx.doi.org/10.1158/1078-0432.CCR-05-0877] [PMID: 16243803]

[48]   Condeelis J, Pollard JW. Macrophages: obligate partners for tumor cell migration, invasion, and metastasis. Cell 2006; 124(2): 263-6.

[http://dx.doi.org/10.1016/j.cell.2006.01.007] [PMID: 16439202]

[49] Murdoch C, Giannoudis A, Lewis CE. Mechanisms regulating the recruitment of macrophages into hypoxic areas of tumors and other ischemic tissues. Blood 2004; 104(8): 2224-34.
[http://dx.doi.org/10.1182/blood-2004-03-1109] [PMID: 15231578]

[50] Sinha P, Clements VK, Bunt SK, Albelda SM, Ostrand-Rosenberg S. Cross-talk between myeloid-derived suppressor cells and macrophages subverts tumor immunity toward a type 2 response. J Immunol 2007; 179(2): 977-83.
[http://dx.doi.org/10.4049/jimmunol.179.2.977] [PMID: 17617589]

[51] Bronte V, Serafini P, Mazzoni A, Segal DM, Zanovello P. L-arginine metabolism in myeloid cells controls T-lymphocyte functions. Trends Immunol 2003; 24(6): 301-5.
[http://dx.doi.org/10.1016/S1471-4906(03)00132-7] [PMID: 12810105]

[52] Meredith MM, Liu K, Darrasse-Jeze G, *et al.* Expression of the zinc finger transcription factor zDC (Zbtb46, Btbd4) defines the classical dendritic cell lineage. J Exp Med 2012; 209(6): 1153-65.
[http://dx.doi.org/10.1084/jem.20112675] [PMID: 22615130]

[53] Satpathy AT, Kc W, Albring JC, *et al.* Zbtb46 expression distinguishes classical dendritic cells and their committed progenitors from other immune lineages. J Exp Med 2012; 209(6): 1135-52.
[http://dx.doi.org/10.1084/jem.20120030] [PMID: 22615127]

[54] De Larco JE, Wuertz BRK, Furcht LT. The potential role of neutrophils in promoting the metastatic phenotype of tumors releasing interleukin-8. Clin Cancer Res 2004; 10(15): 4895-900.
[http://dx.doi.org/10.1158/1078-0432.CCR-03-0760] [PMID: 15297389]

[55] Youn JI, Gabrilovich DI. The biology of myeloid-derived suppressor cells: The blessing and the curse of morphological and functional heterogeneity. Eur J Immunol 2010; 40(11): 2969-75.
[http://dx.doi.org/10.1002/eji.201040895] [PMID: 21061430]

[56] Sugimoto H, Mundel TM, Kieran MW, Kalluri R. Identification of fibroblast heterogeneity in the tumor microenvironment. Cancer Biol Ther 2006; 5(12): 1640-6.
[http://dx.doi.org/10.4161/cbt.5.12.3354] [PMID: 17106243]

[57] Tomasek JJ, Gabbiani G, Hinz B, Chaponnier C, Brown RA. Myofibroblasts and mechano-regulation of connective tissue remodelling. Nat Rev Mol Cell Biol 2002; 3(5): 349-63.
[http://dx.doi.org/10.1038/nrm809] [PMID: 11988769]

[58] Willis BC, duBois RM, Borok Z. Epithelial origin of myofibroblasts during fibrosis in the lung. Proc Am Thorac Soc 2006; 3(4): 377-82.
[http://dx.doi.org/10.1513/pats.200601-004TK] [PMID: 16738204]

[59] Brittan M, Hunt T, Jeffery R, *et al.* Bone marrow derivation of pericryptal myofibroblasts in the mouse and human small intestine and colon. Gut 2002; 50(6): 752-7.
[http://dx.doi.org/10.1136/gut.50.6.752] [PMID: 12010874]

[60] Spaeth EL, Dembinski JL, Sasser AK, *et al.* Mesenchymal stem cell transition to tumor-associated fibroblasts contributes to fibrovascular network expansion and tumor progression. PloS one 2009; 4(4): 1-11.
[http://dx.doi.org/10.1371/journal.pone.0004992]

[61] Orimo A, Gupta PB, Sgroi DC, *et al.* Stromal fibroblasts present in invasive human breast carcinomas promote tumor growth and angiogenesis through elevated SDF-1/CXCL12 secretion. Cell 2005; 121(3): 335-48.
[http://dx.doi.org/10.1016/j.cell.2005.02.034] [PMID: 15882617]

[62] Nieman KM, Kenny HA, Penicka CV, *et al.* Adipocytes promote ovarian cancer metastasis and provide energy for rapid tumor growth. Nat Med 2011; 17(11): 1498-503.
[http://dx.doi.org/10.1038/nm.2492] [PMID: 22037646]

[63] Carmeliet P, Jain RK. Molecular mechanisms and clinical applications of angiogenesis. Nature 2011; 473(7347): 298-307.

[http://dx.doi.org/10.1038/nature10144] [PMID: 21593862]

[64]   Armulik A, Genové G, Betsholtz C. Pericytes: developmental, physiological, and pathological perspectives, problems, and promises. Dev Cell 2011; 21(2): 193-215.
[http://dx.doi.org/10.1016/j.devcel.2011.07.001] [PMID: 21839917]

[65]   Cooke VG, LeBleu VS, Keskin D, *et al.* Pericyte depletion results in hypoxia-associated epithelial-t-
-mesenchymal transition and metastasis mediated by met signaling pathway. Cancer Cell 2012; 21(1): 66-81.
[http://dx.doi.org/10.1016/j.ccr.2011.11.024] [PMID: 22264789]

[66]   Swartz MA, Lund AW. Lymphatic and interstitial flow in the tumour microenvironment: linking mechanobiology with immunity. Nat Rev Cancer 2012; 12(3): 210-9.
[http://dx.doi.org/10.1038/nrc3186] [PMID: 22362216]

[67]   Alitalo K. The lymphatic vasculature in disease. Nat Med 2011; 17(11): 1371-80.
[http://dx.doi.org/10.1038/nm.2545] [PMID: 22064427]

[68]   Gamble D, Jaiswal PK, Lutz I, Johnston KD. The use of ceramics in total hip arthroplasty. Ortho & Rheum 2017; 4(3): 555636.

[69]   McEntire BJ, Bal BS, Rahaman MN, Chevalier J, Pezzotti G. Ceramics and ceramic coatings in orthopaedics. J Eur Ceram Soc 2015; 35(16): 4327-69.
[http://dx.doi.org/10.1016/j.jeurceramsoc.2015.07.034]

[70]   Nickoli MS, Hsu WK. Ceramic-based bone grafts as a bone grafts extender for lumbar spine arthrodesis: a systematic review. Global Spine J 2014; 4(3): 211-6.
[http://dx.doi.org/10.1055/s-0034-1378141] [PMID: 25083364]

[71]   Mohseni E, Zalnezhad E, Bushroa AR. Comparative investigation on the adhesion of hydroxyapatite coating on Ti–6Al–4V implant: A review paper. Int J Adhes Adhes 2014; 48: 238-57.
[http://dx.doi.org/10.1016/j.ijadhadh.2013.09.030]

[72]   Dearnaley G, Arps JH. Biomedical applications of diamond-like carbon (DLC) coatings: A review. Surf Coat Tech 2005; 200(7): 2518-24.
[http://dx.doi.org/10.1016/j.surfcoat.2005.07.077]

[73]   Prakash L. Ceramics in arthroplasty, arthritis and orthopaedics. Res Arthritis Bone Study 2018; 1(1): 1-4.

[74]   Abbasi Z, Bahrololoom ME, Shariat MH, Bagheri RA. Bioactive glasses in dentistry: a review. J Dent Biomater 2015; 2(1): 1-9.

[75]   Shadjou N, Hasanzadeh M. Graphene and its nanostructure derivatives for use in bone tissue engineering: Recent advances. J Biomed Mater Res A 2016; 104(5): 1250-75.
[http://dx.doi.org/10.1002/jbm.a.35645] [PMID: 26748447]

[76]   Pascotto R, Pini N, Aguiar FHB, Lima DANL, Lovadino JR, Terada RSS. Advances in dental veneers: materials, applications, and techniques. Clin Cosmet Investig Dent 2012; 4: 9-16.
[http://dx.doi.org/10.2147/CCIDE.S7837] [PMID: 23674920]

[77]   Özkurt Z, Kazazoğlu E. Zirconia dental implants: a literature review. J Oral Implantol 2011; 37(3): 367-76.
[http://dx.doi.org/10.1563/AAID-JOI-D-09-00079] [PMID: 20545529]

[78]   Najeeb S, Khurshid Z, Zafar M, *et al.* Modifications in glass ionomer cements: nano-sized fillers and bioactive nanoceramics. Int J Mol Sci 2016; 17(7): 1134-48.
[http://dx.doi.org/10.3390/ijms17071134] [PMID: 27428956]

[79]   Makeeva IM, Polyakova MA, Avdeenko OE, Paramonov YO, Kondrat'ev SA, Pilyagina AA. Effect of long term application of toothpaste Apadent Total Care Medical nano-hydroxyapatite. Stomatologia (Mosk) 2016; 95(4): 34-6.
[http://dx.doi.org/10.17116/stomat201695434-36] [PMID: 27636759]

[80]   Hegde R, Thakkar J. Comparative evaluation of the effects of casein phosphopeptide-amorphous calcium phosphate (CPP-ACP) and xylitol-containing chewing gum on salivary flow rate, pH and buffering capacity in children: An *in vivo* study. J Indian Soc Pedod Prev Dent 2017; 35(4): 332-7.
[http://dx.doi.org/10.4103/JISPPD.JISPPD_147_17] [PMID: 28914246]

[81]   Utneja S, Nawal RR, Talwar S, Verma M. Current perspectives of bio-ceramic technology in endodontics: calcium enriched mixture cement - review of its composition, properties and applications. Restor Dent Endod 2015; 40(1): 1-13.
[http://dx.doi.org/10.5395/rde.2015.40.1.1] [PMID: 25671207]

[82]   Catalu CT, Istrate SL, Voinea LM, Mitulescu C, Popescu V, Ciuluvică R. Ocular implants-methods of ocular reconstruction following radical surgical interventions. Rom J Ophthalmol 2018; 62(1): 15-23.
[http://dx.doi.org/10.22336/rjo.2018.3] [PMID: 29796430]

[83]   Jordan DR, Klapper SR. Controversies in Enucleation Technique and Implant Selection: Whether to Wrap, Attach Muscles, and Peg? Oculoplastics and Orbit 2010: 195-209.

[84]   Baino F, Potestio I. Orbital implants: State-of-the-art review with emphasis on biomaterials and recent advances. Mater Sci Eng C 2016; 69: 1410-28.
[http://dx.doi.org/10.1016/j.msec.2016.08.003] [PMID: 27612842]

[85]   Tian T, Wu C, Chang J. Preparation and *in vitro* osteogenic, angiogenic and antibacterial properties of cuprorivaite (CaCuSi$_4$O$_{10}$, Cup) bioceramics. RSC Advances 2016; 6(51): 45840-9.
[http://dx.doi.org/10.1039/C6RA08145B]

[86]   Han P, Wu C, Chang J, Xiao Y. The cementogenic differentiation of periodontal ligament cells *via* the activation of Wnt/β-catenin signalling pathway by Li$^+$ ions released from bioactive scaffolds. Biomaterials 2012; 33(27): 6370-9.
[http://dx.doi.org/10.1016/j.biomaterials.2012.05.061] [PMID: 22732362]

[87]   Wu C, Zhou Y, Lin C, Chang J, Xiao Y. Strontium-containing mesoporous bioactive glass scaffolds with improved osteogenic/cementogenic differentiation of periodontal ligament cells for periodontal tissue engineering. Acta Biomater 2012; 8(10): 3805-15.
[http://dx.doi.org/10.1016/j.actbio.2012.06.023] [PMID: 22750735]

[88]   Wu C, Zhou Y, Xu M, *et al.* Copper-containing mesoporous bioactive glass scaffolds with multifunctional properties of angiogenesis capacity, osteostimulation and antibacterial activity. Biomaterials 2013; 34(2): 422-33.
[http://dx.doi.org/10.1016/j.biomaterials.2012.09.066] [PMID: 23083929]

[89]   Zhou C, Yu B, Yang X, *et al.* Lipid-coated nano-calcium-phosphate (LNCP) for gene delivery. Int J Pharm 2010; 392(1-2): 201-8.
[http://dx.doi.org/10.1016/j.ijpharm.2010.03.012] [PMID: 20214964]

[90]   Jones JR. Review of bioactive glass: From Hench to hybrids. Acta Biomater 2013; 9(1): 4457-86.
[http://dx.doi.org/10.1016/j.actbio.2012.08.023] [PMID: 22922331]

[91]   Hessel CM, Pattani VP, Rasch M, *et al.* Copper selenide nanocrystals for photothermal therapy. Nano Lett 2011; 11(6): 2560-6.
[http://dx.doi.org/10.1021/nl201400z] [PMID: 21553924]

[92]   Wang X, Lv F, Li T, *et al.* Electrospun micropatterned nanocomposites incorporated with Cu$_2$S nanoflowers for skin tumor therapy and wound healing. ACS Nano 2017; 11(11): 11337-49.
[http://dx.doi.org/10.1021/acsnano.7b05858] [PMID: 29059516]

[93]   Yu Q, Han Y, Tian T, *et al.* Chinese sesame stick-inspired nano-fibrous scaffolds for tumor therapy and skin tissue reconstruction. Biomaterials 2019; 194: 25-35.
[http://dx.doi.org/10.1016/j.biomaterials.2018.12.012] [PMID: 30572284]

[94]   Ma H, Luo J, Sun Z, *et al.* 3D printing of biomaterials with mussel-inspired nanostructures for tumor therapy and tissue regeneration. Biomaterials 2016; 111: 138-48.
[http://dx.doi.org/10.1016/j.biomaterials.2016.10.005] [PMID: 27728813]

[95]   Wust P, Hildebrandt B, Sreenivasa G, *et al.* Hyperthermia in combined treatment of cancer. Lancet Oncol 2002; 3(8): 487-97.
[http://dx.doi.org/10.1016/S1470-2045(02)00818-5] [PMID: 12147435]

[96]   Habash RWY. Therapeutic hyperthermia. Handb Clin Neurol 2018; 157: 853-68.
[http://dx.doi.org/10.1016/B978-0-444-64074-1.00053-7] [PMID: 30459045]

[97]   Aspasio RD, Borges R, Marchi J. Biocompatible glasses for cancer treatment. Biocompatible Glasses 2016; pp. 249-65.
[http://dx.doi.org/10.1007/978-3-319-44249-5_10]

[98]   Cheng Y, Weng S, Yu L, Zhu N, Yang M, Yuan Y. The role of hyperthermia in the multidisciplinary treatment of malignant tumors. Integr Cancer Ther 2019; 18
[http://dx.doi.org/10.1177/1534735419876345] [PMID: 31522574]

[99]   Sohail A, Ahmad Z, Bég OA, Arshad S, Sherin L. A review on hyperthermia *via* nanoparticle-mediated therapy. Bull Cancer 2017; 104(5): 452-61.
[http://dx.doi.org/10.1016/j.bulcan.2017.02.003] [PMID: 28385267]

[100]   Ruiz-Hernández E, Serrano MC, Arcos D, Vallet-Regí M. Glass–glass ceramic thermoseeds for hyperthermic treatment of bone tumors. J Biomed Mater Res A 2006; 79A(3): 533-43.
[http://dx.doi.org/10.1002/jbm.a.30889] [PMID: 16788969]

[101]   Lin Y, Xiao W, Liu X, Bal BS, Bonewald LF, Rahaman MN. Long-term bone regeneration, mineralization and angiogenesis in rat calvarial defects implanted with strong porous bioactive glass (13–93) scaffolds. J Non-Cryst Solids 2016; 432: 120-9.
[http://dx.doi.org/10.1016/j.jnoncrysol.2015.04.008]

[102]   Bretcanu O, Miola M, Bianchi CL, *et al. In vitro* biocompatibility of a ferrimagnetic glass-ceramic for hyperthermia application. Mater Sci Eng C 2017; 73: 778-87.
[http://dx.doi.org/10.1016/j.msec.2016.12.105] [PMID: 28183672]

[103]   Baeza A, Arcos D, Vallet-Regí M. Thermoseeds for interstitial magnetic hyperthermia: from bioceramics to nanoparticles. J Phys Condens Matter 2013; 25(48): 484003-14.
[http://dx.doi.org/10.1088/0953-8984/25/48/484003] [PMID: 24200980]

[104]   Goya GF, Berquó TS, Fonseca FC, Morales MP. Static and dynamic magnetic properties of spherical magnetite nanoparticles. J Appl Phys 2003; 94(5): 3520-8.
[http://dx.doi.org/10.1063/1.1599959]

[105]   Kozissnik B, Bohorquez AC, Dobson J, Rinaldi C. Magnetic fluid hyperthermia: Advances, challenges, and opportunity. Int J Hyperthermia 2013; 29(8): 706-14.
[http://dx.doi.org/10.3109/02656736.2013.837200] [PMID: 24106927]

[106]   Luderer AA, Borrelli NF, Panzarino JN, *et al.* Glass-ceramic-mediated, magnetic-field-induced localized hyperthermia: response of a murine mammary carcinoma. Radiat Res 1983; 94(1): 190-8.
[http://dx.doi.org/10.2307/3575874] [PMID: 6856765]

[107]   Laurent S, Dutz S, Häfeli UO, Mahmoudi M. Magnetic fluid hyperthermia: Focus on superparamagnetic iron oxide nanoparticles. Adv Colloid Interface Sci 2011; 166(1-2): 8-23.
[http://dx.doi.org/10.1016/j.cis.2011.04.003] [PMID: 21601820]

[108]   Hilger I, Frühauf K, Andrä W, Hiergeist R, Hergt R, Kaiser WA. Heating potential of iron oxides for therapeutic purposes in interventional radiology. Acad Radiol 2002; 9(2): 198-202.
[http://dx.doi.org/10.1016/S1076-6332(03)80171-X] [PMID: 11918373]

[109]   Maier-Hauff K, Ulrich F, Nestler D, *et al.* Efficacy and safety of intratumoral thermotherapy using magnetic iron-oxide nanoparticles combined with external beam radiotherapy on patients with recurrent glioblastoma multiforme. J Neurooncol 2011; 103(2): 317-24.
[http://dx.doi.org/10.1007/s11060-010-0389-0] [PMID: 20845061]

[110]   Hervé K, Douziech-Eyrolles L, Munnier E, *et al.* The development of stable aqueous suspensions of

PEGylated SPIONs for biomedical applications. Nanotechnology 2008; 19(46): 465608-15.
[http://dx.doi.org/10.1088/0957-4484/19/46/465608] [PMID: 21836255]

[111] Colilla M, González B, Vallet-Regí M. Mesoporous silicananoparticles for the design of smart delivery nanodevices. Biomater Sci 2013; 1(2): 114-34.
[http://dx.doi.org/10.1039/C2BM00085G] [PMID: 32481793]

[112] Goon IY, Lai LMH, Lim M, Munroe P, Gooding JJ, Amal R. Fabrication and dispersion of gold-shel--protected magnetite nanoparticles: systematic control using polyethyleneimine. Chem Mater 2009; 21(4): 673-81.
[http://dx.doi.org/10.1021/cm8025329]

[113] Yu MK, Jeong YY, Park J, *et al.* Drug-loaded superparamagnetic iron oxide nanoparticles for combined cancer imaging and therapy *in vivo.* Angew Chem Int Ed 2008; 47(29): 5362-5.
[http://dx.doi.org/10.1002/anie.200800857] [PMID: 18551493]

[114] He X, Wu X, Cai X, *et al.* Functionalization of magnetic nanoparticles with dendritic-linear-brush-like triblock copolymers and their drug release properties. Langmuir 2012; 28(32): 11929-38.
[http://dx.doi.org/10.1021/la302546m] [PMID: 22799877]

[115] Lee DE, Koo H, Sun IC, Ryu JH, Kim K, Kwon IC. Multifunctional nanoparticles for multimodal imaging and theragnosis. Chem Soc Rev 2012; 41(7): 2656-72.
[http://dx.doi.org/10.1039/C2CS15261D] [PMID: 22189429]

[116] Bardhan R, Chen W, Perez-Torres C, *et al.* Nanoshells with targeted simultaneous enhancement of magnetic and optical imaging and photothermal therapeutic response. Adv Funct Mater 2009; 19(24): 3901-9.
[http://dx.doi.org/10.1002/adfm.200901235]

[117] Vallet-Regí M, Ruiz-González L, Izquierdo-Barba I, González-Calbet JM. Revisiting silica based ordered mesoporous materials: medical applications. J Mater Chem 2006; 16(1): 26-31.
[http://dx.doi.org/10.1039/B509744D]

[118] Liong M, Lu J, Kovochich M, *et al.* Multifunctional inorganic nanoparticles for imaging, targeting, and drug delivery. ACS Nano 2008; 2(5): 889-96.
[http://dx.doi.org/10.1021/nn800072t] [PMID: 19206485]

[119] Meng H, Xue M, Xia T, *et al.* Use of size and a copolymer design feature to improve the biodistribution and the enhanced permeability and retention effect of doxorubicin-loaded mesoporous silica nanoparticles in a murine xenograft tumor model. ACS Nano 2011; 5(5): 4131-44.
[http://dx.doi.org/10.1021/nn200809t] [PMID: 21524062]

[120] Li L, Tang F, Liu H, *et al. In vivo* delivery of silica nanorattle encapsulated docetaxel for liver cancer therapy with low toxicity and high efficacy. ACS Nano 2010; 4(11): 6874-82.
[http://dx.doi.org/10.1021/nn100918a] [PMID: 20973487]

[121] Vivero-Escoto JL, Slowing II, Lin VSY. Tuning the cellular uptake and cytotoxicity properties of oligonucleotide intercalator-functionalized mesoporous silica nanoparticles with human cervical cancer cells Hela. Biomaterials 2010; 31(6): 1325-33.
[http://dx.doi.org/10.1016/j.biomaterials.2009.11.009] [PMID: 19932923]

[122] Ravanbakhsh M, Labbaf S, Karimzadeh F, Pinna A, Houreh AB, Nasr-Esfahani MH. Mesoporous bioactive glasses for the combined application of osteosarcoma treatment and bone regeneration. Mater Sci Eng C 2019; 104: 109994-04.
[http://dx.doi.org/10.1016/j.msec.2019.109994] [PMID: 31500021]

[123] Dorozhkin SV. Bioceramics of calcium orthophosphates. Biomaterials 2010; 31(7): 1465-85.
[http://dx.doi.org/10.1016/j.biomaterials.2009.11.050] [PMID: 19969343]

[124] Salinas AJ, Vallet-Regí M. Evolution of ceramics with medical applications. Z Anorg Allg Chem 2007; 633(11-12): 1762-73.
[http://dx.doi.org/10.1002/zaac.200700278]

[125] Anselme K. Osteoblast adhesion on biomaterials. Biomaterials 2000; 21(7): 667-81.
[http://dx.doi.org/10.1016/S0142-9612(99)00242-2] [PMID: 10711964]

[126] Yuan H, van Blitterswijk CA, de Groot K, de Bruijn JD. Cross-species comparison of ectopic bone formation in biphasic calcium phosphate (BCP) and hydroxyapatite (HA) scaffolds. Tissue Eng 2006; 12(6): 1607-15.
[http://dx.doi.org/10.1089/ten.2006.12.1607] [PMID: 16846356]

[127] Tang J, Howard CB, Mahler SM, Thurecht KJ, Huang L, Xu ZP. Enhanced delivery of siRNA to triple negative breast cancer cells *in vitro* and *in vivo* through functionalizing lipid-coated calcium phosphate nanoparticles with dual target ligands. Nanoscale 2018; 10(9): 4258-66.
[http://dx.doi.org/10.1039/C7NR08644J] [PMID: 29436549]

[128] Arami H, Khandhar A, Liggitt D, Krishnan KM. *In vivo* delivery, pharmacokinetics, biodistribution and toxicity of iron oxide nanoparticles. Chem Soc Rev 2015; 44(23): 8576-607.
[http://dx.doi.org/10.1039/C5CS00541H] [PMID: 26390044]

[129] Mitragotri S, Lahann J. Physical approaches to biomaterial design. Nat Mater 2009; 8(1): 15-23.
[http://dx.doi.org/10.1038/nmat2344] [PMID: 19096389]

[130] Olton DYE, Close JM, Sfeir CS, Kumta PN. Intracellular trafficking pathways involved in the gene transfer of nano-structured calcium phosphate-DNA particles. Biomaterials 2011; 32(30): 7662-70.
[http://dx.doi.org/10.1016/j.biomaterials.2011.01.043] [PMID: 21774979]

[131] Liu Y, Liu , He F, *et al.* An efficient calcium phosphate nanoparticle-based nonviral vector for gene delivery. Int J Nanomedicine 2011; 6: 721-7.
[http://dx.doi.org/10.2147/IJN.S17096] [PMID: 21556346]

[132] Klesing J, Wiehe A, Gitter B, Gräfe S, Epple M. Positively charged calcium phosphate/polymer nanoparticles for photodynamic therapy. J Mater Sci Mater Med 2010; 21(3): 887-92.
[http://dx.doi.org/10.1007/s10856-009-3934-7] [PMID: 19924519]

[133] White JE, Day DE. Rare earth aluminosilicate glasses for *in vivo* radiation delivery. Key Eng Mater 1994; 94-95: 181-208.
[http://dx.doi.org/10.4028/www.scientific.net/KEM.94-95.181]

[134] Giammarile F, Bodei L, Chiesa C, *et al.* Therapy, Oncology and Dosimetry Committees. EANM procedure guideline for the treatment of liver cancer and liver metastases with intra-arterial radioactive compounds. Eur J Nucl Med Mol Imaging 2011; 38(7): 1393-406.
[http://dx.doi.org/10.1007/s00259-011-1812-2] [PMID: 21494856]

[135] Hench LL, Thompson I. Twenty-first century challenges for biomaterials. J R Soc Interface 2010; 7(Suppl 4) (Suppl. 4): S379-91.
[PMID: 20484227]

[136] Bose S, Tarafder S. Calcium phosphate ceramic systems in growth factor and drug delivery for bone tissue engineering: A review. Acta Biomater 2012; 8(4): 1401-21.
[http://dx.doi.org/10.1016/j.actbio.2011.11.017] [PMID: 22127225]

[137] Sokolova V, Epple M. Inorganic nanoparticles as carriers of nucleic acids into cells. Angew Chem Int Ed 2008; 47(8): 1382-95.
[http://dx.doi.org/10.1002/anie.200703039] [PMID: 18098258]

[138] Zhang M, Kataoka K. Nano-structured composites based on calcium phosphate for cellular delivery of therapeutic and diagnostic agents. Nano Today 2009; 4(6): 508-17.
[http://dx.doi.org/10.1016/j.nantod.2009.10.009]

[139] Khalifehzadeh R, Arami H. The CpG molecular structure controls the mineralization of calcium phosphate nanoparticles and their immunostimulation efficacy as vaccine adjuvants. Nanoscale 2020; 12(17): 9603-15.
[http://dx.doi.org/10.1039/C9NR09782A] [PMID: 32314980]

[140] Chen WY, Lin MS, Lin PH, Tasi P-S, Chang Y, Yamamoto S. Studies of the interaction mechanism between single strand and double-strand DNA with hydroxyapatite by microcalorimetry and isotherm measurements. Colloids Surf A Physicochem Eng Asp 2007; 295(1-3): 274-83.
[http://dx.doi.org/10.1016/j.colsurfa.2006.09.013]

[141] Mostaghaci B, Loretz B, Haberkorn R, Kickelbick G, Lehr C-M. One-step synthesis of nanosized and stable amino-functionalized calcium phosphate particles for DNA transfection. Chem Mater 2013; 25(18): 3667-74.
[http://dx.doi.org/10.1021/cm401886u]

[142] Mostaghaci B, Susewind J, Kickelbick G, Lehr CM, Loretz B. Transfection system of amino-functionalized calcium phosphate nanoparticles: *in vitro* efficacy, biodegradability, and immunogenicity study. ACS Appl Mater Interfaces 2015; 7(9): 5124-33.
[http://dx.doi.org/10.1021/am507193a] [PMID: 25692576]

[143] Bisso S, Mura S, Castagner B, Couvreur P, Leroux JC. Dual delivery of nucleic acids and PEGylated-bisphosphonates *via* calcium phosphate nanoparticles. Eur J Pharm Biopharm 2019; 142: 142-52.
[http://dx.doi.org/10.1016/j.ejpb.2019.06.013] [PMID: 31220571]

[144] Tobin LA, Xie Y, Tsokos M, *et al.* Pegylated siRNA-loaded calcium phosphate nanoparticle-driven amplification of cancer cell internalization *in vivo*. Biomaterials 2013; 34(12): 2980-90.
[http://dx.doi.org/10.1016/j.biomaterials.2013.01.046] [PMID: 23369215]

[145] Zhang M, Li J, Xing G, *et al.* Variation in the internalization of differently sized nanoparticles induces different DNA-damaging effects on a macrophage cell line. Arch Toxicol 2011; 85(12): 1575-88.
[http://dx.doi.org/10.1007/s00204-011-0725-y] [PMID: 21881955]

[146] Xiaoyu M, Xiuling D, Chunyu Z, *et al.* Polyglutamic acid-coordinated assembly of hydroxyapatite nanoparticles for synergistic tumor-specific therapy. Nanoscale 2019; 11(32): 15312-25.
[http://dx.doi.org/10.1039/C9NR03176F] [PMID: 31386744]

[147] Shamsi M, Majidi Zolbanin J, Mahmoudian B, *et al.* A study on drug delivery tracing with radiolabeled mesoporous hydroxyapatite nanoparticles conjugated with 2DG/DOX for breast tumor cells. Nuclear Medicine Review 2018; 21(1): 32-6.
[http://dx.doi.org/10.5603/NMR.a2018.0008] [PMID: 29319137]

[148] Wu Y, Gu W, Xu ZP. Enhanced combination cancer therapy using lipid-calcium carbonate/phosphate nanoparticles as a targeted delivery platform. Nanomedicine (Lond) 2019; 14(1): 77-92.
[http://dx.doi.org/10.2217/nnm-2018-0252] [PMID: 30543136]

[149] Abdullayev E, Lvov Y. Halloysite clay nanotubes for controlled release of protective agents. J Nanosci Nanotechnol 2011; 11(11): 10007-26.
[http://dx.doi.org/10.1166/jnn.2011.5724] [PMID: 22413340]

[150] Vergaro V, Abdullayev E, Lvov YM, *et al.* Cytocompatibility and uptake of halloysite clay nanotubes. Biomacromolecules 2010; 11(3): 820-6.
[http://dx.doi.org/10.1021/bm9014446] [PMID: 20170093]

[151] Price RR, Gaber BPY, Lvov Y. In-vitro release characteristics of tetracycline HCl, khellin and nicotinamide adenine dineculeotide from halloysite; a cylindrical mineral. J Microencapsul 2001; 18(6): 713-22.
[http://dx.doi.org/10.1080/02652040010019532] [PMID: 11695636]

[152] Sun L, Mills DK. Halloysite nanotube-based drug delivery system for treating osteosarcoma. Annu Int Conf IEEE Eng Med Biol Soc 2014; 2920-3.
[PMID: 25570602]

[153] Li K, Zhang Y, Chen M, *et al.* Enhanced antitumor efficacy of doxorubicin-encapsulated halloysite nanotubes. Int J Nanomedicine 2017; 13: 19-30.
[http://dx.doi.org/10.2147/IJN.S143928] [PMID: 29296083]

[154] Yendluri R, Lvov Y, de Villiers MM, *et al.* Paclitaxel encapsulated in halloysite clay nanotubes for

intestinal and intracellular delivery. J pharma Sci. 2017; 106: 3131-9.

[155]   Gaharwar AK, Mihaila SM, Swami A, *et al.* Bioactive silicate nanoplatelets for osteogenic differentiation of human mesenchymal stem cells. Adv Mater 2013; 25(24): 3329-36.
[http://dx.doi.org/10.1002/adma.201300584] [PMID: 23670944]

[156]   Xu F, Liu M, Li X, *et al.* Loading of indocyanine green within polydopamine-coated laponite nanodisks for targeted cancer photothermal and photodynamic therapy. Nanomaterials (Basel) 2018; 8(5): 347-63.
[http://dx.doi.org/10.3390/nano8050347] [PMID: 29783745]

[157]   Razavi M, Fathi M, Savabi O, Tayebi L, Vashaee D. Improvement of *in vitro* behavior of an Mg alloy using a nanostructured composite bioceramic coating. J Mater Sci Mater Med 2018; 29(10): 159.
[http://dx.doi.org/10.1007/s10856-018-6170-1] [PMID: 30350229]

[158]   Ren H, Zhu C, Li Z, Yang W, Song E. Emodin-loaded magnesium silicate hollow nanocarriers for anti-angiogenesis treatment through inhibiting VEGF. Int J Mol Sci 2014; 15(9): 16936-48.
[http://dx.doi.org/10.3390/ijms150916936] [PMID: 25250911]

[159]   Yu L, Chen Y, Lin H, Gao S, Chen H, Shi J. Magnesium-engineered silica framework for pH-accelerated biodegradation and DNAzyme-triggered chemotherapy. Small 2018; 14(35): 1800708.
[http://dx.doi.org/10.1002/smll.201800708] [PMID: 30070076]

[160]   Sun TW, Zhu YJ, Chen F, *et al.* Superparamagnetic yolk–shell porous nanospheres of iron oxide@magnesium silicate: synthesis and application in high-performance anticancer drug delivery. RSC Advances 2016; 6(105): 103399-411.
[http://dx.doi.org/10.1039/C6RA21492D]

[161]   Vallet-Regí M. Bio-ceramics with clinical applications. John Wiley & Sons; 2014.
[http://dx.doi.org/10.1002/9781118406748]

[162]   Comparison of Bioactive Glass and Beta-Tricalcium Phosphate as Bone Graft Substitute. Available at: https://clinicaltrials.gov/ct2/show/study/NCT00841152?term=Bioceramics&cond=Cancer&draw=2&rank=1

[163]   Lindfors NC, Koski I, Heikkilä JT, Mattila K, Aho AJ. A prospective randomized 14-year follow-up study of bioactive glass and autogenous bone as bone graft substitutes in benign bone tumors. J Biomed Mater Res B Appl Biomater 2010; 94: 157-64.
[http://dx.doi.org/10.1002/jbm.b.31636] [PMID: 20524190]

[164]   Complete Twelve Month Bone Remodeling With a Bi-phasic Injectable Bone Substitute in Benign Bone Tumors. Available at: https://clinicaltrials.gov/ct2/show/study/NCT02567084?term=Ceramic&cond=Cancer&draw=2&rank=2

[165]   Available at: https://www.bostonscientific.com/en-US/products/cancer-therapies/therasphere--90-glass-microspheres.html

[166]   Available at: https://clinicaltrials.stanford.edu/browse-all-trials.html?ctid=NCT03295006

[167]   Wu C, Zhou Y, Fan W, *et al.* Hypoxia-mimicking mesoporous bioactive glass scaffolds with controllable cobalt ion release for bone tissue engineering. Biomaterials 2012; 33(7): 2076-85.
[http://dx.doi.org/10.1016/j.biomaterials.2011.11.042] [PMID: 22177618]

[168]   Hench LL. The story of Bioglass®. J Mater Sci Mater Med 2006; 17(11): 967-78.
[http://dx.doi.org/10.1007/s10856-006-0432-z] [PMID: 17122907]

[169]   Hench LL. An introduction to bioceramics. World Scientific 1993.
[http://dx.doi.org/10.1142/2028]

[170]   Vallet-Regí M, Balas F, Arcos D. Mesoporous materials for drug delivery. Angew Chem Int Ed 2007; 46(40): 7548-58.
[http://dx.doi.org/10.1002/anie.200604488] [PMID: 17854012]

[171]   Rana KS, Souza LP, Isaacs MA, Raja FNS, Morrell AP, Martin RA. Development and

characterization of gallium-doped bioactive glasses for potential bone cancer applications. ACS Biomater Sci Eng 2017; 3(12): 3425-32.
[http://dx.doi.org/10.1021/acsbiomaterials.7b00283] [PMID: 33445381]

[172] Yu Q, Chang J, Wu C. Silicate bioceramics: from soft tissue regeneration to tumor therapy. J Mater Chem B Mater Biol Med 2019; 7(36): 5449-60.
[http://dx.doi.org/10.1039/C9TB01467E] [PMID: 31482927]

[173] Pecqueux F, Tancret F, Bouler JM. Young's modulus of macroporous bioceramics: Measurement and numerical simulation. Bioceramics Development and Applications 2010; 1: 1-3.
[http://dx.doi.org/10.4303/bda/D110101]

Frontiers in Nanomedicine, 2023, Vol. 4, 247-271

CHAPTER 7

Advanced Materials and Processing Techniques

Smita S. Bhuyar-Kharkhale[1], Sudhir S. Bhuyar[2], Ajay K. Potbhare[2], Manjiri S. Nagmote[2,*], Nakshatra B. Singh[3] and Ratiram G. Chaudhary[2,*]

[1] *Department of Chemistry, Lemdeo Patil College, Mandhal, India*

[2] *Department of Chemistry, Seth Kesarimal Porwal College of Arts and Science and Commerce Kamptee, India*

[3] *Department of Chemistry and Biochemistry and RDC, Sharda University, Greater Noida, India*

Abstract: Advanced materials and processing techniques are the backbone of the smart industry. The smart industry could not be run without a furnish of raw materials. Further, the raw materials could become advanced materials by employing good processing technology. Owing to this, new advanced materials exhibit compelling properties and applications in various fields. In the present chapter, we have provided insight into the current development of advanced materials comprising different fabrication methodologies and their incorporation with nanofillers, as well as their advanced processing techniques. Moreover, advanced materials' applications have been emphasized in different fields, like tissue engineering, biomedical, agriculture sector, and pesticides.

Keywords: Advanced materials, Agricultural, Biomedical devices, Fertilizers, Pesticides, Processing techniques, Tissue engineering.

INTRODUCTION

The unprecedented and rapid technological development started in the 19th century, especially in the fabrication and processing of advanced materials. Nowadays, advanced materials, particularly nanomaterials (NMs), have become one of the most significant generic materials due to their potential applications in different sectors [1]. The materials considered as advanced materials of 21st century are advanced nanoceramics, smart polymers, graphene-based nanomaterials, and nanocomposites (NCs). The advanced nanomaterials are shown in Fig. (**1**).

* **Corresponding authors Manjiri S. Nagmote & Ratiram G. Chaudhary:** Department of Chemistry, Seth Kesarimal Porwal College, Kamptee, India; E-mails: manjirinagmote@gmail.com and chaudhary_rati@yahoo.com

Felipe López-Saucedo (Ed.)

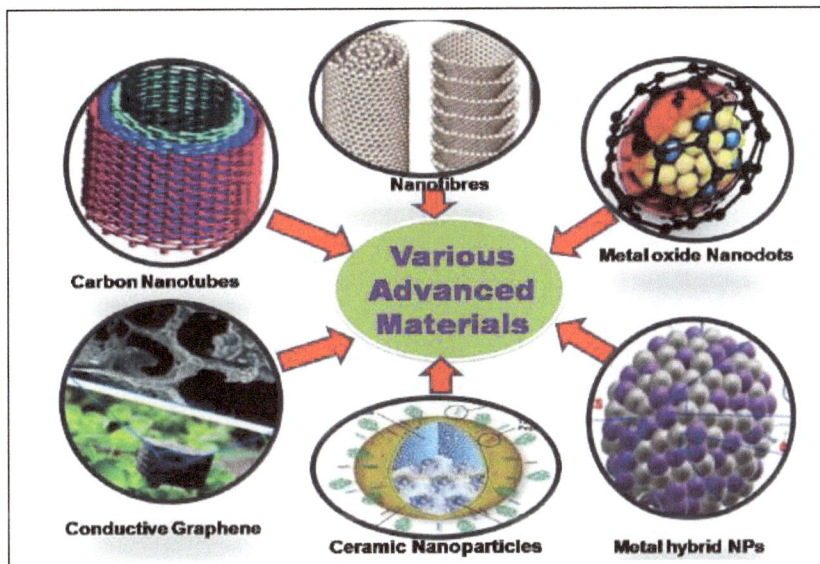

Fig. (1). Advanced nanomaterials.

Basically, advanced materials show higher strength, improved density ratios, better hardness, superior thermal behaviour and electrical, optical, and chemical properties. For instance, the graphene-based NCs are one of the best advanced materials, which have been used for numerous applications. Besides, many researchers have developed different types of advance manufacturing processing techniques to enhance materials' efficiency and productivity [2 - 4]. Material processing is the heart of material science and engineering, which involves fundamental principles responsible for the processing of all types of materials. It is a multi-step process involving chemical, thermal, and physical processes.

Polymer NCs are one of the most the significantly advanced materials as they possess toughness, stiffness, corrosion resistance, and lightness [6 - 8]. These are used in aviation industries because of their advanced features, such as tailor-made mechanical properties, design flexibility, lower weight, anti-corrosion, and better fatigue performance [9]. The material processing technique for graphene-reinforced polymer composites is shown in Fig. (**2**). A well-known aviation industry, Boeing, uses around 80% composites for its single Dremliner-787. Each Dreamliner requires 35 metric tons of carbon fiber reinforced plastic (CFRP) out of twenty-three metric tons of composites made of carbon [10]. Likewise, carbon is the prominent element having new forms of carbon-based materials, such as fullerene [11], carbon nanotubes (CNTs) [12], and graphene [13]. Graphene is popular among carbon-based materials. Graphene has gained significant interest in materials science, and it has been employed in numerous applications, like sensors, electrodes, energy storage devices, various types of solar cells, *etc.* [14, 15]. Moreover, fibber-reinforced NCs are also important materials and can be manufactured by resin transfer moulding (RTM) and vacuum-assisted resin transfer moulding (VARTM) [16 - 19]. Different types of glasses and their composites with superior properties have been prepared using different techniques [20, 21]. The current trend in materials science is to see beyond conventional and well-known materials, which are new having better properties. It is high time that conventional processing techniques for materials must change. It is of utmost importance that processing techniques must change the properties of advanced materials to a great extent. From the beginning, the processing methods should be integrated into the design and development. There are a number of challenges faced by the industry tycoon to achieve excellence in each process to design, develop, produce, and distribute material products to customers. Excellent functional capabilities must be integrated. Therefore, keeping this perspective, the present chapter highlights a recent development in advanced materials' processing techniques and their important applications in various sectors.

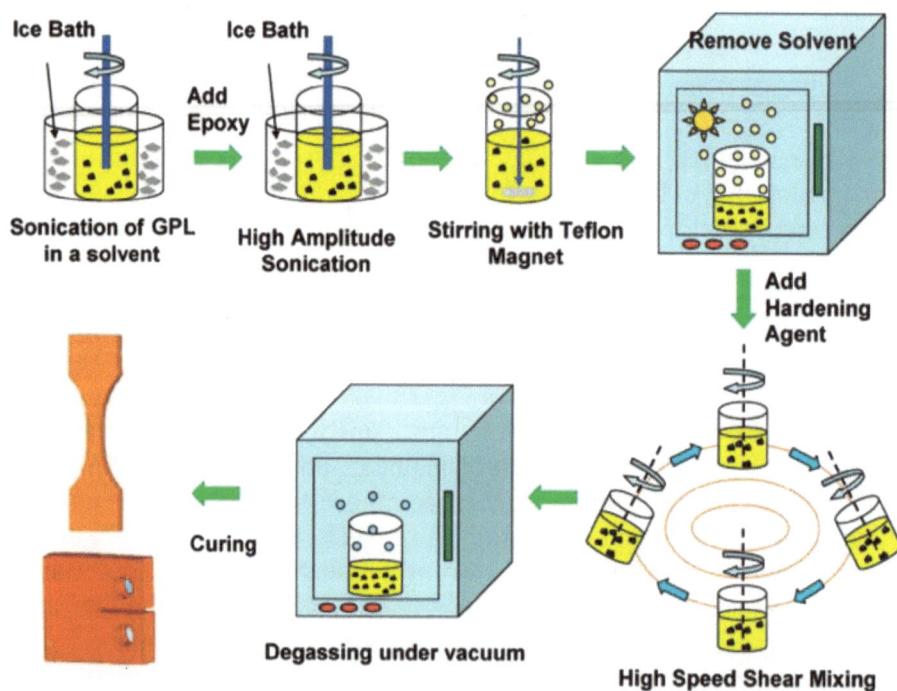

Fig. (2). Solution mixing for the manufacture of graphene-reinforced polymer composites (GRPCs). Reproduced with permission from the American Chemical Society [5].

ADVANCED TECHNIQUES FOR MATERIALS' MANUFACTURING

The following techniques are used for the production of advanced materials:

Solvent/Solution Processing Technique

The solvent/solution technique is purely based on a solvent system, equally combined polymer, and pre-polymer in a suitable solvent, and nanofillers in various solvents, such as water, toluene, or chloroform. While polymer matrix and nanofiller particles are dispersed in the same or different solvents properly, both solutions are intermixed by a powerful agitation, like magnetic stirring, refluxing, high shear type of mixing, or by regular sonication. The polymer chains intercalate and displace the solvent within the interlayer of the silicate. Upon solvent removal, the intercalated/exfoliated structure remains, resulting in polymer NCs [22]. This method is also recognized as the intercalation of polymer or pre-polymer from solution used to prepare NCs with layered silicates. Usually, high aspect ratio in nanoparticles dispersion is poor, and the quality of dispersion can be enhanced by using a proper surface-active agent.

After homogeneous dispersion of nanoparticles (NPs), the polymer solution is poured into a suitable mold after complete evaporation of solvent, leaving behind the film or sheet-like structure of desired NCs, and the selection of solvent in this method is completely dependent on the solubility of polymers. The environmental friendliness of this method depends on the use of less toxic solvents, like ethanol, water, *etc.* Due to the proper interaction between solvent and polymer, this method is a good option for the dispersion of NPs in polymers. This is the trouble-free method to get good NCs. If some consideration is taken for the control of the dissolvable polymer, it can be disposed of subsequently [23 - 26]. Paul *et al.* [27] synthesized polystyrene-clay NCs by solution intercalation techniques. Indeed, they initially prepared laponite aqueous dispersion stock, which was only miscible with hydrophilic polymers in ion-exchange reactions; they made use of some alkylammonium ions of cetyltrimethyl ammonium bromide (CTAB) to provide layered silicate miscible with hydrophobic-PS. Later, in the laponite aqueous solution, CTAB solution was added in a stepwise manner with dispersion. To obtain an optimum level of absorption, the mixture was stirred at 50 °C for 24 h. Afterward, the reaction mixer was centrifuged at 10,000 rpm for the entire removal of halide ions by continuous washing with deionised water. This gives a white precipitate of organically modified laponite (O-laponite). In the next step, chloroform was used as the solvent, and an intercalation technique where PS was a base matrix, was used for the dispersion of O-laponite. Composites of PS and O-laponite were obtained by continuous magnetic stirring for 48 h and complete removal of chloroform after three days at room temperature. Similarly, Jin *et al.* conducted experiments to prepare multiwall-carbon nanotube composites using a polymer matrix of polyhydroxyaminoether (PHAE) as a precursor by sonication process [28].

Melt Blending Technique

The melt blending technique is more economical, beneficial, and straightforward, which could be used for the preparation of polymer NCs, like HDPE, thermoplastic polymers, polypropylene polycarbonate, polystyrene, *etc.* The ternary melt blends of poly (lactic acid)/poly(vinyl alcohol)-chitosan are shown in Fig. (**3**). The melt blending technique is cost-effective, environment-friendly, and good for bulk production to meet the industrial demand. The beauty of this technique is that it does not require any solvent. Hence, it is the most adaptable technique for NCs production. Under this process, NCs can be processed using ordinary devices; if we provide mixing conditions, then there will be no limitation for the melt blending technique. The major parameters taken into consideration for proper mixing are screw length, design of screw, the position of mixing, processing temperature, shear, and residence time for melt [29, 30]. Another important parameter for getting better dispersion of nanofillers is the well-balanced screw configuration (*e.g.,* feed zone, mixing zone, and metering zone) and nature of the extruder [31].

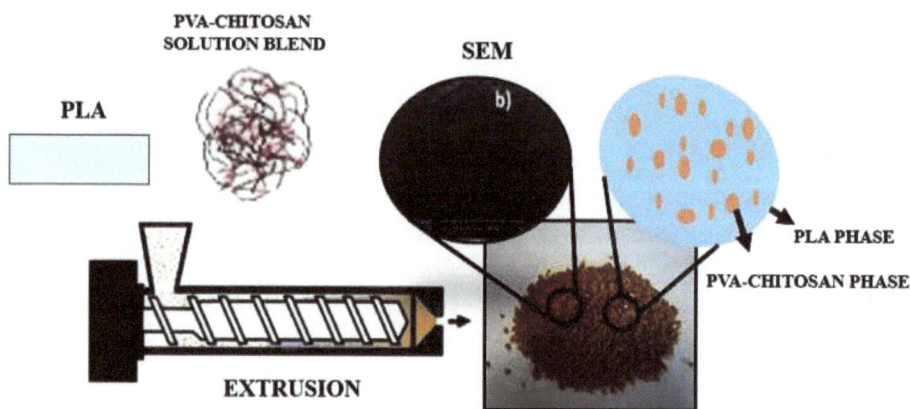

Fig. (3). Ternary melt blends of poly(lactic acid)/poly(vinyl alcohol)-chitosan.

The major disadvantages of this technique are the high ratio of nanofillers, the high viscosities of the material, and the problem of dispersion [32]. The conventional method used for mixing thermoplastics with graphene, which comes under this technique, is extrusion and injection molding [33 - 35]. Moreover, the polymer changes its viscosity with the addition of NPs into the polymer matrix because of the high shear condition. Therefore, all processing conditions are properly optimized for the entire range of polymers/nanoparticles [36]. When this method is used for the preparation of clay-based NCs, the molecules of polymer can penetrate into the available interlayer space sandwiched between clay particles. The process of diffusion is ready to peel the clay layers away. The intercalated or exfoliated structure achievement can be completely attained by processing conditions and the compatibility present between the clay and polymer matrix. The single-screw and twin-screw extruders have been used by Cho and Paul for the preparation of Nylon 6/clay NCs [37]. Adak *et al.* [38] prepared functionalized-graphene NCs (FGN) films produced by solution master-batching and later melt mixing, followed by compression molding with different concentrations of graphene, where nanocomposite films enhanced barrier properties of helium gas at a large level with an increase in the concentration of graphene. Secondly, tensile strength and stiffness of the nanocomposite films enhanced drastically with the increase in graphene concentration. Similarly, Noorunnisa *et al.* [39] fabricated composites of graphene nanoplatelets and linear low-density polyethylene under various extrusion conditions, where they investigated an effect of screw speed, graphene nanoplatelets content and feeder speed on the multiple properties, like electrical, thermal, and mechanical properties. Moreover, they also reported an insertion of graphene nanoplatelets into the polymer matrix, and therefore, increased the thermal stability and conductivity. Various polymer NCs have been fabricated by melt-blending techniques, as presented in Table **1** [40 - 46].

Table1. The various polymer NCs fabricated by melt blending technique.

S. No.	Advanced Materials	Properties	References
1	Poly(MA-co-NIPA)-graphene CPs	Improve dielectric constant	[40]
2	Poly(vinylalcohol) (PVA)/graphene	Thermal and mechanical properties	[41]
3	Poly(vinyl chloride) (PVC)/rGO NCs	Improvement in dielectric permittivity	[42]
4	Polypropylene/graphene nanoplatelet composites	Step up mechanical and thermal properties	[43]
5	Polypropylene/graphene nanoplatelet composites	Improve rigidity and photo-oxidation behaviour of PP	[44]
6	Polypropylene/poly(methyl methacrylate)/graphene composites	Improvement of electrical conductivity/storage modulus	[45]
7	Isotacticpolypropylene/graphene composites	Greatly improved thermal stability of composites	[46]

In-situ Polymerization Technique

Presently, *in-situ* polymerization is a comprehensive polymer synthetic-technique, which has been generally exploited in the synthesis of various types of polymer NCs [47, 48]. The schematic illustration of the *in-situ* polymerization processing technique is given in Fig. (**4**). This technique was employed for the first time for the synthesis of nylon-6 based clay NCs [49]. In-situ polymerization methods are straightforward, cost-effective, eco-friendly, and useful techniques for the fabrication of polymer NCs, and therefore, they are employed in various mixed dispersion and polymerization reactions [50]. Moreover, they are very similar techniques to *in-situ* graphene-based NCs [51 - 54]. The polymerization reaction may be initiated within an intercalated sheet by heat, radiation, suitable initiator diffusion or fixed catalyst by cation-exchange process [55]. The best examples of polymerization techniques are thermoplastic and thermosetting, which can be treated by this process using various nanofillers (*e.g.,* metallic oxide, ceramic, polymers, silicate, *etc.*). One of the most excellent applications of thermoplastic NCs is 'Toyota Motor Corporation' which had made its first successful attempt to fabricate thermoplastic NCs montmorillonite (MMT) using amino acid as compatibilizer. The modified MMT of ε-caprolactam monomer was swollen at 100 °C, followed by polymerization to get polyamide/MMT NCs [56]. The various permutation combinations of different nanofillers and compatibilizers have been used to get longer chains for intercalation in polymers. Depending on the quantity of montmorillonite introduced, exfoliated (<15 wt %) or intercalated structures (15 to 70 wt %) may be obtained. Later, ε-caprolactam intercalative polymerization could be understood with no modification of the MMT surface. Different kinds of NCs have been prepared by *in-situ* polymerization using various kinds of polymer matrices of thermoplastic. The polymethylmethacrylate (PMMA), graphene/polyurethane and other NCs can be fabricated by this processing technique, as shown in Table **2** [56 - 63].

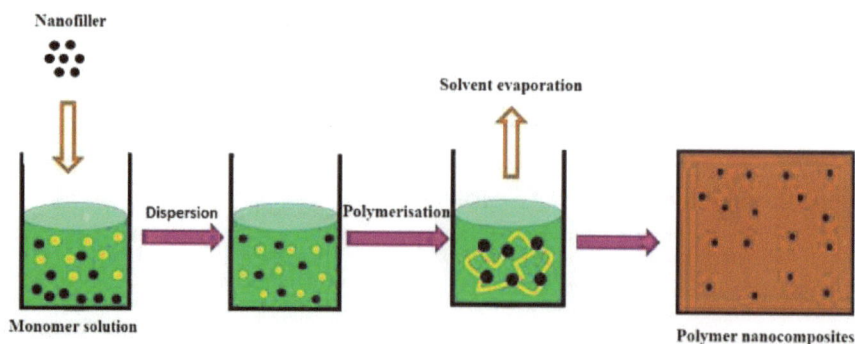

Fig. (4). Schematic illustration for the *in-situ* polymerization processing technique.

Table 2. Various polymer NCs fabricated by *in-situ* polymerization technique [49].

S. No.	Advanced Materials	Properties	References
1	Polystyrene sulphonate NCs	Improvement in electrocatalytic activity	[56]
2	Graphene/polyaniline NCs	Great enhancement in electrical property and better flexibility	[57]
3	Graphene/polyurethane NCs	High electrical and good thermal property	[58]
4	Graphene/epoxy NCs	Enhanced thermal, mechanical, and electrical properties	[59]
5	Graphene/polystyrene NCs	High electrical conductivity and good thermal property	[60]
6	Graphene/PVA NCs	Strongly enhanced water vapour barrier and mechanical properties	[61]
7	Polyethylene terephthalate (PET)-Clay NCs	Improvement in heat distortion temperature and modulus property	[62]
8	Polymethylmethacrylate (PMMA) NCs	Improvement in conductivity	[63]

The melt blending technique is not suitable for the preparation of thermosetting NCs [64, 65]. Therefore, for the preparation of exfoliated and intercalated NCs, organoclays have been used as nanofillers, where exfoliation is significantly affected by organoclays. For example, when comparably intra-layer and extra-layer polymerization reactions take place, enough amount of heat is produced by the curing process, which diminishes the attractive forces between the silicate lamellae resulting in an exfoliated NCs structure. On the other hand, resin will cure before if the extra-layer polymerization is faster and the intra-layer resin produces enough amount of curing heat to drive the clay to exfoliate, and subsequently, exfoliation will not be reached. The photo-reticulation by UV light is another important technique that is used for the preparation of thin insulating coating used for electrical application. Davidson and Sangermano [66, 67] reported NCs of cycloaliphatic epoxy resin doped by nanoclays filler using the photo-curing technique, where the reaction mixture was photo-cured by UV light. UV curing technique is considered as an environmentally friendly technique due to its energy-saving and solvent-free process. Today, UV technique is tremendously used because of its unique characteristics, where rapid conversion of liquid monomer into a solid film takes place with tailor-made physical, chemical, and mechanical properties. This technique may be used for both types of polymers, *i.e.,* thermoplastic and thermosetting with various types of nanofillers. Most recently, polymer extrusion has been directly done by using this technique; therefore, it is also considered the most productive industrial technique.

Sol-gel Technique

Sol-gel is one of the best techniques employed for the preparation of organic-inorganic based-material polymer NCs [68, 69]. This technique works on the principle of relatively low temperature and molecular precursor. In this technique, various organic and inorganic molecular masses can be condensed by maintaining the homogeneity of NPs. Using this technique, glasses and ceramics can be fabricated using various organic/inorganic composites, porous composites, and polycrystals. The preparation of carbon nanotube/TiO$_2$ composites *via* the sol-gel technique is shown in Fig. (**5**) [69].

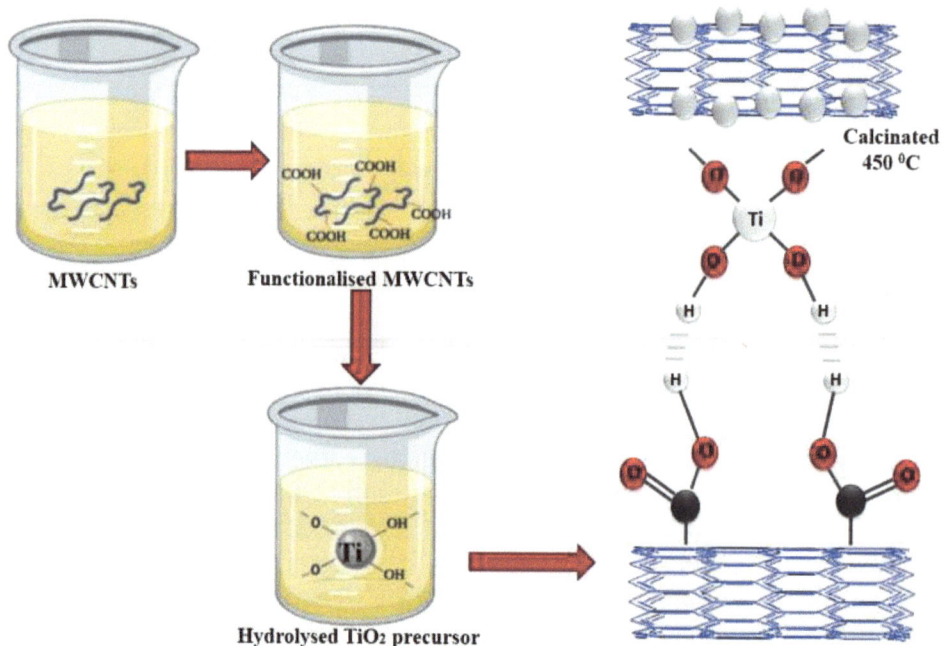

Fig. (5). The preparation of carbon nanotube/TiO$_2$ composites *via* sol-gel technique [69].

Basically, metal alkoxide M(OR)$_n$ is to be dispersed homogeneously in water or other solvents, like alcohol, ammonia, acid, *etc.* Further, alkoxide is to be hydrolysed with water, and then converted to alcohol and metal hydroxide. The different metals, such as Ba, Cu, Al, Si, Ti, Zr, Ge, V, W, Y, *etc.*, and some other alkoxides, like silicon alkoxides, TEOS (tetraethoxysilane) and TMOS (tetramethoxysilane), can be used.

The polymerization and hydrolysis reaction of TEOS is given below [70].

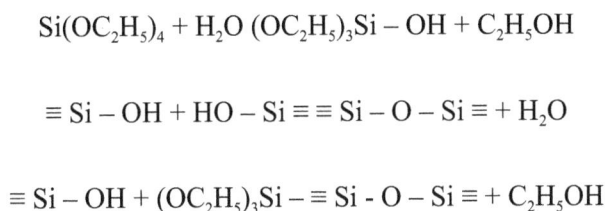

$$Si(OC_2H_5)_4 + H_2O \ (OC_2H_5)_3Si - OH + C_2H_5OH$$

$$\equiv Si - OH + HO - Si \equiv \ \equiv Si - O - Si \equiv + H_2O$$

$$\equiv Si - OH + (OC_2H_5)_3Si - \equiv Si - O - Si \equiv + C_2H_5OH$$

The various industries are manufacturing smart sol-gel-mediated compounds by condensation of tetraethoxysilane or acidification of sodium silicates functionalised by silicic acid.

PROCESSING AND APPLICATIONS OF MATERIALS IN DIFFERENT SECTORS

Materials' Processing in Tissue Engineering

Tissue engineering (TE) is utilized for the improvement, restoration and maintenance of impaired parts of the human body by virtue of active biomolecules. Thus, the field of TE involves outstanding scope; however, artificial skin and cartilage are yet best outputs of TE [71]. TE has come up as an emerging multidisciplinary field where failure and damaged tissues are re-functionalised by artificially or naturally developed tissues. The need for tissue transplantation can be minimized by the application of engineered tissue. The injured tissues and organs are replaced by artificial scaffold materials. Generally, TE scaffold materials are made up of biomaterials, metal oxide, metal alloys, polymers, bio-ceramics, and composites. The engineered biomaterials NCs play a pivotal role in the regeneration and restoration of damaged and failing tissues. The regeneration of accidental or any other disease-caused tissue damage is made highly possible with the blessings of TE. Nonetheless, improvement, restoration, and maintenance of damaged tissue or whole organ may also be carried out by utilizing TE. The surgical procedure is also common to recover damaged tissues due to the limited auto-repair capabilities of tissues. However, the use of scaffold materials has been proven to be the best alternative for surgical procedures [71]. The scaffolds' materialistic tissue frameworks are designed for desired cell interaction, particularly in medical sciences. These highly porous 3-D scaffold tissue frames are incorporated into cellular structures to functionalise them. Materials utilised for scaffolding in TE are chitosan, hydroxyapatite, collagen, hydrogel, nanostructured material, and polymer proteins. Hence, to design suitable scaffolds for TE, biocompatible, biodegradable, porous, and mechanically strong materials need to be taken into consideration for biomedical applications [72].

Biomaterials

Biomaterials are synthetically or naturally existing materials used for designing scaffolds or tissue constructs for all grafting or auto-grafting in the human body. The chemically synthesized biomaterials can be metals, polymers, ceramics, and composites. It seems to be more advantageous to use composite scaffolds made of biomaterials along with ceramics, synthetic or natural polymers, thus becoming common in preference. Moreover, ceramic scaffolding is composed of hydroxyl-apatite and tri-calcium phosphate, which possess mechanical hardness, low elasticity, and hard brittle surface, as well as are biocompatible with neighboured bone due to similar chemical compositions [73, 74]. But ceramic scaffolding preference is restricted because of brittleness, uncontrolled degrading rates of hydroxyl-apatite, and difficulties in reshaping and remoulding [75 - 77]. Instead of some hydroxyapatite, synthetic polymers have been proven to have controlled degradation rates [78 - 81]. Similarly, bio-originated materials, such as collagen, a variety of proteoglycans, and chitosan, have become the next efficient option to fabricate tissue constructs in TE due to their superior cellular adhesion and growth, unlike synthetic polymers. To resolve the functioning problems of single-phase materials and boost the bio-capacity of materials, composite/metal oxide or multi-phased biomaterials can be doped with other materials to produce scaffold nanocomposites [82 - 86]. In addition, the use of collagen-based scaffolds is a classic approach that enhances the biomechanical properties of tissue framework for bone restoration [87].

Metals and Metal Alloys

The use of metal scaffolds with good biocompatibility in orthopaedic fixations is very common in medical sciences. These metal scaffolds can be framed from stainless steel, cobalt alloys, and titanium alloys. Inefficiency in the biological perception of metallic surfaces can be overcome by surface coating, resulting in good biomechanical characteristics of metallic scaffolds [88]. Due to physical and mechanical characteristics, Ta, Mg and Mg, Ti, and Ni-Ti alloys have been proven to have the best scope in TE. The biomechanical properties of metallic alloys are superior to single metallic, polymeric, or ceramic-based scaffolds. As the metals are less porous in nature, they restrict cell proliferation; therefore, metallic scaffolds are replaced by bio-ceramics due to their good porosity and biodegradability [89].

Polymers

Polymer is also a good resource for tissue constructs due to its high surface area, biodegradation, mechanical properties, compatibility with tissues, and well-interconnected porous structure. Compared to metal and metal alloys, degradation rates of natural polymers are very less. The best examples of natural polymers are collagen and chitin, which possess high biocompatibility; however, they exhibit low biomechanical properties. However, compared to collagen, organometallic polymers and chitosan/organic-based materials exhibit superior biocompatibility due to high porosity but poor showing mechanical properties [90 - 92]. These limitations can be resolved by synthetic polymers, like polylactic acid, polyglycolic acid, *etc.*, having good porosity, cell compatibility, and biomechanical characteristics [93].

Bioceramics

Ceramic biomaterials are utilized to restore and regenerate infected or damaged cellular structures. These bio-inert materials can be amorphous and crystalline; chemically, they may be made of zirconia, alumina, and hydroxy-apatite [94]. Ceramic implants are applied in hard tissue regeneration due to their good cellular compatibility [95]. The scaffold ceramic materials are immensely used for tissue regeneration because they exhibit friendly metabolic behaviours in the human body without releasing any toxicity. So, amongst various methods and resources, ceramic scaffolds framing calcium-based ceramics and their composite scaffolds can be preferred as they are biocompatible and non-inflammatory [96].

Composites

Composite scaffolds are implemented in TE; specifically, hard tissue restoration can be done using a blend of natural or synthetic polymers along with metals and ceramics. The progression of the composite framework in TE is fascinating since its characteristics can be altered or modified as per the needs of the host body [97]. The ceramic materials can be inert, bioactive, or semi-inert, exhibiting enhanced osteo-conductivity [98, 99]. To overcome fragility of composites of bio-ceramics glasses, polymer/ceramics-based composites have been conceptualised, whereas ceramic or glass-polymer composites have been engineered to achieve superior biomechanical characteristics.

Material Processing in Biomedical Devices

To achieve smooth, advanced, and more precise medical technology, the design and manufacturing of medical devices are a priority for increasing the safe and standard life of human being. Therefore, it is an urgent need in the medical field to formulate innovative materials with improved functionalities as well as superior mechanical, smart physical-chemical, and electrical properties. The three basic properties of materials, like fine structure, good properties, and smart processing, collaboratively establish the performance of biomedical devices. Nowadays, requirements for smart devices are significantly increasing day by day due to the spread of diverse pandemic diseases, increasing diseases in population and thousands of organ implantations in a day. Moreover, implantable biomedical devices have been proven to be a blessing for partially resolving vital organ failures. However, research and development are being carried out to design non-invasive smart devices with superior sensitivity and minimal side effects. The different kinds of biomaterials, like metals, ceramics, polymers, and composites, are being utilized in the manufacturing of implantable and smart medical devices. The implantable and smart biomedical devices using smart materials are given in Fig. (**6**).

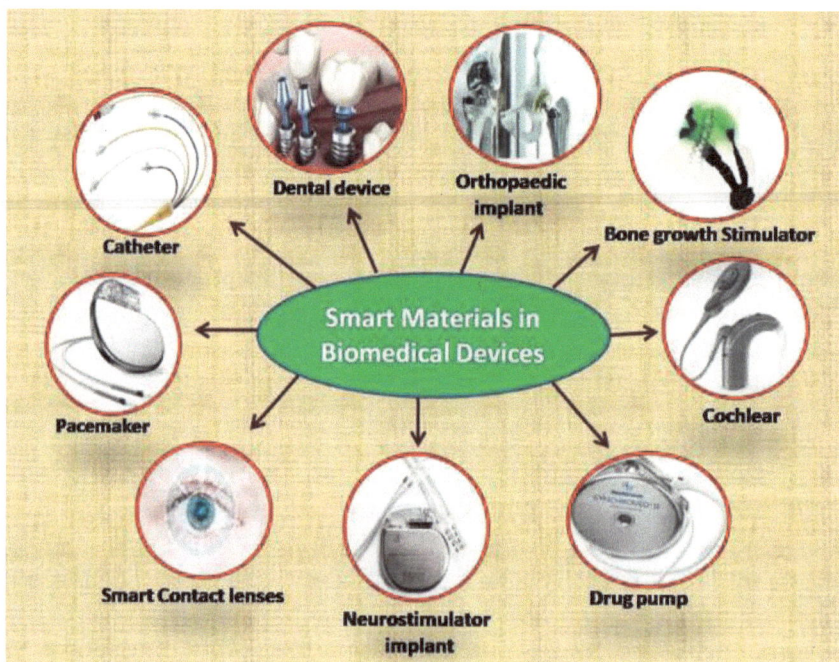

Fig. (6). Implantable and smart biomedical devices using smart materials.

Metals

Metals are extremely ductile and pliable, with high compressive and tensile strengths. The electrical and thermal conductivity are both high. The properties of metals can be altered by preparing their alloys. Due to the easy oxidation of metals, stainless steel alloy comprising iron, carbon, and chromium is a preferred choice in designing medical instruments and devices. Researchers working on metal-containing materials always aim to create novel alloys and straightforward processing methods to improve metal characteristics for applications in medicine. One of the better alloys is titanium because its modulus of elasticity is similar to that of bone rather than steel. New titanium alloys, particularly nickel-free alloys, are actively being investigated. Moreover, bio-absorption or bio-adsorption are potential applications of metals. Polymers are bioabsorbable, but magnesium and iron could make alloys with good properties of bioabsorption. Therefore, materials engineers are striving to design metals with lesser sensitivity to magnetism. Metal-containing compounds can resist absorbing or interacting with proteins, viruses, and other biological molecules due to surface modifications.

Ceramics

Ceramic is also one of the important materials, which include glass, clay, and concrete, used usually in oxide forms. They involve carbides, nitrides, or silicides also. They are mechanically rigid and brittle, with very little flexibility. They have a high compressive strength while having a low tensile and shear strength. Ceramics have low electrical conductivity in general, yet they can act as semiconductors, and a few can even become superconductors at high temperatures. But chemically, they are inactive. Thus, ceramics are said to be good insulators and play a significant role in the production of medical devices. Also, aluminium oxide is the most frequently used ceramic biomaterial in medical devices. Zirconium dioxide is also more popular. It has greater strength than aluminium oxide when stabilized with yttrium oxide, allowing the material to achieve the same strength as aluminium oxide even at small sizes. Moreover, piezoelectric ceramics materials are progressively replacing metal sensors in several medical devices.

Polymers

Rubber is the best example of a polymer, commonly used everywhere. In general, rubbers are light in weight and flexible, as well as being relatively inexpensive. More than 70% of the polymers used in medical devices manufacturing are thermoplastics, which can be molded to exact tolerance. While metals interfere with medical scanning technologies, such as MRIs, polymers do not have the same problem. In the manufacturing of medical devices, sterilizable, contamination-resistant, and low-toxic polymers are preferred. Processed polymers have a wide range of mechanical properties, which can also be altered for new uses.

In the advancement of medical devices, polymers are playing a significant role in developing 3D printing. It is now possible to produce device components using 3D printers due to recent technological breakthroughs. There are two common polymers for printing: acrylonitrile butadiene styrene and polylactic acid. Additionally, 3D printing has simplified the prototyping of medical devices, resulting in faster development cycles and cost saving. The prototypes can be printed using polymers, even if the final medical equipment is made of metals or ceramics.

Composite

A composite is one of the most recently developed materials used in designing medical devices. The composite is a blend of materials from two or more materials. The composites offer materials the desired properties while suppressing their undesirable properties. By the integration of metallic fibers, polymer-metals composite can preserve the lightweight and molding ability of plastics with increased strength. The composite's macroscopic layer is the place where two materials are combined. Because many tissues in the human body, such as skin, bones, muscles, and teeth, are composite materials, compatibility in replicating these tissues is the most vital aspect while designing and developing the composite scaffold or implant.

Agricultural Applications

There is a need for raw materials in intensive and extensive-scale commercial farming, like any other industry. The farmers need to minimise the impact of excessive synthetic fertilizers as they are hazardous to the environment and humans too. The quality of crop production may be decreased due to the use of hazardous chemicals that reduce the soil's fertility. However, farmers must maximise fertilizers' efficiency by using nitrogen-based biofertilizers. Several biomaterials can increase the fertility of the soil by adding biodegradable organic matter. The biodegradable organic matter enriches the soil by enhancing the components, such as carbon and nitrogen. High pH of the soil can result in the loss of nitrogen by 50% of the fertilisers to the atmosphere. Dispersants help and stabilise the particles and chelate, such as calcium gluconate assists with plants' uptake of nutrients. Agricultural production is generally seen as a source of raw materials.

Fertilisers and Trace Elements

Fertilizers are natural or synthetic substances that enhance soil fertility or nourish plants. The fertilizers may improve soil effectiveness by increasing water retention and filtering surplus liquid, and there can be organic and non-organic fertilizers. Natural processes can only be used to create organic fertilizers. Nevertheless, non-organic fertilizers are created to suit the nutritional needs of plants. Crops require nutrients to grow and be harvested for nutritious food, and fertilizers play a key role in giving those nutrients to crops. Fertilizers are plant nutrients given to agricultural areas to replenish essential elements, like nitrogen, phosphorus, and potassium, along with trace elements, including boron, molybdenum, manganese, copper, zinc, and iron, which are found naturally in the soil, but when soil nutrients are lacking or in short supply, the plants suffer from nutrient shortage and stop developing. So, soil improvements can be made by adding fertilizers to the soil. Urea, ammonium nitrate, and calcium nitrate are nitrogen-supplying fertilizers, while diammonium phosphate and monoammonium phosphate are common phosphorus providers. Similarly, potassium nitrate and potassium chloride are fertilizers that enrich the soil with potassium. While, synthetic fertilizers mainly use nitrogen, potassium, and phosphorus. Secondly, natural gas can combine with nitrogen in the air to form anhydrous ammonia. This can be applied directly to crops. Likewise, potassium is also used in fertilizers, such as potassium nitrate, potassium sulphate, and potassium chloride. Trace elements used as fertilizers include calcium, sulphur, and magnesium.

Nowadays, the use of plastics in agriculture has helped farmers to increase crop productivity while improving food quality and reducing their ecological imprint. These include polyolefin polymers (polypropylene (PP)) and ethylene vinyl acetate copolymer (EVA). Less often used plastics include polyvinyl chloride, polycarbonate, and polymethyl-methacrylate (PMMA). Greenhouses, low tunnels, mulching, plastic reservoirs and irrigation systems, crop collection, handling, and transport are all made possible by these polymeric plastics. Used plastics are frequently washed to remove sand, herbs, and pesticides before being ground and extruded into pellets once they are retrieved from the fields. Chemical recycling and energy recovery are complementary solutions with distinct technologies available when mechanical recycling is not feasible [100, 101].

Application in Pesticides

Pesticides are used to control pests and weeds. The pesticides are prepared from phosphoric acid compounds. An active component is combined with inert substances to form pesticides. The active component kills the pests, while the inert ingredients make spraying and coating the target plant easier. Active chemicals are distilled from natural sources, but they are now mostly produced in a laboratory or on pilot level. Pesticides contain elements, like chlorine, oxygen, sulphur, phosphorus, nitrogen, and bromine. The pesticides are classified as herbicides, insecticides, and fungicides (Fig. 7).

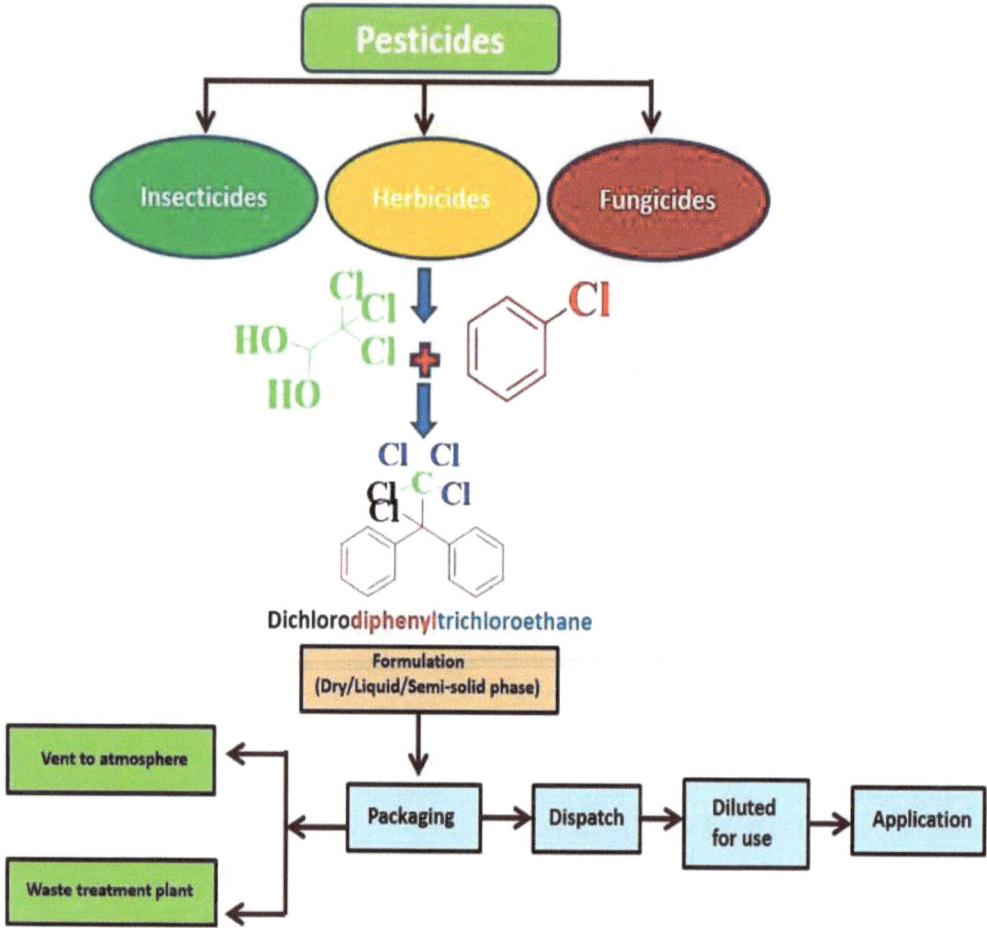

Fig. (7). Schematic process of pesticide manufacturing.

Several biopowder-based materials can be used to control the pest. Vegetable materials, such as pulverized nut shells or corn cobs, clays, diatomite, talc, or calcium carbonate, are commonly used as dust pesticides. Different types of materials are used for manufacturing various pesticides. For example, for DDT, chloral and chlorobenzene are used as raw materials. In the production of pesticides, an active substance is manufactured in a chemical factory, then either formulated there or transported to a formulator that prepares the powder or semi-solid form by mixing active material with inert powder. The pesticide is then sent to a farmer or other qualified applicator, who dilutes it before applying it to the crops (Fig. **7**).

CONCLUSION AND FUTURE ASPECTS

In this article, various manufacturing techniques have been discussed, and out of all, the melt blending technique is found superior because of cost, straightforwardness, environment friendliness, and is suited for bulk production to meet the industrial demand. In-situ polymerisation technique is also good for the fabrication of thermosetting and thermoplastic NCs. Each technique has its own importance for the fabrication of NCs. Further, it has been revealed that various processing techniques are significantly useful for advanced materials' applications, like TE, biomedical agents, fertilizers, and pesticides. Overall, different advanced materials and processing techniques are very constructive to fabricate future smart materials benign to the environment to fulfill the requirement and living standards of human beings.

CONSENT FOR PUBLICATION

Not applicable.

CONFLICT OF INTEREST

The author declares no conflict of interest, financial or otherwise.

ACKNOWLEDGEMENT

The authors are thankful to the Principal of Seth Kesarimal Porwal College Kamptee for his immense support in permitting us to use the Departmental Library and internet facility during the scientific writing of the chapter.

REFERENCES

[1]   Kaounides LC. The Revolution in Materials Science and Engineering: Strategic Implications for Developing Countries in the 1990s. United Nations Industrial Development Organization 1991.

[2]   Jones RM. Mechanics of composite materials. CRC Press 2014.

[3]   Lubin G. Handbook of composites. Springer Science & Business Media 2013.

[4]   Guo N, Leu MC. Additive manufacturing: technology, applications and research needs. Front Mech Eng 2013; 8(3): 215-43.
      [http://dx.doi.org/10.1007/s11465-013-0248-8]

[5]   Rafiee MA, Rafiee J, Wang Z, Song H, Yu ZZ, Koratkar N. Enhanced mechanical properties of nanocomposites at low graphene content. ACS Nano 2009; 3(12): 3884-90.
      [http://dx.doi.org/10.1021/nn9010472] [PMID: 19957928]

[6]   Hale DK. The physical properties of composite materials. J Mater Sci 1976; 11(11): 2105-41.
      [http://dx.doi.org/10.1007/PL00020339]

[7]   Yan DX, Ren PG, Pang H, Fu Q, Yang MB, Li ZM. Efficient electromagnetic interference shielding of lightweight graphene/polystyrene composite. J Mater Chem 2012; 22(36): 18772-4.
      [http://dx.doi.org/10.1039/c2jm32692b]

[8]   Sahmaran M, Li VC, Andrade C. Corrosion resistance performance of steel-reinforced engineered cementitious composite beams. ACI Mater J 2008; 105: 243.

[9]   Dinca I, Ban C, Stefan A, Pelin G. Nanocomposites as advanced material for aerospace industry. INCAS Bull 2012; 4: 73-83.

[10]   Available from: http://en.wikipedia.org/wiki/Boeing_787_Dreamliner

[11]   Kroto HW, Heath JR, O'Brien SC, Curl RF, Smalley RE. C60: Buckminsterfullerene. Nature 1985; 318(6042): 162-3.
[http://dx.doi.org/10.1038/318162a0]

[12]   Iijima S, Ichihashi T. Single-shell carbon nanotubes of 1-nm diameter. Nature Publishing group 1993; 336: 603-5.

[13]   Novoselov KS, Geim AK, Morozov SV, *et al.* Two-dimensional gas of massless Dirac fermions in graphene. Nature 2005; 438(7065): 197-200.
[http://dx.doi.org/10.1038/nature04233] [PMID: 16281030]

[14]   Tomar R, Abdala AA, Chaudhary RG, Singh NB. Photocatalytic degradation of dyes by nanomaterials. Mater Today Proc 2020; 29: 967-73.
[http://dx.doi.org/10.1016/j.matpr.2020.04.144]

[15]   Umekar MS, Bhusari GS, Potbhare AK, *et al.* Bioinspired reduced graphene oxide based nanohybrids for photocatalysis and antibacterial applications. Curr Pharm Biotechnol 2021; 22(13): 1759-81.
[http://dx.doi.org/10.2174/1389201022666201231115826] [PMID: 33390112]

[16]   Fielding JC, Chen C, Borges J. Vacuum infusion process for nanomodified aerospace epoxy resins. In SAMPE Symposium & Exhibition 2004.

[17]   Haque A, Hossain F, Dean D, Shamsuzzoha M. S2-Glass/vinyl ester polymer nanocomposites: Manufacturing, structures, thermal and mechanical properties. J Adv Mater 2005; 37: 16-27.

[18]   Haque A, Shamsuzzoha M, Hussain F, Dean D. S2-glass/epoxy polymer nanocomposites: manufacturing, structures, thermal and mechanical properties. J Compos Mater 2003; 37(20): 1821-37.
[http://dx.doi.org/10.1177/002199803035186]

[19]   Chowdhury FH, Hosur MV, Jeelani S. Studies on the flexural and thermomechanical properties of woven carbon/nanoclay-epoxy laminates. Mater Sci Eng A 2006; 421(1-2): 298-306.
[http://dx.doi.org/10.1016/j.msea.2006.01.074]

[20]   Roy S, Lu H, Narasimhan K, Hussain F. Characterization and modeling of strength enhancement mechanisms in a polymer/clay nanocomposite. AIAA Conference Proceedings. Texas. 2005.
[http://dx.doi.org/10.2514/6.2005-1853]

[21]   Hussain F, Roy S, Narasimhan K, Vengadassalam K, Lu H. E-Glass—Polypropylene pultruded nanocomposite: manufacture, characterization, thermal and mechanical properties. Journal of Thermoplastic Composite Materials 2007; 20(4): 411-34.
[http://dx.doi.org/10.1177/0892705707079604]

[22]   Gurses A, Introduction to Polymer-Clay Nanocomposites. Singapore: Panstanford Publishing 2016.

[23]   Silva PSRC, Tavares MIB. Solvent Effect on the Morphology of Lamellar Nanocomposites Based on HIPS. Mater Res 2015; 18(1): 191-5.
[http://dx.doi.org/10.1590/1516-1439.307314]

[24]   Monteiro MSSB, Rodrigues CL, Neto RPC, Tavares MIB. The structure of polycaprolactone-clay nanocomposites investigated by ^{1}H NMR relaxometry. J Nanosci Nanotechnol 2012; 12(9): 7307-13.
[http://dx.doi.org/10.1166/jnn.2012.6431] [PMID: 23035469]

[25]   Soares IL, Chimanowsky JP, Luetkmeyer L, Silva EO, Souza DHS, Tavares MIB. Evaluation of the influence of modified TiO_2 particles on polypropylene composites. J Nanosci Nanotechnol 2015; 15(8): 5723-32.

[http://dx.doi.org/10.1166/jnn.2015.10041] [PMID: 26369145]

[26]   Antonio de Pádua C B C, Maria Inês Bruno T, Emerson Oliveira S, Soraia Z, Zaioncz S. The effect of montmorillonite clay on the crystallinity of poly(vinyl alcohol) nanocomposites obtained by solution intercalation and *in situ* polymerization. J Nanosci Nanotechnol 2015; 15(4): 2814-20.
[http://dx.doi.org/10.1166/jnn.2015.9233] [PMID: 26353498]

[27]   Paul PK, Hussain SA, Bhattacharjee D, Pal M. Preparation of polystyrene–clay nanocomposite by solution intercalation technique. Bull Mater Sci 2013; 36(3): 361-6.
[http://dx.doi.org/10.1007/s12034-013-0498-4]

[28]   Jin L, Bower C, Zhou O. Alignment of carbon nanotubes in a polymer matrix by mechanical stretching. Appl Phys Lett 1998; 73(9): 1197-9.
[http://dx.doi.org/10.1063/1.122125]

[29]   Paul DR, Robeson LM. Polymer nanotechnology: Nanocomposites. Polymer (Guildf) 2008; 49(15): 3187-204.
[http://dx.doi.org/10.1016/j.polymer.2008.04.017]

[30]   Wang K, Liang S, Du R, Zhang Q, Fu Q. The interplay of thermodynamics and shear on the dispersion of polymer nanocomposite. Polymer (Guildf) 2004; 45(23): 7953-60.
[http://dx.doi.org/10.1016/j.polymer.2004.09.053]

[31]   Tan LJ, Zhu W, Zhou K. Recent progress on polymer materials for additive manufacturing. Adv Funct Mater 2020; 30(43): 2003062.
[http://dx.doi.org/10.1002/adfm.202003062]

[32]   Singh V, Joung D, Zhai L, Das S, Khondaker SI, Seal S. Graphene based materials: Past, present and future. Prog Mater Sci 2011; 56(8): 1178-271.
[http://dx.doi.org/10.1016/j.pmatsci.2011.03.003]

[33]   Lee JH, Kim SK, Kim NH. Effects of the addition of multi-walled carbon nanotubes on the positive temperature coefficient characteristics of carbon-black-filled high-density polyethylene nanocomposites. Scr Mater 2006; 55(12): 1119-22.
[http://dx.doi.org/10.1016/j.scriptamat.2006.08.051]

[34]   Kalaitzidou K, Fukushima H, Drzal LT. A new compounding method for exfoliated graphite–polypropylene nanocomposites with enhanced flexural properties and lower percolation threshold. Compos Sci Technol 2007; 67(10): 2045-51.
[http://dx.doi.org/10.1016/j.compscitech.2006.11.014]

[35]   Wang WP, Pan CY. Preparation and characterization of polystyrene/graphite composite prepared by cationic grafting polymerization. Polymer (Guildf) 2004; 45(12): 3987-95.
[http://dx.doi.org/10.1016/j.polymer.2004.04.023]

[36]   Sinha Ray S, Okamoto M. Polymer/layered silicate nanocomposites: a review from preparation to processing. Prog Polym Sci 2003; 28(11): 1539-641.
[http://dx.doi.org/10.1016/j.progpolymsci.2003.08.002]

[37]   Utracki LA. Polymeric nanocomposites: compounding and performance. J Nanosci Nanotechnol 2008; 8(4): 1582-96.
[http://dx.doi.org/10.1166/jnn.2008.18225] [PMID: 18572559]

[38]   Adak B, Joshi M, Butola BS. Polyurethane/functionalized-graphene nanocomposite films with enhanced weather resistance and gas barrier properties. Compos, Part B Eng 2019; 176: 107303.
[http://dx.doi.org/10.1016/j.compositesb.2019.107303]

[39]   Noorunnisa Khanam P, AlMaadeed MA, Ouederni M, *et al.* Melt processing and properties of linear low density polyethylene-graphene nanoplatelet composites. Vacuum 2016; 130: 63-71.
[http://dx.doi.org/10.1016/j.vacuum.2016.04.022]

[40]   Torğut G. Fabrication, characterization of poly(MA-co-NIPA)-graphene composites and optimization the dielectric properties using the response surface method (RSM). Polym Test 2019; 76: 312-9.

[http://dx.doi.org/10.1016/j.polymertesting.2019.03.035]

[41]   Rattanakot J, Potiyaraj P. Poly(Lactic Acid)/Poly(Vinyl Alcohol)/Graphene Nanocomposites. Key Eng Mater 2018; 773: 10-4.
[http://dx.doi.org/10.4028/www.scientific.net/KEM.773.10]

[42]   Akhina H, Mohammed Arif P, Gopinathan Nair MR, Nandakumar K, Thomas S. Development of plasticized poly (vinyl chloride)/reduced graphene oxide nanocomposites for energy storage applications. Polym Test 2019; 73: 250-7.
[http://dx.doi.org/10.1016/j.polymertesting.2018.10.015]

[43]   Ajorloo M, Fasihi M, Ohshima M, Taki K. How are the thermal properties of polypropylene/graphene nanoplatelet composites affected by polymer chain configuration and size of nanofiller? Mater Des 2019; 181: 108068.
[http://dx.doi.org/10.1016/j.matdes.2019.108068]

[44]   Mistretta MC, Botta L, Vinci AD, Ceraulo M, La Mantia FP. Photo-oxidation of polypropylene/graphene nanoplatelets composites. Polym Degrad Stabil 2019; 160: 35-43.
[http://dx.doi.org/10.1016/j.polymdegradstab.2018.12.003]

[45]   You F, Li X, Zhang L, Wang D, Shi CY, Dang ZM. Polypropylene/poly(methyl methacrylate)/graphene composites with high electrical resistivity anisotropy *via* sequential biaxial stretching. RSC Advances 2017; 7(10): 6170-8.
[http://dx.doi.org/10.1039/C6RA28486H]

[46]   Zhao S, Chen F, Huang Y, Dong JY, Han CC. Crystallization behaviors in the isotactic polypropylene/graphene composites. Polymer (Guildf) 2014; 55(16): 4125-35.
[http://dx.doi.org/10.1016/j.polymer.2014.06.027]

[47]   Kuilla T, Bhadra S, Yao D, Kim NH, Bose S, Lee JH. Recent advances in graphene based polymer composites. Prog Polym Sci 2010; 35(11): 1350-75.
[http://dx.doi.org/10.1016/j.progpolymsci.2010.07.005]

[48]   Lang MA, Guo-jian W, Jin-feng DA. Preparation and properties of reduced graphene oxide/polyimide composites produced by *in-situ* polymerization and solution blending methods. N Carbon Mater 2016; 31: 129-34.

[49]   Okada A, Usuki A. Twenty years of polymer-clay nanocomposites. Macromol Mater Eng 2006; 291(12): 1449-76.
[http://dx.doi.org/10.1002/mame.200600260]

[50]   Pavlidou S, Papaspyrides CD. A review on polymer–layered silicate nanocomposites. Prog Polym Sci 2008; 33(12): 1119-98.
[http://dx.doi.org/10.1016/j.progpolymsci.2008.07.008]

[51]   Shrirame TS, Khan JS, Umekar MS, *et al.* Graphene-Polymer Nanocomposites for Environmental Remediation of Organic Pollutants, Metal Nanocomposites for Energy and Environmental Applications. Singapore: Springer 2022; pp. 321-49.
[http://dx.doi.org/10.1007/978-981-16-8599-6_14]

[52]   Umekar M, Chaudhary R, Bhusari G, Potbhare A. Fabrication of zinc oxide-decorated phytoreduced graphene oxide nanohybrid *via* Clerodendrum infortunatum. Emerging Materials Research 2021; 10(1): 75-84.
[http://dx.doi.org/10.1680/jemmr.19.00175]

[53]   Chaudhry AU, Lonkar SP, Chudhary RG, Mabrouk A, Abdala AA. Thermal, electrical, and mechanical properties of highly filled HDPE/graphite nanoplatelets composites. Mater Today Proc 2020; 29(3): 704-8.
[http://dx.doi.org/10.1016/j.matpr.2020.04.168]

[54]   Chaudhary RG, Potbhare AK, Chouke PB, Rai AR. Graphene-based materials and their Nanocomposites with Metal Oxides : Biosynthesis, Electrochemical, Photocatalytic and Antimicrobial

Applications Magnetic Oxides and Composites II. Material Research Foundation 2020; Vol. 83: pp. 79-116.
[http://dx.doi.org/10.21741/9781644900970-4]

[55]   Alexandre M, Dubois P. Polymer-layered silicate nanocomposites: preparation, properties and uses of a new class of materials. Mater Sci Eng Rep 2000; 28(1-2): 1-63.
[http://dx.doi.org/10.1016/S0927-796X(00)00012-7]

[56]   Usuki A, Kawasumi M, Kojima Y, Okada A, Kurauchi T, Kamigaito O. Swelling behavior of montmorillonite cation exchanged for ω-amino acids by ε-caprolactam. J Mater Res 1993; 8(5): 1174-8.
[http://dx.doi.org/10.1557/JMR.1993.1174]

[57]   Baruah B, Kumar A, Umapathy GR, Ojha S. Enhanced electrocatalytic activity of ion implanted rGO/PEDOT:PSS hybrid nanocomposites towards methanol electro-oxidation in direct methanol fuel cells. J Electroanal Chem (Lausanne) 2019; 840: 35-51.
[http://dx.doi.org/10.1016/j.jelechem.2019.03.053]

[58]   Miao J, Li H, Qiu H, Wu X, Yang J. Graphene/PANI hybrid film with enhanced thermal conductivity by *in situ* polymerization. J Mater Sci 2018; 53(12): 8855-65.
[http://dx.doi.org/10.1007/s10853-018-2112-z]

[59]   Wang X, Hu Y, Song L, Yang H, Xing W, Lu H. *In situ* polymerization of graphene nanosheets and polyurethane with enhanced mechanical and thermal properties. J Mater Chem 2011; 21(12): 4222-7.
[http://dx.doi.org/10.1039/c0jm03710a]

[60]   Bao C, Guo Y, Song L, Kan Y, Qian X, Hu Y. *In situ* preparation of functionalized graphene oxide/epoxy nanocomposites with effective reinforcements. J Mater Chem 2011; 21(35): 13290-8.
[http://dx.doi.org/10.1039/c1jm11434d]

[61]   Hu H, Wang X, Wang J, *et al.* Preparation and properties of graphene nanosheets–polystyrene nanocomposites *via in situ* emulsion polymerization. Chem Phys Lett 2010; 484(4-6): 247-53.
[http://dx.doi.org/10.1016/j.cplett.2009.11.024]

[62]   Ma J, Li Y, Yin X, *et al.* Poly(vinyl alcohol)/graphene oxide nanocomposites prepared by *in situ* polymerization with enhanced mechanical properties and water vapor barrier properties. RSC Advances 2016; 6(55): 49448-58.
[http://dx.doi.org/10.1039/C6RA08760D]

[63]   Ke Y, Long C, Qi Z. Crystallization, properties, and crystal and nanoscale morphology of PET-clay nanocomposites. J Appl Polym Sci 1999; 71(7): 1139-46.
[http://dx.doi.org/10.1002/(SICI)1097-4628(19990214)71:7<1139::AID-APP12>3.0.CO;2-E]

[64]   Okamoto M, Morita S, Kotaka T. Dispersed structure and ionic conductivity of smectic clay/polymer nanocomposites. Polymer (Guildf) 2001; 42(6): 2685-8.
[http://dx.doi.org/10.1016/S0032-3861(00)00642-X]

[65]   Lan T, Pinnavaia TJ. Clay-reinforced epoxy nanocomposites. Chem Mater 1994; 6(12): 2216-9.
[http://dx.doi.org/10.1021/cm00048a006]

[66]   Becker O, Varley R, Simon G. Morphology, thermal relaxations and mechanical properties of layered silicate nanocomposites based upon high-functionality epoxy resins. Polymer (Guildf) 2002; 43(16): 4365-73.
[http://dx.doi.org/10.1016/S0032-3861(02)00269-0]

[67]   Davidson RS. Exploring the science, technology and applications of UV and EB curing. London, UK: SITA Technology Ltd. 1998.

[68]   Sangermano M, Bongiovanni R, Malucelli G, Priola A. New developments in cationic photopolymerization: process and properties.Horizons in polymer research. New York: Nova Science Publisher Inc. 2006.

[69]   Tanaka T, Montanari GC, Mulhaupt R. Polymer nanocomposites as dielectrics and electrical

insulation-perspectives for processing technologies, material characterization and future applications. IEEE Trans Dielectr Electr Insul 2004; 11(5): 763-84.
[http://dx.doi.org/10.1109/TDEI.2004.1349782]

[70]   Leila Bazli, Mostaf Shiravic Arman. A review of carbon nanotube/TiO$_2$ composite prepared via sol-gel method, J Compos Compd. 2019; 1-9.

[71]   Kumar P, Sindhu A. Materials for tissue engineering. Advances in Animal Biotechnology and its Applications 2018; 357-70.
[http://dx.doi.org/10.1007/978-981-10-4702-2_20]

[72]   Potbhare AK, Umekar MS, Chouke PB, *et al.* Bioinspired graphene-based silver nanoparticles: Fabrication, characterization and antibacterial activity. Mater Today Proc 2020; 29: 720-5.
[http://dx.doi.org/10.1016/j.matpr.2020.04.212]

[73]   Hench LL. Bioceramics. J Am Ceram Soc 1998; 81(7): 1705-28.
[http://dx.doi.org/10.1111/j.1151-2916.1998.tb02540.x]

[74]   Ambrosio AMA, Sahota JS, Khan Y, Laurencin CT. A novel amorphous calcium phosphate polymer ceramic for bone repair: I. Synthesis and characterization. J Biomed Mater Res 2001; 58(3): 295-301.
[http://dx.doi.org/10.1002/1097-4636(2001)58:3<295::AID-JBM1020>3.0.CO;2-8]   [PMID: 11319744]

[75]   Wang M. Developing bioactive composite materials for tissue replacement. Biomaterials 2003; 24(13): 2133-51.
[http://dx.doi.org/10.1016/S0142-9612(03)00037-1] [PMID: 12699650]

[76]   Van Landuyt P, Li F, Keustermans JP, Streydio JM, Delannay F, Munting E. The influence of high sintering temperatures on the mechanical properties of hydroxylapatite. J Mater Sci Mater Med 1995; 6(1): 8-13.
[http://dx.doi.org/10.1007/BF00121239]

[77]   Dorozhkin SV. Calcium orthophosphate (CaPO$_4$) scaffolds for bone tissue engineering applications. Journal of Biotechnology and Biomedical Science 2018; 1(3): 25-93.
[http://dx.doi.org/10.14302/issn.2576-6694.jbbs-18-2143]

[78]   Lu L, Peter SJ, Lyman MD, *et al.* *In vitro* and *in vivo* degradation of porous poly(DL-lactic-co-glycolic acid) foams. Biomaterials 2000; 21(18): 1837-45.
[http://dx.doi.org/10.1016/S0142-9612(00)00047-8] [PMID: 10919687]

[79]   Oh S, Kang SG, Kim ES, Cho SH, Lee JH. Fabrication and characterization of hydrophilic poly(lactic-co-glycolic acid)/poly(vinyl alcohol) blend cell scaffolds by melt-molding particulate-leaching method. Biomaterials 2003; 24(22): 4011-21.
[http://dx.doi.org/10.1016/S0142-9612(03)00284-9] [PMID: 12834596]

[80]   Rowlands AS, Lim SA, Martin D, Cooper-White JJ. Polyurethane/poly(lactic-co-glycolic) acid composite scaffolds fabricated by thermally induced phase separation. Biomaterials 2007; 28(12): 2109-21.
[http://dx.doi.org/10.1016/j.biomaterials.2006.12.032] [PMID: 17258315]

[81]   Liu H, Slamovich EB, Webster TJ. Less harmful acidic degradation of poly(lactic-co-glycolic acid) bone tissue engineering scaffolds through titania nanoparticle addition. Int J Nanomedicine 2006; 1(4): 541-5.
[http://dx.doi.org/10.2147/nano.2006.1.4.541] [PMID: 17722285]

[82]   Kim SS, Sun Park M, Jeon O, Yong Choi C, Kim BS. Poly(lactide-co-glycolide)/hydroxyapatite composite scaffolds for bone tissue engineering. Biomaterials 2006; 27(8): 1399-409.
[http://dx.doi.org/10.1016/j.biomaterials.2005.08.016] [PMID: 16169074]

[83]   Huang YX, Ren J, Chen C, Ren TB, Zhou XY. Preparation and properties of poly(lactide-c--glycolide) (PLGA)/ nano-hydroxyapatite (NHA) scaffolds by thermally induced phase separation and rabbit MSCs culture on scaffolds. J Biomater Appl 2008; 22(5): 409-32.

[http://dx.doi.org/10.1177/0885328207077632] [PMID: 17494961]

[84]   Chaudhary RG, Potbhare AK, Aziz SKT, Umekar MS, Bhuyar SS, Mondal A. Phytochemically fabricated reduced graphene Oxide-ZnO NCs by *Sesbania bispinosa* for photocatalytic performances. Mater Today Proc 2021; 36: 756-62.
[http://dx.doi.org/10.1016/j.matpr.2020.05.821]

[85]   Mishra RK, Verma K, Chaudhary RG, Lambat T, Joseph K. An efficient fabrication of polypropylene hybrid nanocomposites using carbon nanotubes and PET fibrils. Mater Today Proc 2020; 29: 794-800.
[http://dx.doi.org/10.1016/j.matpr.2020.04.753]

[86]   Wu W, Feng X, Mao T, *et al.* Engineering of human tracheal tissue with collagen-enforced poly-lactic-glycolic acid non-woven mesh: A preliminary study in nude mice. Br J Oral Maxillofac Surg 2007; 45(4): 272-8.
[http://dx.doi.org/10.1016/j.bjoms.2006.09.004] [PMID: 17097777]

[87]   O'Brien FJ. Biomaterials & scaffolds for tissue engineering. Mater Today 2011; 14(3): 88-95.
[http://dx.doi.org/10.1016/S1369-7021(11)70058-X]

[88]   Liu C, Xia Z, Czernuszka JT. Design and development of three-dimensional scaffolds for tissue engineering. Chem Eng Res Des 2007; 85(7): 1051-64.
[http://dx.doi.org/10.1205/cherd06196]

[89]   Gahlawat SK, Duhan JS, Salar RK, Siwach P, Kumar S, Kaur P. Advances in Animal Biotechnology and its Applications. 1st ed. Springer 2018; pp. 1-418.
[http://dx.doi.org/10.1007/978-981-10-4702-2]

[90]   Bueno EM, Glowacki J. Cell-free and cell-based approaches for bone regeneration. Nat Rev Rheumatol 2009; 5(12): 685-97.
[http://dx.doi.org/10.1038/nrrheum.2009.228] [PMID: 19901916]

[91]   Zohora FT, Yousuf A, Anwarul M. Biomaterials as porous scaffolds for tissue engineering applications: a review. Eur Sci J 2014; 10: 186-209.

[92]   Chandraprabha MN, Krishna RH, Samrat K, Pradeepa K, Patil NC, Sasikumar M. Biogenic collagen-nano zno composite membrane as potential wound dressing material: structural characterization, antibacterial studies and *in vivo* wound healing studies. J Inorg Organomet Polym Mater 2022; 32(9): 3429-44.
[http://dx.doi.org/10.1007/s10904-022-02351-8]

[93]   Naughton GK, Tolbert WR, Grillot TM. Emerging developments in tissue engineering and cell technology. Tissue Eng 1995; 1(2): 211-9.
[http://dx.doi.org/10.1089/ten.1995.1.211] [PMID: 19877929]

[94]   Imran Z, Imran Z, Farooq U, Leghari A, Ali H. Bioactive glass: a material for the future. World J Dent 2012; 3(2): 199-201.
[http://dx.doi.org/10.5005/jp-journals-10015-1156]

[95]   Luz GM, Mano JF. Preparation and characterization of bioactive glass nanoparticles prepared by sol–gel for biomedical applications. Nanotechnology 2011; 22(49): 494014.
[http://dx.doi.org/10.1088/0957-4484/22/49/494014] [PMID: 22101770]

[96]   Apelt D, Theiss F, El-Warrak AO, *et al.* In vivo behavior of three different injectable hydraulic calcium phosphate cements. Biomaterials 2004; 25(7-8): 1439-51.
[http://dx.doi.org/10.1016/j.biomaterials.2003.08.073] [PMID: 14643619]

[97]   Rezwan K, Chen QZ, Blaker JJ, Boccaccini AR. Biodegradable and bioactive porous polymer/inorganic composite scaffolds for bone tissue engineering. Biomaterials 2006; 27(18): 3413-31.
[http://dx.doi.org/10.1016/j.biomaterials.2006.01.039] [PMID: 16504284]

[98]   Hench LL. Bioceramics: from concept to clinic. Am Ceram Soc Bull 1993; 72: 93-8.

[99]   Lakes RS, Park J. Biomaterials: an introduction. Springer Science Business Media 1992.

[100]   Maraveas C. Environmental sustainability of plastic in agriculture. Agriculture 2020; 10(8): 310.
        [http://dx.doi.org/10.3390/agriculture10080310]

[101]   Scarascia-Mugnozza G, Sica C, Russo G. Plastic materials in European agriculture: actual use and
        perspectives. J Agric Eng 2012; 42(3): 15-28.
        [http://dx.doi.org/10.4081/jae.2011.28]

CHAPTER 8

Regulators of Biomedical Devices

Umut Beylik[1] and **Erhan Akdoğan**[2,3,*]

[1] *Department of Health Management, Faculty of Gulhane Health Sciences, University of Health Sciences, Turkey*

[2] *Health Institutes of Turkey, İstanbul, Turkey*

[3] *Department of Mechatronics Engineering, Faculty of Mechanical Engineering, Yıldız Technical University, İstanbul, Turkey*

Abstract: Regulators of medical devices in the world regulate global competition in the medical device sector, and on the other hand, they play a decisive role in security, performance, and access issues. Technological development has also increased in the medical device industry, and medical devices have become more important in the diagnosis and treatment of health services. However, it is important that legal regulations must be implemented correctly and effectively in order to prevent public health or unethical behaviors. In this context, the regulations of the United States of America (USA) and the European Union (EU), the leaders in the sector, along with their high markets are discussed. In addition, medical device regulations in Japan, China, and Brazil, which have an important position in technological development and competition and have high potential, are also included. Considering the urgency and possible consequences of healthcare services, it is necessary to consider the fund and the regulations of the medical device sector separately in individual, national and global dimensions, from macro to micro. In addition to the safety, cost, and effectiveness of medical devices, it is important to discuss the conformity assessment, approval system processes, and how long it takes for a medical device to be put on the market. Considering the rapid technology change, regulations should be made to carry out the licensing and approval processes effectively and quickly in medical device regulations.

Keywords: Biomedical devices, Classification, Comparison, Medical devices, Regulations, Regulatory, Safety.

INTRODUCTION

The medical device sector has an important place in the health sector where a lot of effort and the latest technology are used extensively. The reasons, such as the

* **Corresponding author Erhan Akdoğan:** Health Institutes of Turkey, İstanbul, Turkey; and Department of Mechatronics Engineering, Faculty of Mechanical Engineering, Yıldız Technical University, İstanbul, Turkey; E-mail: erhan.akdogan@tuseb.gov.tr

Felipe López-Saucedo (Ed.)

increase in the elderly population, the spread of chronic diseases, the development of science day by day, and the high purchasing power have led to the development of the medical device industry. Globalization, countries' development of their own technologies, and countries' desire to export have accelerated the spread of this sector in the world. It seems that global organizations dominate the medical device industry. All over the world, it seems that the end products of medical devices are of high cost. For this reason, medical device production and distribution to the market are mostly carried out by big organizations, and small and medium-sized organizations operating in the medical device sector are usually purchased by big organizations or strategic partnerships are established with them, and activities are carried out at a global level [1].

It is seen that more than 60% of medical device production in the world is foreign trade. This situation reveals that the medical device industry has become global. Therefore, international organizations are included in this sector and they take the competition to a higher level. Of course, some countries do not go into privatization and try to get stronger in their local markets [2].

All processes that medical devices go through until they become available to the market are aimed at ensuring the safety of products and minimizing the risk of error and damage. Therefore, regulations developed for medical devices in the world aim to protect public health and ensure the use of safe products [3].

In this chapter, firstly, medical device regulations are discussed from a general perspective, and then the USA and EU regulations are discussed in detail. Also, medical device regulations in Japan, China, and Brazil, which have an important role and potential in the medical device sector, are briefly discussed, and the subject is evaluated in detail with the results, discussions, and recommendations section.

MEDICAL DEVICE REGULATIONS OVERVIEW

Medical device regulations are made with the participation of all partners. Representatives of the leading countries, scientific committees, universities, medical device manufacturers producing worldwide, notified institutions that check whether the products are at a sufficient level in terms of quality and safety, and international testing and certification institutions and working groups determine the regulations and universal standards in line with the needs [4].

Thanks to international trade agreements, such as the Customs Union and the free-enterprise policy, the circulation of medical devices between countries has become easier, as is the case with many product groups. Within the scope of the agreements made, countries have updated their current legislation and developed common legislation. As a result of the common laws and rules applied, a product in any country can enter the market and can be exported without any obstacles. For example, a medical device produced and certified in accordance with the requirements of the medical device directives within the borders of the EU can be placed on the market after the necessary customs notifications are made in other Union member countries [5].

The desire of multinational organizations to reduce production and transfer costs, which emerged after the initiation of intercontinental trade, enabled medical device organizations to establish production facilities in different continents. Thanks to the new facilities, although organizations manage to reduce production and transfer costs, the whole world is prevented from turning into a single medical device market due to reasons, such as the absence of a common legislation or audit program covering all countries, including Russia, China, Japan, and America. Today, working groups and boards consisting of representatives of the leading countries of the sector develop projects on this issue [6].

The International Medical Device Regulators Forum (IMDRF) is a voluntary working group that brings together medical device regulators from various countries of the world and aims to accelerate the joint work of international medical device regulators. IMDRF consists of medical device regulatory authorities and representatives of Australia, Brazil, Canada, China, the EU, Japan, Russia, Singapore, and the USA. The World Health Organization (WHO) and the Asia-Pacific Economic Cooperation - Life Sciences Innovation Forum - Regulatory Harmonization Steering Committee (APEC-LSIF-RHSC) are official observers. Among the aims of the organization are to effectively respond to industry challenges, establish a regulatory model for medical devices, protect public health, and maximize safety.

There are different medical device risk groups in the world. These groups are commonly called classification systems. Although there are differences in the classification of medical devices according to countries, in general, products are classified according to the place of use, purpose of use, indication, device components and risks [4]. The first classification system was prepared with a system called "Nomenclature regulation system". In this system, products are classified as low, medium, and high, according to the increasing risk level. The Nomenclature classification system is still used in the USA and Japan. In these countries, device classes are determined according to the products called generic and with a certain risk class [7]. EU, in 1993, developed a Medical Device Legislation, and Medical Device Directives brought a new perspective to the classification of products. They have established general rules that can be applied to products by identifying potential hazards to the human body. According to the new Medical Device Directives, medical devices are classified within the framework of these rules. Since the EU and some other countries develop systems based on the accuracy of the manufacturer's declaration, manufacturers and notified bodies, the risk classes of medical devices are determined within the framework of the classification rules in the directives [7].

MEDICAL DEVICE REGULATIONS IN THE WORLD

In this section, information is given about the medical device regulations of the USA, EU, Japan, and China, which are active in the medical device sector in the world.

United States of America

The organization operating as a controller and regulator in the US is the Food and Drug Administration (FDA). The FDA organization works under the Federal Food, Drug and Cosmetics Act under the US Department of Health and Human Services [4]. In order for medical devices to be placed in the USA, FDA approval is required. Before a medical device can be certified in the USA, it must be shown that the device is not only safe but also effective. In the USA, since the FDA is the only authority system, the market entry processes of all medical devices can go through a costly and lengthy evaluation process [8]. The FDA assigns devices to one of three regulatory classes according to their intended use, depending on whether they are invasive or implantable and pose risk to users.

Table **1** shows the US classification of medical devices and regulatory requirements for approval and market surveillance. As also shown in Table **1**, the device class determines the level of evidence and evaluation required to show safety and effectiveness [9].

In the USA, manufacturers should follow the steps below before putting their devices on the market [7]:

- Classify the device.
- Choosing the right pre-market distribution.
- Preparing FDA-approved information for pre-market presentation.
- Send pre-market presentation to FDA and interact with FDA staff during the review.
- Complete the organization registration and device list.

Table 1. USA classification of medical devices and regulatory requirements for approval and market surveillance.

Classification	Description	Premarket Requirements	General Time to Clearance/Approval	Postmarket Requirements
Class I	These devices are typically simple in design and manufacture and have a history of safe use. Device examples here are tongue depressors, crutches, and scalpels. *No to negligible risk.*	Subject to the least regulatory control, most Class I devices are exempt from premarket notification and/or good manufacturing practices regulation, although some general controls apply (device registration and listing, labeling regulations).	Varies.	Reports of device safety and performance problems are mandatory for manufacturers but voluntary for providers and users. They use the MAUDE (Manufacturer and User Facility Device Experience database), MedSun (Medwatch adverse event reporting program), and Medical Device Surveillance Network (network of facilities collecting data on device-related problems). Postmarket studies are required for certain devices, particularly those in Classes II and III.
Class II	These devices are more complicated and are associated with a higher level of risk than Class I technologies and include endoscopes, infusion pumps, and condoms. *Low risk.*	Most Class II devices are required to clear premarket notification 510(k) requirements. In rare cases, clinical studies are required for a 510(k) submission. In addition, these devices may be subject to other special controls, such as special labeling requirements and mandatory postmarket surveillance.	6 to 12 months.	
Class III	Devices belonging to this category usually support or sustain human life, are of substantial importance in preventing impairment of human health, or present a potential, unreasonable risk of illness or injury to the patient. Such devices include coronary stents, defibrillators, and tissue grafts. *Medium and high risk.*	These devices have the most stringent requirements. Typically, the information is insufficient to ensure safety and effectiveness solely through general or special controls. Therefore, a premarket application is required for Class III devices, which includes evidence from prospective, randomized control trials.	12+ months.	

Resource [9, 10]:

Unless exempted, manufacturers must apply one of the following types of pre-market applications for medical devices [7]:

- 510(k) Pre-Market Notification - applicable to some Class I and mostly Class II devices.
- Pre-Market Approval - mostly applicable to Class III.
- For automatic Class III identification, De Novo Evaluation - may be applied to new low to medium-risk devices for which there is no previously approved device.
- Humanitarian Device Exemption (HDE) - Applicable to class III devices for patients with rare diseases or conditions, meaning less than 4,000 people may be affected per year.

The examinations of medical devices by the FDA are made according to different applications and the risk class. In this context, general controls check whether medical devices are manufactured under a quality assurance program. Special controls apply to Class II devices. Pre-market approval is the highest level of regulatory control in the USA and applies to Class III devices [9].

The basic legal provisions of the US FD&C Act, which provides the regulatory framework for medical devices in the USA, are named General Controls. It includes the following regulatory controls [7]:

- Decreasing the quality of the goods/cheating.
- Misbranding.
- Organization registration and device list.
- Prohibited devices.
- Notification of risk, repair, replacement, and payback.
- Records and reports.
- Restricted devices.
- Quality System Regulation (QSR), Good Manufacturing Practices (GMP).

In order for a medical device to be allowed to enter the market by the FDA, the device must first pass general controls, such as registration, listing, labeling, and Good Manufacturing Practice (GMP). For Class I medical devices to be placed on the market, it is sufficient to pass the general controls [11].

In the USA, one of the important practices for medical devices is the Good Manufacturing Practice (GMP). 21 CFR 820 section has established the legal infrastructure of the GMP quality system for medical devices. Quality system regulation contains requirements, such as design, production, medical packaging, storage, and maintenance services. Manufacturers must establish and monitor quality systems for maintenance, repair, and service activities to ensure that their products consistently comply with applicable requirements and specifications. Quality systems for products (food, pharmaceuticals, biological materials, and devices) that are subject to control by the FDA are known as current good manufacturing practices (CGMPs). However, there is no obligation to establish a GMP quality system for all medical devices and the list of these low-risk products is determined and announced by the FDA [12].

The Premarket Notification-510 (k) process has been developed for market entry of Class II medical devices. Class II medical device manufacturers submit applications to the FDA promising that their devices are substantially equivalent in terms of effectiveness and safety to another device that was previously legally available on the market and called a "predicate" [13]. In order for a medical device to be considered substantially equivalent to a predicate medical device, the medical device must have the same intended use and/or the same/different technological features as the predicate device. This technological difference should indicate new performance and effectiveness evaluations of the device. Generally, animal and test tests are sufficient to prove equivalence with the predicate device [14].

For Class II devices, they are included in the so-called Special Controls applied and are generally applied specifically to the device and include the following [7]:

- Performance standards and guidelines,
- Pre-market data requirements,
- Private labeling requirements,
- Post-market inspection requirements, and
- Patient records.

For Class III or for an unmatched medical device on the market, FDA requests pre-market approval from medical device manufacturers to decide whether it is sufficient in terms of safety and performance related to medical devices. Compared to the 510(k) process, the pre-market approval application requires more extensive research and submission of evidence on the effectiveness and performance of the device. In practice, this standard is achieved through small clinical studies in a selected group of patients. Studies often do not include randomized designs, and the FDA generally does not allow manufacturers to collect long-term efficacy data [15].

Section 226 of the Food and Drug Administration Amendments Act of 2007 (FDAAA) and section 614 of the Food and Drug Administration Safety and Innovation Act of 2012 (FDASIA) amended the Federal Food, Drug, and Cosmetic Act to add section 519(f), which directs FDA to publish regulations establishing a unique device identification system for medical devices. The Unique Device Identification/UDI Recommended Rule was published on July 10th, 2012, followed by a change published on November 19th, 2012, that changed the implementation timeline for some devices. In developing it, different perspectives as possible were included (manufacturers, global regulatory organizations, clinical communities, and patient advocacy organizations). On September 24th, 2013, FDA published a final rule establishing a unique device identification system (the UDI Rule). UDI initiatives also continue globally. The European Commission published a framework for the UDI System in April 2013; The International Forum of Medical Device Regulators (IMDRF) UDI Working Group published a guidance document on UDI in December 2013 (FDA, 2014). Later, the Medical Device Regulation dated 5th April, 2017, and numbered 2017/745 and the *in vitro* Medical Diagnostic Devices Regulation dated 5th April, 2017, and numbered 2017/746 were published in the EU Official Journal, and related UDI rules have also entered into force in the EU [9].

European Union

EU medical device regulations have been rearranged for various reasons, especially the poly implant prothese scandal and the defective medical implants. Research and transition process from the previous regulations to the new regulation is ongoing. The reasons for the changes in EU medical device legislation are listed below [16]:

1. Notified Bodies:
    a. Failure to avoid the danger and dysfunction posed by the business perspective of notified bodies. There is no state sanction in the conformity assessment process of high-risk medical devices, and this decision is left to notified bodies with commercial concerns.
    b. Lack of clinical knowledge and expertise of notified bodies.
    c. The freedom to choose a notified body allows manufacturers to choose the body that is more suitable for them. In fact, a device rejected by one notified body can be approved by another.
2. Criticisms of clinical evaluation methodology:
    a. No requirement for clinical efficacy; keeping innovation ahead of clinical effectiveness.
    b. The situation caused by confidentiality requirements in accessing clinical

data and a situation that is not in line with the principles of transparency of the Declaration of Helsinki.

   c. No clinical research obligation, even for new technologies. A literature review or data based on recalled or prohibited medical devices or equivalency can be considered clinical data.
   d. No pre-marketing requirement for clinical evidence of efficacy for high-risk devices.
3. Unavailability of data and a lack of public database:
   a. Clinical database.
   b. Documentation applications.
   c. Information about the number of different medical devices.
4. Failure to ensure traceability.
5. Inadequacy of conformity assessment processes.
6. Lack of a common medical device nomenclature.
7. Systemic deficiencies in vigilance activities and failure to prevent incomplete reporting.
8. Legislation guides do not serve as an international guideline for the planning of pre-market clinical trials.
9. The conditions of placing medical devices on the market are much easier compared to drugs.
10. Inability to adapt to new technologies.
11. Lack of manufacturer's clinical knowledge and expertise.

Substituted by Corrigendum to Regulation (EU) 2017/745 of the European Parliament and of the Council of 5th April, 2017, on medical devices, amending Directive 2001/83/EC, Regulation (EC) No 178/2002 and Regulation (EC) No. 1223/2009 and repealing Council Directives 90/385/EEC and 93/42/EEC have been published in the Official Journal of the EU L 117 on 5th May, 2017.

This latest medical device regulation, previous regulations still in effect, and transitional regulations are described in this section.

Previous Regulations Still Applied (Transition Period)

Medical devices within the EU are currently regulated by the following 3 directives [17]:

1) Council Directive 90/385/EEC on Active Implantable Medical Devices (1990):

Devices covered by the Council Directive 90/385/EEC of 20th June, 1990, on active implantable medical devices must meet the relevant provisions of this directive. This directive covers all activities related to the design, manufacture, placing on the market, putting into service, use, and inspection of active implantable (implantable) medical devices [18].

2) Council Directive 93/42/EEC on Medical Devices (1993):

Devices covered by the Council Directive 93/42/EEC of 14th June, 1993, on medical devices must comply with those specified in this directive. This directive covers all activities related to the design, manufacture, placing on the market, putting into service, use and inspection of medical devices. Medical Devices Directive 93/42/EEC also regulates the obligations of the competent authorities of the member states regarding medical devices [19].

3) Council Directive 98/79/EC on *In Vitro* Diagnostic Medical Devices (IVDMD) (1998):

For *in vitro* medical diagnostic devices, "Directive 98/79/EC of the European Parliament and of the Council of 27th October, 1998" is valid. This directive covers all activities related to the design, manufacture, placing on the market, putting into service, use and inspection of *in vitro* medical diagnostic devices [18].

In the EU, every marketed medical device must carry a Conformité Européenne (CE) mark indicating that it conforms to relevant directives of the EC Medical Device Directives of the EU. A device with a CE mark can be marketed in any EU member state. Medical devices that are non-implantable and considered low risk are "self-marked" meaning that the manufacturer itself simply certifies compliance and applies a CE mark. High-risk devices must undergo a more extensive outside review. Through a complex system of legislation, high-risk medical device approval applications can be filed in any member state and reviewed by a notified body established within that state and authorized by that state's Competent Authority or health agency to assess and assure conformity with requirements of the relevant EC directive. Notified bodies are private organizations that contract with manufacturers to supply these certifications for a fee, and there are currently around 76 notified bodies in the EU. Once the notified body agrees that the device meets the requirements for conformity, the notified body issues a CE mark, and the device can then be marketed in EU member states [8].

European medical device database, EUDAMED (Europäische Datenbank für Medizinprodukte, originally in German), is a web-based application developed by the European Commission. Its purpose is to exchange legal information on the implementation of medical device directives between the European Commission's Directorate General for Health and Consumers, the EU member states, and the competent authorities of the member states (European Commission, 2012). Since the EUDAMED system only aims at the official exchange of information between the European Commission and the competent authorities of the member states, there is no public access to this system [8, 19].

The legal source of the EUDAMED system involves three main directives of the EU, which contain legal regulations on medical devices. These directives are [18]:

- Council Directive 90/385/EEC on Active Implantable Medical Devices.
- Council Directive 93/42/EEC on Medical Devices.
- 12 of Council Directive 98/79/EC on *In Vitro* Diagnostic Medical Devices (IVDMD).

In the Commission Decision on 19th April, 2010, regarding EUDAMED, the purpose of the EUDAMED system is to strengthen market surveillance and control by providing the competent authorities with quick access to information on manufacturers, authorized representatives, medical devices, certificates, vigilance (warning) data and clinical trials, and to ensure uniform application of medical device directives, especially in matters related to the registration system [18]. According to article 10b of Council Directive 90/385/EEC dated 20th June, 1990, on active implantable medical devices, it is specified as what information the European Data Bank will contain [19].

Mandatory fields related to the data that the member states of the EU must transfer to the EUDAMED system are listed in the Commission Decision on 19th April, 2010, related to EUDAMED. Accordingly, member states are obliged to define the following mandatory fields in the EUDAMED system. The mandatory information in question is [18]:

Within the scope of the 93/42/EEC Medical Device Directive:

For the manufacturer or authorized representatives; for medical devices with name, address, country, and contact information; internationally accepted classification code, device name or generic name.

Within the scope of the 90/385/EEC Active Implantable Medical Devices Directive:

For certificates; certificate number, type of certificate, date of certificate issuance, expiry date, manufacturer and notified body information, the general scope of the certificate and details of the devices, status of the certificate and, the reasons for the notified body.

Within the scope of 98/79/EC *In Vitro* Medical Diagnostic Devices Directive:

- For all IVD manufacturers or their authorized representatives: Name, address, country, and contact information.
- For certificates: Certificate number, certificate type, certificate issuance date, expiry date, manufacturer, and authorized representative information, notified body information, the general scope of certificate and details of devices, the status of certificate and reasons of the notified body. For events within the scope of the warning system; competent authority reference number, manufacturer and authorized representative information, manufacturer contact information, manufacturer reference/Field Safety Corrective Action (FSCA) information, device information with lot number, serial number, and software version, notified body information, market known devices' all investigations, background information and description, outcome, recommendations, activities performed and definitions.

New Regulations

The new regulations entered into force in all EU member states as of 25th May, 2017. Transition periods are given for the implementation of both regulations. In general, this transition period is three years for the Medical Device Regulation and five years from the effective date for the *In Vitro* Diagnostic Medical Devices Regulation. Special transition periods are also foreseen for the implementation of some provisions of both regulations [20].

In Table **2**, in order to ensure the safety of medical devices, new EU regulations and old regulations have been compared.

Table 2. A comparison of new EU regulations and old EU regulations.

Old Regulations	New Regulations
Regulations on medical devices date back to the 1990s and do not reflect current technological development.	New regulations that consider technological development and initiate development.
Control of high-risk devices, such as implants, depends on the national notified body.	Control of high-risk devices, such as implants, also includes independent expert committees within the EU.
Clinical studies conducted by more than one Member State are subject to a large number of national evaluations.	Clinical studies carried out by more than one Member State are coordinated and evaluated in one place.
Most aesthetic products, such as color contact lenses, are organized into "general products".	Many aesthetic products regulated as medical devices are subject to more detailed controls.

(Table 2) cont.....

Old Regulations	New Regulations
Before placing on the market, only one out of five *in vitro* medical diagnostic devices is checked by the notified body.	Prior to release, four out of five *in vitro* medical diagnostic devices are checked by the notified body.
The European database contains limited information on non-public medical devices.	The European database contains extensive information on medical devices, most of which are publicly available.
There is variable and often limited information regarding devices implanted in patients.	There is an "implant card" for implanted devices so that more information can be obtained.
There may be financial losses arising from medical devices. For example, compensation is not guaranteed if the manufacturer goes bankrupt.	There is a financial mechanism. In possible cases, compensation is available.
Multiple registration procedures may be required for medical devices.	Manufacturers only need to register their devices at the EU level once.

Resource [20].

Japan

Japan is one of the largest shareholders in the medical device industry and the international medical device market. According to official figure published by the Ministry of Health, Labor, and Social Security, there are approximately 34 billion dollars spent on medical devices in Japan. In addition, due to the aging population of Japan, it is stated in these studies that the use of medical devices has increased. With the effect of all these and Japan being one of the leading countries in technology, there is an established supervisory and regulatory system in the field of medical devices in Japan. Medical devices regulated by the Ministry of Health, Labor, and Social Welfare of Japan, are subject to the regulatory agency of the Medicines and Medical Devices Agency located in the Ministry. Medical devices were regulated by the Pharmaceuticals and Medical Devices Act, which came into force in 2014, and a risk-based classification system and classification systems were determined [21].

Medical device classes, based on Japanese Medical Device Nomenclature (JMDN) codes and medical device names whose generic names are determined, arc classified according to risk levels. In Japan, medical devices are classified into four risk classes, class I, class II, class III, and class IV, and determined with reference to the Global Harmonization Task Force classification rules. Before applying to Japanese institutions, a manufacturer determines the type and risk class of the device according to the code of the JMDN. The risk code is also used to determine the regulatory controls that should be applied to the device [7]. However, a quality management system for medical devices is implemented in Japan. The quality management system is applied to the products to be supplied to the market and is subject to control by the Pharmaceuticals and Medical Devices Agency. The quality management system applied in Japan complies with the ISO13485 standard applied in Europe and the inspections carried out by the FDA in the USA [22].

China

China, one of the countries that has shown rapid development in terms of production in the 2000s, has also shown remarkable progress in the field of medical devices. Since medical device trade is carried out worldwide, organizations operating in China have to follow all world regulations. However, there are also local legal regulations in the country. Supervisory and regulatory activities regarding medical devices are carried out by the Food and Drug Administration of China. Following the laws published by the institution, manufacturers determine the risk class of their devices using the Medical Devices Classification Search System (MDCSS) and the China Medical Device Regulatory Database (CMDRD).

In general, the medical device regulations in China are similar to the US FDA. Medical devices and *in vitro* diagnostic medical devices are classified as Class I, Class II, and Class III, and the control methods applied to the highest-risk class Class III group products differ from other risk classes and are more comprehensive [23].

Brazil

Medical equipment is inspected in Brazil by the Brazilian National Health Inspection Agency called ANVISA. Coordinating ministerial decisions regarding the adoption and financing of health technologies in the public health care system is a complex process. Therefore, there are problems in paying the costs of health technologies adopted by public health institutions with the decisions of the Ministry and due to the lack of coordination [24].

When inspections are done properly, the use of unsafe and inefficient medical devices is prevented. At the same time, ANVISA's authorization processes can create societal demand for certain products. Lawsuits are being filed against federal, state, and local authorities that use products not licensed by ANVISA. Concern always persists that incompatibilities between institutions may create results that encourage the demand for expensive equipment when there are cheaper and more suitable solutions. Brazilian medical device regulations are shown in Table **3** [24].

Table 3. Brazilian medical device regulations.

National Regulatory Authority	ANVISA
Key laws/regulations	RDC 185/200158 (only for Brazil)
Risk classification system	There is a quadruple risk classification system similar to EU Directive 93/42/EEC
What is necessary for registrations and import?	• RDC 27/2011 for electrical equipment
	• Technical file based on RDC 185/2011, Annex 3, Part A/B/C
	• Free sale certificate from the manufacturer country
	• Risk management in compliance with ISO 14971 for all implants
Seller's records/certification	Work allowance permit is required for the organization to sell, distribute, and import products in Brazil.
Quality system requirements	Inspection of class III and IV medical device manufacturers twice a year by ANVISA and according to BGMP'59
What for clinical tests?	Clinical tests are necessary for high-risk products and innovation
New regulation	RDC 16/201360

Resource [24].

COMPARISON OF THE EUROPEAN UNION AND THE UNITED STATES OF AMERICA

In the USA, the FDA is the authorized body to regulate the processes related to the placing of devices on the market. In other words, unlike the EU system, the process is managed through a public institution. Similar to the EU system, devices are classified according to risk groups, but the classification rules are not as detailed as in the EU and the classification is made by the FDA [25].

As can be seen in Fig. (1), the approval of medical devices in both the EU and the USA shares some similarities. The FDA assigns devices to 3 main regulatory classes: low risk or Class I, moderate risk or Class II, and high risk or Class III. In the USA, a Class I device requires merely a Premarket Notification without clinical trials, whereas Class III devices require clinical trials and/or other evidence unless they are not substantially different from an already-marketed Class III device. If they are similar to a previously approved predicate device, they can usually forgo clinical testing or undergo only limited clinical investigations. About 75% of Class II devices in the USA require some form of clinical trials to demonstrate their safety and that they perform as expected, although the level of evidence required for approval is often less rigorous than that for new drug approval. Randomized controlled trials, for example, are uncommon because of difficulties in randomization and blinding, and many devices are approved based on small observational studies [8].

In Europe, the Council of Europe's New Approach Directives defined the "Essential Requirements" applicable to all countries to ensure the safety and performance of devices. The EU Commission assigns devices to 4 classes. Class I or low-risk devices are only required to "declare" to the National Competent Authority in their country that they comply with the Basic Requirements. For example, in the UK, the Competent Authority is the Medicines and Healthcare Products Regulatory Agency [8]. Intermediate and high-risk devices (Class IIa, IIb, and III) require clinical and/or non-clinical evidence to support approval. As in the US, if a device is shown to be substantially similar to an already approved device, data from the validated device can be used to support validation, and new clinical trials may not be needed [8]. There are fundamental differences in the US and European approaches to medical device regulations. These differences help explain why the US and Europe have adopted different regulatory processes and evidence requirements for devices [26]. For example, the FDA was established to promote and protect public health through the regulation of medicinal products, whereas notified bodies in the European system were developed as part of a wider initiative to strengthen innovation and industry policy across Europe. Notified bodies are, therefore, not designed to function as public health agencies. Instead, the protection of public health depends on Competent Authorities, whose roles vary greatly between member states [10].

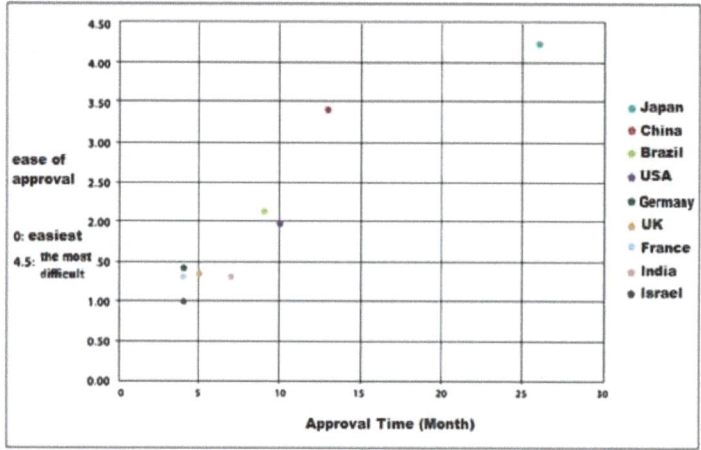

Fig. (1). Countries' average approval times-ease of approval situations [25].

Globally, the largest share of medical devices is investigated and approved in the USA and in the EU. Although the regulatory processes in the USA and Europe share common goals and have many similarities, the different history of device regulation in both regions contributes to significant regulatory dissimilarities. Whereas the FDA was founded as a centralized consumer protection agency, the current European systems were driven by a need to standardize commercial rules across the European member states. As a result, the FDA is sometimes seen as overplaying safety concerns at the cost of commercial enterprise, whereas the European systems are sometimes characterized as being primarily concerned with preserving commercial interests to the detriment of patient safety. Despite assertions that drugs are approved more slowly in the USA, analysis indicates that approved drugs reach the public more quickly in the USA than in Europe. Whether there is a true "device lag" between Europe and the USA is less clear. Nevertheless, device safety concerns and device failures on both sides of the "pond" lead both the USA and EU to seek greater cooperation and to explore tightening regulations regarding device approvals. Legislative efforts in both the USA and EU are currently underway to promote transparency and mutual standardization of medical device approval processes [8]. A comparison of device approval processes in the USA and EU is shown in Table **4**.

Table 4. Comparison of device class approved processes in the USA and EU.

United States of America	European Union
Class I (low risk): Premarket notification process does not require clinical trials. Class II (intermediate risk): 25% can undergo premarket notification process, 75% require clinical evidence.	Class I (low risk): "Self declare" to the Competent Authority of a state of the EU and can be marketed throughout the EU.
Class II (intermediate risk): 75% require clinical evidence. Class III (high risk): De Nevo devices: The application reclassifies a device that was automatically classified (as a new application) as a Class III device and as a Class II device. Less stringent clinical evidence will generally be required. Class III device with predicates: If substantial similarity to previous "predicate" devices, generally, clinical trials are not needed.	Class II IIb (intermediate risk) and Class III (high risk): Device with predicates: If substantial similarity to previous "predicate" devices, it generally does not need clinical evidence.
Class III device without predicates: Clinical trials to show safety and efficacy.	Class IIa, IIb, III devices without predicates: Clinical evidence to show safety that the device performs as planned.
Application for FDA Approval.	Decentralized Approval Process. Application to any of the notified bodies of any EU state: notified body examines the application to assure compliance with EC regulations. If the device meets regulatory requirements, a CE is applied, and the device can be marketed throughout Europe.

Reference [8].

The FDA assigns devices to 3 main regulatory classes: low risk or Class I, moderate risk or Class II, and high risk or Class III (Table **5**). In the USA, a Class I device requires merely a Premarket Notification without clinical trials, whereas Class III devices require clinical trials and/or other evidence unless they are not substantially different from an already-marketed Class III device.

Table 5. Risk classification of medical devices in the USA and EU.

United States of America	European Union
Class I: low risk of illness or injury, e.g., gauze, toothbrushes.	Class I: low risk; e.g., sterile dressings, gloves.
Class II: moderate risk of illness or injury, e.g., suture, needles.	Class IIa: low-medium risk; e.g., surgical blades, suction equipment. Class IIb: medium to high risk; e.g., ventilators, some implants, radiotherapy equipment.
Class III: significant risk of illness or injury, e.g., pacemakers, implantable defibrillators.	Class III: high risk; e.g., drug-eluting cardiac stents, pacemakers, implantable defibrillators.

Resource [8].

According to the medical device regulations in the USA and EU countries, Average Approval Times - Approval Ease Situations are shown in Fig. (**1**). According to the Fig. (**1**), these time periods in Japan and China are longer compared to the USA and EU countries and the USA, and Israel seems to be in the best position.

CONCLUSION AND DISCUSSION

All legal regulations and policies regarding medical devices aim to ensure the development and competitiveness of the innovative structure of the medical device industry, while at the same time aiming to protect public health and patient safety at the highest level. Rules regarding the safety and performance of medical devices were published in the EU in the 1990s as three main medical device directives. Within the scope of the inadequacies and implementation problems of these regulations, important updates were made in the regulations in 2017, and transition processes were defined for compliance with these regulations.

Regulation controls innovation in a given area. However, technological progress and changes have an impact on regulation. This interaction must be perceived as a success in changing regulation [27]. Regulatory change fully responds to changes in technical, social, and economic conditions surrounding them. In the case of medical devices and the legislative changes currently under discussion in Regulation (EU) 2017/745 of the European Parliament and of the Council as of 5th April, 2017, on medical devices, this is to ensure greater safety for patients. At the same time, it means an increased economic burden for medical device manufacturers in the form of conformity assessment costs. It is currently a question of whether this change will place a significant burden on small and medium-sized enterprises, which are the main innovators in the field of medical devices, to the point where production is reduced, or whether it will, as with so many examples of successfully implemented regulation, it has become a powerful stimulus for further innovation [28]. With the arrival of the new European Commission Medical Device Regulation 2017/745 (https://eur-lex.europa.eu), there will be further and more demanding requirements for manufacturers to enter the European market. However, it will always be crucial to correctly classify a risky medical device, regardless of the country of origin of the manufacturer or the medical device [27].

When the medical device regulations are evaluated within the scope of the approval of the medical devices to be put on the market, it is seen that Israel, the USA, Germany, France, Netherlands, and the UK, are ahead of the others, but China is behind, and Japan is far behind. As a result, it is one of the important issues that medical device regulations are aimed at the rapid use of effective and reliable medical devices in the market. Eliminating unnecessary processes by defining adequate and effective approval mechanisms in the relevant regulations, regulations are among the issues to be considered. Reconsideration of these processes, especially in Japan, where there are too many technological organizations, makes us think that it may face the situation of not being able to evaluate the infrastructure and not being able to compete. Also, among the issues, it should be considered that countries in this situation may result in the withdrawal of organizations from the medical device sector.

General recommendations for medical device regulations are listed below:

- In the globalizing world, the establishment of global and national medical device inventory systems in harmony with each other and an authorized global authority organization can ensure that the management of medical devices is carried out more effectively and safely.
- For medical supplies, tools, and equipment that are not included in the definition of medical devices but do not have a device feature, it is thought that separating the sectors from each other by making separate regulations will make the systems more manageable.
- Considering the training of human resources and especially the effects of health services on public health, national and international regulations can be made for the employment of trained manpower in the public sector in countries without advanced technology.

- To protect competition in the sector and to prevent monopolization, market intervention mechanisms can be defined in accordance with each other.
- Because the subject has the feature of improving public health and providing access to quality health services rather than individual health, regulations encouraging public and private sector medical device research centers are recommended in national regulations.

In the medical device industry, it is necessary to harmonize and disseminate good practice examples all over the world by updating the internationally published guidelines with the specific features of medical devices, envisaging that they should be published and the processes become standardized. Also, it is required to develop models suitable for medical device evaluations, to support the developed models with clinical and economic evaluation processes, to collect the data to be used in the evaluations, to analyze the collected data, to interpret the evaluation results correctly and to implement them [11]. Innovation in medical devices, due to their potential benefits, is the most prominent area to ensure the efficient distribution of resources and the right to health access. In order to prevent problems related to the evaluation of medical devices, it is suggested that medical device evaluations should be considered as an iterative process and that decisions about medical devices should be reviewed with the acquisition of new evidence. Handling medical device evaluations as an iterative process and adapting the evaluated medical device to the health system in line with this process will both alleviate the tension between the effective distribution of resources and the right of access to health, and that the need to develop additional evidence in the process will encourage technological development. The medical device will be evaluated at the stage of collecting additional evidence. It is also important to consider that long-term data on the clinical and economic effectiveness of the evaluated medical device can be accessed without endangering the health of the patient, which is extremely important. Finally, it is recommended that all countries comply with these regulations and that new medical devices are released to the market by removing the inherent risks, because a healthy competitive environment is imperative, considering the mission of the private sector.

CONSENT FOR PUBLICATION

Not applicable.

CONFLICT OF INTEREST

The author declares no conflict of interest, financial or otherwise.

ACKNOWLEDGEMENT

Declared none.

REFERENCES

[1] Maliyamu, M. Current situation of the medical devices industry in foreign trade: the case of turkey. dokuz eylul university institute of social sciences master's thesis, 2019.

[2] Koç, ZZ, Competitive analysis of the medical device industry according to the five power models: the case of istanbul province. duzce university institute of social sciences master thesis, 2020.

[3] Available from: https://www.ema.europa.eu/en/human-regulatory/overview/medical-devices

[4] Barstugan D. Investigation of imported and domestic medical devices in Turkey market in terms of product safety, cost and benefit, İstanbul University Master's Thesis, İstanbul 2019.

[5] European Commission, Evaluation of the Eudamed, European Commission 2012. http://ec.europa.eu/DocsRoom/documents/12981/attachments/1/translations/en/renditio ns/native

[6] Technology Development Foundation of Turkey. Medical Device Sector and Strategy Proposal in the World and in Turkey. Ankara 2013.

[7] Theisz V. Medical Device Regulatory Practices International Perspective, Romanya. CRC Press 2015. [http://dx.doi.org/10.1201/b18817]

[8] Van Norman GA. Drugs and Devices. JACC Basic Transl Sci 2016; 1(5): 399-412. [http://dx.doi.org/10.1016/j.jacbts.2016.06.003] [PMID: 30167527]

[9] Bahçeci S. The Evaluation of Medical Device Registration System in Turkey 2018.

[10] Sorenson C, Drummond M. Improving medical device regulation: the United States and Europe in perspective. Milbank Q 2014; 92(1): 114-50. [http://dx.doi.org/10.1111/1468-0009.12043] [PMID: 24597558]

[11] Yıldız T. Evaluation of Health Technologies in Medical Devices. Journal of Social Security 2017; 7(13): 116-46.

[12] Available from: https://www.fda.gov/AboutFDA/Transparency/Basics/ucm194877.htm

[13] Fraser A G, Daubert J C, Werf F, *et al.* Clinical evaluation of cardiovascular devices: principles, problems, and proposals for European regulatory reform: Report of a policy conference of the European Society of Cardiology. Eur Heart J 2011; 32(13): 1673–86.

[14] Eldessoukı R. Therapeutic and Diognastic Device Regulations, Therapeutic and Diognastic Device Outcomes Research. USA: ISPOR 2011; pp. 41-3.

[15] Taylor RS, Iglesias CP. Assessing the clinical and cost-effectiveness of medical devices and drugs: are they that different? Value Health 2009; 12(4): 404-6. [http://dx.doi.org/10.1111/j.1524-4733.2008.00476_2.x] [PMID: 19138305]

[16] Tatlı E. Changes brought by the European Medical Device Regulation No. 2017/745 and Measuring the Adaptation Level of Medical Device Manufacturers in Turkey to New Regulations, Ege University Graduate School of Sciences. Department of Biomedical Technologies, Master's Thesis, İzmir 2019.

[17] European Commission, Regulatory Framework. Available from: https://ec.europa.eu/growth/sectors/medical-devices/regulatory-framework_en

[18] Official Journal of the European Union, Commission Decision of 19 April 2010 on the European Databank on Medical Devices. (Eudamed). Eur Union 2010.

[19] European Commission, Market surveillance and vigilance, 2018. Available from: http://ec.europa.eu/growth/sectors/medical-devices/marketsurveillance/ (Access date: 19/04/2021).

[20]   TİTCK, Announcement on EU's New Medical Device Regulations, Turkish Medicines and Medical Devices Agency, 2018. Available from: http://www.titck.gov.tr/Haberler/HaberGetir?id=983

[21]   International Trade Administration, Top Markets Report Medical Devices Country Case Study I, 2016. Available from: https://www.trade.gov/topmarkets/pdf/Medic al_Devices_Japan.pdf

[22]   Available from: https://www.emergogroup.com/services/japan/quality-management-systemcompliance

[23]   Available from: https://www.emergogroup.com/resources/china-process-chart

[24]   Kiper M. Medical Device Sector and Strategy Proposal in the World and in Turkey. Turkey Technology Development Foundation. Ankara 2018.

[25]   National Institute for Public Health and the Environment. Comparison of market authorization systems of medical devices in USA and Europe. Bilthoven: Netherlands National Institute for Public Health and the Environment 2015.

[26]   Kramer DB, Xu S, Kesselheim AS. Regulation of medical devices in the United States and European Union. N Engl J Med 2012; 366(9): 848-55.
        [http://dx.doi.org/10.1056/NEJMhle1113918] [PMID: 22332952]

[27]   Available from: https://www.oecd.org/regreform/ (Access date: 23/03/2021).

[28]   Lukas. P, Ladislav, H, Petra, M, Martin, A, and Marek, P, Medical Devices: Regulation, Risk Classification, and Open Innovation. J Open Innov 2020; 6(42): 1-13.

SUBJECT INDEX

A

Accumulation 6, 159, 164, 187
　effective tumor 159
　liposome 164
Acid 51, 55, 123, 124, 152, 159, 229
　folic 159
　gallic 123, 124
　glutamic 51
　hyaluronic 55, 152, 229
　methacrylic 152
Acute 185, 210
　lymphocytic leukaemia 210
　myelogenous leukemia (AML) 185
Alzheimer's disease 6
Angiogenesis 15, 40, 53, 173, 212, 216, 218
　factor-mediated 173
　inhibitors 218
Angiostrongylus cantonensis 117
Anti-acanthamoebic effects 124
Anti-amoebic activity 124
Antiamoebic effect 123
Antibacterial activity 52
Anticancer 148, 153, 157, 162, 163, 167, 170,
　　171, 172, 173, 174, 175, 177, 183, 185,
　　188
　activity 148, 157, 163, 167, 170, 172, 173,
　　174, 175, 177, 183, 185
　agents 153, 162, 171, 172, 175, 188
　delivery applications 170
Anticancer drugs 150, 151, 164, 169, 171,
　　179, 183, 187, 232
　liposomal 164
Antigen detection methods 126
Antiparasitic agent 122
Anti-parasitic medications 116
Antiseptic technique 77
Anti-Toxocara 118, 136
　effect 136
Anti-Toxoplasma 128, 129
　effects 128, 129
　medications 128

Antitumor 148, 154, 155, 157, 165, 166, 167,
　　168, 170
　activity 155, 165, 166, 167, 168, 170
　effect 148, 154, 157, 165
ANVISA's authorization processes 285
Apoptosis 30, 45, 166, 182, 211
　inducing 182
Arthroplasty 220

B

Bacterial artificial chromosome (BAC) 39
Bioactive glass ceramics 222, 234
Bioceramics 209, 219, 226, 227, 234
　biodegradable 219
　fabricating 234
　mesoporous 209, 226, 227
Biomimetic sensors 88
Bone 52, 54, 75, 96, 100, 106, 165, 227
　cancer 75, 96, 100, 106, 165, 227
　marrow homing peptide (BMHP) 54
　repair periodontal tissue regeneration 52
Bovine serum albumin (BSA) 153, 173
Brachytherapy 217, 222, 227
Breast cancer 161, 162, 164, 167, 172, 173,
　　184, 185, 215, 216, 217, 224
　malignant 215
　metastatic 161, 185

C

Cancer
　associated fibroblasts (CAFs) 153, 164, 174,
　　176, 184, 185, 210, 213, 214, 216, 217,
　　223, 229
　cervical 174
　colon 174, 229
　gastic 184
　gastric 174, 185
　malignant 210
　melanoma skin 223
　neck 217

www.ingramcontent.com/pod-product-compliance
Lightning Source LLC
Chambersburg PA
CBHW050811220326
41598CB00006B/175